# DETERMINANTS IN THE EVOLUTION OF THE EUROPEAN CHEMICAL INDUSTRY, 1900–1939

# Chemists and Chemistry

VOLUME 16

*A series of books devoted to the examination of the history and development of chemistry from its early emergence as a separate discipline to the present day. The series will describe the personalities, processes, theoretical and technical advances which have shaped our current understanding of chemical science.*

*The titles published in this series are listed at the end of this volume.*

# DETERMINANTS IN THE EVOLUTION OF THE EUROPEAN CHEMICAL INDUSTRY, 1900–1939

## New Technologies, Political Frameworks, Markets and Companies

*Edited by*


Anthony S. Travis
*Sidney M. Edelstein Center,*
*Hebrew University,*
*Jerusalem, Israel*

Harm G. Schröter
*University of Konstanz,*
*Konstanz, Germany*

Ernst Homburg
*Rijksuniversiteit Limburg,*
*Maastricht, The Netherlands*

and

Peter J.T. Morris
*Science Museum,*
*London, U.K.*



KLUWER ACADEMIC PUBLISHERS
DORDRECHT / BOSTON / LONDON

A C.I.P. Catalogue record for this book is available from the Library of Congress.



ISBN 0-7923-4890-7



Published by Kluwer Academic Publishers,
P.O. Box 17, 3300 AA Dordrecht, The Netherlands.

Sold and distributed in North, Central and South America
by Kluwer Academic Publishers,
101 Philip Drive, Norwell, MA 02061, U.S.A.

In all other countries, sold and distributed
by Kluwer Academic Publishers,
P.O. Box 322, 3300 AH Dordrecht, The Netherlands.

*Printed on acid-free paper*



Printed in the Netherlands

# Contents

# Foreword

The editors wish to thank the European Science Foundation for its support of the programme on the Evolution of Chemistry in Europe, 1789-1939, as well as for sponsoring the publication of this volume. Through the subdivision of this initiative that deals specifically with chemical industry it has been possible for historians of science, technology, business and economics to share often widely differing viewpoints and develop consensus across disciplinary and cultural boundaries.

The contents of this volume are based on the third of three workshops that have considered the emergence of the modern European chemical industry prior to 1939, the first held in Liège (1994), the second in Maastricht (1995), and the third in Strasbourg (1996).

All contributors and participants are thanked for their participation in often lively and informative debates. The generous hospitality of the European Science Foundation and its staff in Strasbourg is gratefully acknowledged.

# Introduction

Emerging chemical knowledge and the development of chemical industry, and particularly the interaction between them, offer rich fields of study for the historian. This is reflected in the contents of the three workshops dealing with the emergence of chemical industry held under the aegis of the European Science Foundation's Evolution of Chemistry in Europe, 1789-1939, programme. The first workshop focused mainly on science for industry, 1789-1850, and the second on the two-way traffic between science and industry, 1850-1914. The third workshop, dealing with the period 1900-1939, covers similar issues, but within different, and wider, contexts.

World War I acted as a catalyst in the development of chemical know how, especially as applied to the needs of industry and national defence. Roy MacLeod demonstrates that the catalytic effect did more than enlarge on previous structures. It represented the beginning of a new type of relationship between science and industry. Moreover, the war not only disrupted international trade, but created, among other things, a new bureaucracy, one that promoted the organisation of, and control over, different, and often novel, markets. This facilitated the spread of various forms of cooperation, particularly cartels and trade and research associations, all of which flourished during the interwar years. At the same time the war created an acute awareness of the need for strong national chemical industries, since it had become abundantly clear that this branch of industry was a critical component of state and military power. Consequently, state intervention became a major issue during the two decades following the cessation of hostilities, decades marked by considerable international tensions, economic instability, and concerns over raw materials. Lothar Meinzer in his contribution takes up the issue of how BASF fared in dealing with both French and German authorities. The foundation of new chemical enterprises was fostered by nearly every European government during and after the war. Núria Puig, Helge Kragh, Timo Myllyntaus and Rolf Petri delineate for Spain, Denmark, Finland and Italy, respectively, how European latecomers to modern chemical manufacture attempted by various means and with different degrees of success to catch up.

During the interwar period, market growth in chemical products was much slower than before 1914. Companies therefore adopted various strategies aimed at strengthening competitiveness. One dominant feature was the perceived need to step up research and development (R&D). However, in many states a gap between academic science and chemical industry still had to be bridged. How two bridges were crossed is described by Anders Lundgren for Sweden and Geert J. Somsen for the Netherlands.

Several of these studies demonstrate how industrial success was achieved. The German chemical industry, which was active in R&D even before World War I, turned, for a time at least, to basic questions and fundamental science. How this happened during the 1920s and 30s is considered by Carsten Reinhardt and Peter Morris. The Swiss were also noted for strong R&D programmes, and, despite the fact that a relevant national patent system was not introduced until early in the 20th century, they soon mastered the control of intellectual property rights, as described by Jakob Tanner. Other approaches led to significant successes, as Anthony Travis and Knut Sogner show for Britain and Norway, respectively, where often small and medium sized firms were much more in tune with new needs and changing markets than were the big corporations. Another approach related to rationalisation of production and organisation of professions. Colin Divall and Sean Johnston show that the chemical engineering profession emerged as a unified body much earlier in Britain than in Germany, notwithstanding German successes in applying engineering to chemical manufacture. Stuart Bennett's contribution emphasises how traditional ways of working in the British chemical industry caused it to fall behind the United States in the matter of process control.

The high pressure Haber-Bosch process for synthetic ammonia, and its widespread impact, overshadows developments described in many of the following chapters. High pressure industrial chemistry was nothing less than the "paradigm shift" that thrust the chemical industry into the 20th century. It is thus appropriate to commence our story with an overview of the Haber-Bosch process and its imitators.

# Part 1

## A New Technology for the 20th Century

BASF high pressure converter for ammonia synthesis, with (inset) Carl Bosch (above) and Fritz Haber (below). Photos courtesy BASF (converter and Haber), and Deutsches Museum, Munich (Bosch). Reproduced from the cover of *Chemistry & Industry*, 2 August 1993, with permission.

# HIGH PRESSURE INDUSTRIAL CHEMISTRY: THE FIRST STEPS, 1909-1913, AND THE IMPACT

ANTHONY S. TRAVIS
*Sidney M. Edelstein Center for the History and Philosophy of Science, Technology and Medicine*
*Givat Ram Campus, The Hebrew University*
*Jerusalem 91904, Israel*

**Introduction**

In 1898, William Crookes's Presidential Address to the British Association for the Advancement of Science warned of an impending fertiliser crisis. As an active scientist, spokesman for chemistry (particularly as editor of *Chemical News*), and a one time director of the Native Guano Company, Crookes was well placed to make a strong plea for studies on the fixation of atmospheric nitrogen. Not only might this solve the fertiliser problem, it might also bind academic chemistry and chemical industry closer together. Moreover, Crookes had already become engaged in the enterprise through his work in 1892 on an electric arc process that combined nitrogen with oxygen. With commercialization in mind, this had been taken up by Lord Rayleigh in 1897, and two years later in Manchester by McDougall and Howles. The impetus for this line of investigation was the great success of electrochemical production of aluminium and alkali.[1]

More promising around 1900, however, was the fixation of atmospheric nitrogen by calcium carbide to afford cyanamide, the process of Nikodemus Caro and Adolph Frank, particularly after Caro found that cyanamide applied to the soil reacted with water to give off ammonia. This also relied on electricity, particularly inexpensive

---

1 A comprehensive international survey of the nitrogen industry will be found in Bruno Waeser, *The Atmospheric Nitrogen Industry with Special Consideration of the Production of Ammonia and Nitric Acid*, translated by Ernest Crowley (Philadelphia: P. Blakiston's Son & Co., 1926), 2 vols. A useful outline of electrochemical developments prior to 1940, including a short historical summary, is C.L. Mantell, *Industrial Electrochemistry* (New York: McGraw-Hill, 2nd edition, 1940). See also Ludwig F. Haber, *The Chemical Industry 1900-1930: International Growth and Technological Change* (Oxford: Clarendon, 1971), pp. 76-84. For Crookes and his interest in nitrogen fixation see Christopher Hamlin, "Between knowledge and action: Themes in the history of environmental chemistry," in Seymour H. Mauskopf, ed., *Chemical Sciences in the Modern World* (Philadelphia: University of Pennsylvania Press, 1993), pp. 308-310.



*A.S. Travis et al. (eds.),*
***Determinants in the Evolution of the European Chemical Industry, 1900–1939*, 3–21.**

hydroelectricity, which is why the American Cyanamid Company was set up in Ontario to exploit the power of Niagara Falls. Cyanamide and byproduct ammonia were the chief sources of nitrogen-containing fertiliser until late in World War I.[2]

Electrochemical fixation of nitrogen as its monoxide was realized in Norway, where the electric arc process of Kristian Birkeland and Samuel Eyde was inaugurated in 1903. The nitrogen monoxide formed nitrogen dioxide in the presence of air. The dioxide was reacted with water to give nitric acid, which in turn was converted into the fertiliser calcium nitrate. The yield was low, and the process could be worked commercially only because of the availability of hydroelectricity. The German coal tar dye manufacturer Badische Anilin und Soda Fabrik (BASF), of Ludwigshafen, which had become involved in nitrogen fixation during 1897, and subsequently developed the Schönherr furnace, later had a minor interest in Norsk Hydro (formed by the Norwegians in 1907).

The BASF interest represented a mode of diversification that was not shared by other German dye-manufacturing companies, which preferred pharmaceutical and organic products based on dyestuff intermediates. All these activities, dyes, pharmaceuticals, and now nitrogen fixation, relied increasingly on new trends in academic research. But while dyes and drugs remained firmly in the realms of organic chemistry, nitrogen fixation was related closely to the newly emerging subdiscipline of physical chemistry. BASF's failure to come up with a suitable process in its own laboratories encouraged the company to fund academic research into nitrogen fixation, particularly, from 1908, by Fritz Haber at the Karlsruhe Technische Hochschule.

Haber's work embraced the physical chemistry of gas reactions, following on from the equilibria studies of Le Chatelier and other chemists. Furthermore, Haber, in common with many leading German academic chemists, was well aware of the rewards that might accrue from successful industrial application of laboratory results. Haber's acute awareness of the industrial potential came over in his 1905 book *Thermodynamik technischer Gasreaktionen* (published in English as *Thermodynamics of Technical Gas Reactions*).[3] His credentials and background were well suited to the needs of BASF.

While BASF was at first interested in electrochemical fixation of nitrogen, Haber decided to return to his earlier studies started in 1903 on the catalytic formation of ammonia from its elements, nitrogen and hydrogen. During 1907, the academic work

---

[2] For carbide and cyanamide see Peter J.T. Morris, "The Development of Acetylene Chemistry and Synthetic Rubber by I.G. Farbenindustrie Aktiengesellschaft, 1926-1945," Oxford University D.Phil. thesis, 1982. For a review of nitrogen fixation see Haber, op. cit. (1), pp. 84-97, and Alfred von Nagel, *Stickstoff* (Ludwigshafen: BASF A.G., 1970).

[3] Fritz Haber, *Thermodynamik technischer Gasreaktionen. Sieben Vorträge* (Munich: R. Oldenbourg, 1905), and, translated by Arthur H. Lamb, *Thermodynamics of Technical Gas Reactions* (London: MacMillan, 1908). Haber's reputation in technical electrochemistry derived from his *Grundriss der Technischen Elektrochemie auf theoretischer Grundlage* (Munich: R. Oldenbourg, 1898).

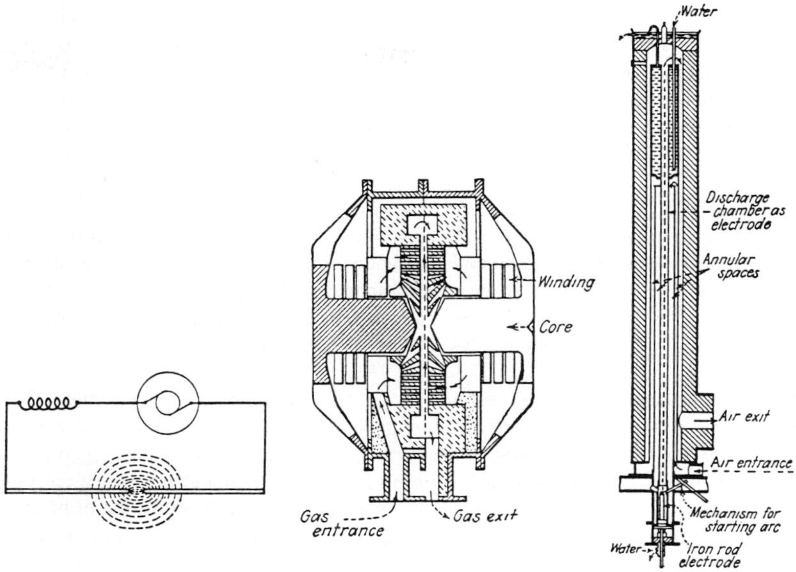


(a) (b) (c)

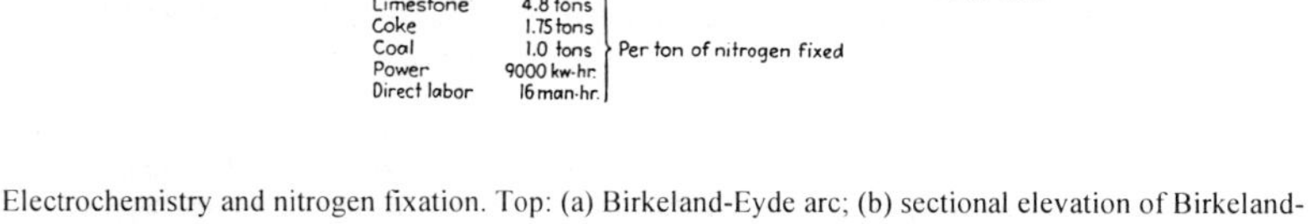


Electrochemistry and nitrogen fixation. Top: (a) Birkeland-Eyde arc; (b) sectional elevation of Birkeland-Eyde furnace, at right angles to arc disc; (c) Schönherr-Hessberger furnace. Below: manufacture of calcium cyanamide, via calcium carbide. (Source: C.L. Mantell, *Industrial Electrochemistry* [New York: McGraw-Hill, 2nd edition, 1940], pages 573, 574, 575, and 521, respectively.)

on ammonia formation brought Haber into conflict with Germany's leading physical chemist, Walther Nernst.[4] Haber's response was to attack the problem with thermodynamics and chemical kinetics, in the form of high pressures, catalysts and elevated temperatures. In this he was aided by his English assistant, Robert Le Rossignol (and a technician named Kirchenbauer). Their results were brought to the attention of BASF later in 1908.[5]

On 2 July 1909 two representatives of BASF visited Haber's laboratory at Karlsruhe. Their brief was to evaluate Haber's apparatus for the high pressure synthesis of ammonia from its elements. Even under optimum conditions the yield was low, around 5 per cent, but Haber had arranged for unreacted hydrogen and nitrogen to be recirculated. Though exothermic, the reaction was carried out at 600°C in order to increase the rate. The preferred catalyst was osmium or uranium.

The benchtop apparatus was dominated by the furnace (later known as converter) in which the reactant gases were preheated by the outgoing reaction mixture. The reactant gases were brought to the desired temperature by means of an electric heating coil, and then, at a pressure of 200 atmospheres, forced into contact with catalyst, when reaction took place. The cooling that occurred moved the equilibrium in the direction of product ammonia. The outgoing mixture of ammonia, hydrogen and nitrogen was dried over soda lime, passed through a heat exchanger, and cooled down in order to liquefy the ammonia and permit its separation from hydrogen and nitrogen. The unreacted gases were then recirculated with a fresh charge. The circulation of gases relied on a small double-acting steel pump, the efficiency of which was restricted by the leather packing on the piston. The apparatus was fitted with a pressure gauge, and the ammonia was determined using Rayleigh's gas interferometer, made by Zeiss.[6]

---

4 For Nernst's participation, including with BASF, see L. Suhling, "Walther Nernst und die Ammoniaksynthese nach Haber und Bosch," in Helmuth Albrecht, ed., *Naturwissenschaft und Technik in der Geschichte. 25 Jahre Lehrstuhl für Geschichte der Naturwissenschaft und Technik am Historischen Institut der Universität Stuttgart* (Stuttgart: Verlag für Geschichte der Naturwissenschaften und der Technik, 1993), pp. 343-356.

5 For Haber see Dietrich Stoltzenberg, *Fritz Haber: Chemiker, Nobelpreisträger, Deutscher, Jude* (Weinheim: VCH, 1994). Nitrogen fixation is covered in chapter 5, "Die Fixierung des Stickstoffs," pp. 133-197. Studies on Haber normally place great emphasis on his controversial gas warfare work. This is often used to introduce a strong political dimension, as for example in the novel by Hermann Heinz Wille, *Der Januskopf: Leben und Wirken des Physikochemikers und Nobelpreisträgers Fritz Haber* (Berlin: Verlag Neues Leben, 1969). See also Ute Deichmann, "Dem Vaterlande - solange es dies wünscht," *Chemie in unserer Zeit*, **30** (1996), 141-149. While several articles and monographs on Haber have appeared in German, there is no full study in the English language, though the following are useful: Morris Goran, *The Story of Fritz Haber* (Oklahama: Norman, 1967), and L.F. Haber, *The Poisonous Cloud: Chemical Warfare in the First World War* (Oxford: Clarendon, 1986). An English language version of Stoltzenberg's book will appear in the American Chemical Society-Chemical Heritage Foundation "History of the Modern Chemical Sciences Series."

6 A comprehensive description of the 1909 apparatus appears in "Journal and patent literature. Ammonia -

The BASF representatives were chemist Carl Bosch and catalyst expert Alwin Mittasch.[7] They were there to find out whether the backing for Haber's nitrogen fixation studies by the head of the managing board, Heinrich von Brunck, had finally paid of. Though the apparatus had been tested and found successful, the initial outcome of the demonstration seemed disappointing; the apparatus refused to work. Carl Bosch left to attend to other business. Mittasch remained in the laboratory and later in the day was rewarded when the apparatus delivered 100 cubic centimetres of ammonia. He was convinced: Haber had overcome the laboratory difficulties involved in the synthesis of ammonia. Patents for the continuous process were filed in Germany and elsewhere, and Haber came to an agreement with BASF over royalties.[8]

The academic foundations of high pressure chemistry. Fritz Haber and colleagues at the Technische Hochschule Karlsruhe in 1909. Haber is seated at the centre of the first row. Seated on the ground is the technician Kirchenbauer. Robert Le Rossignol is second from the left, first row. Friedrich Bergius, pioneer of high pressure coal-to-oil conversion, is standing at extreme left. (Courtesy of BASF Ludwigshafen.)

---

Technical preparation of, from its elements," *Journal of the Society of Chemical Industry*, **32** (1913), 134-138. For the Rayleigh gas interferometer see George Lunge, *Technical Gas Analysis*, revised and rewritten by H.R. Ambler (London: Gurney and Jackson, 1934), pp. 198-199.

7 For Bosch see Karl Holdermann, *Im Banne der Chemie: Carl Bosch, Leben und Werk* (Düsseldorf: Econ Verlag, 1953).

8 The most detailed study of Haber's relationship with BASF is Carsten Reinhardt, "Über Wissenschaft und Wirtschaft. Fritz Habers Zusammenarbeit mit der BASF 1908 bis 1911," in Albrecht, op. cit. (4), *Naturwissenschaft und Technik in der Geschichte*, pp. 287-315. This is also a useful source of references to the German literature on Haber's ammonia work, including historical and historiographical accounts.

The stage was now set for the development of high pressure industrial chemistry; it would emerge as a dominant force throughout the period covered in this volume and beyond. But first the difficulties of scaling up Haber's 0.75m high converter to a pilot plant had to be overcome. That was the formidable challenge confronted by Bosch and colleagues, especially chief engineer Franz Lappe. They had to seek out cheaper catalysts than osmium and uranium, but with similar levels of activity, to build reactors to withstand high temperatures and pressures, and to establish inexpensive sources of nitrogen and hydrogen in adequate amounts and that could be made pure.

**The catalysts**

Alwin Mittasch and Georg Stern undertook catalyst experiments at Ludwigshafen in miniature high pressure tubes designed by Stern. These were heated electrically and air cooled, and designed to accept cartridges containing about two grams of test catalyst. Bosch encouraged the study of iron after examining a recently published atlas of spectra, suggesting that "[A]n element which has such a complicated emission spectrum should be of some use."[9] It did not take long to find catalysts based on iron compounds, some, such as a Swedish magnetite, available just by chance. In 1910, an iron-aluminium catalyst with activity close to that of osmium and uranium was chosen for

Replica of Haber-Le Rossignol apparatus of 1909 at the Deutsches Museum, Munich. At right is a BASF apparatus for testing catalysts, as employed by Mittasch. (Courtesy Deutsches Museum, Munich.)

[9] Quoted in Max Appl, "The Haber-Bosch process and the development of chemical engineering," in W.F. Furter, ed., *A Century of Chemical Engineering* (New York: Plenum, 1982), pp. 29-53, on p. 39. Appl, former head of the BASF ammonia department, provides an excellent overview of the technical development of the Haber-Bosch process. For work on catalysts see also Alwin Mittasch, *Geschichte der Ammoniaksynthese* (Weinheim: Verlag Chemie, 1951).

further development. A stable iron catalyst promoted with aluminium and potassium was found to be successful in 1911, and Mittasch soon added calcium as a third promoter. Related work concerned the careful study of catalyst poisons. By 1912, some 2,500 substances had been examined in 6,500 experiments. In this way Mittasch came up with the catalyst that would be favoured in industrial synthesis of ammonia, and established a strategy for screening catalysts that was later widely adopted. Around thirty pressure tubes saw continuous activity until around 1920, by which time there had been 20,000 test runs.

**The converter**

Carl Bosch took on the task of scaling up the converter. Stainless steel, that would have avoided many problems, was not available. What was available was the type of steel that withstood the great heat and pressures of massive cannons, though cannons were not subjected to hydrogen under such conditions. Explosions of the test converters were frequent, caused by embrittlement as carbon in the iron of the converter combined with

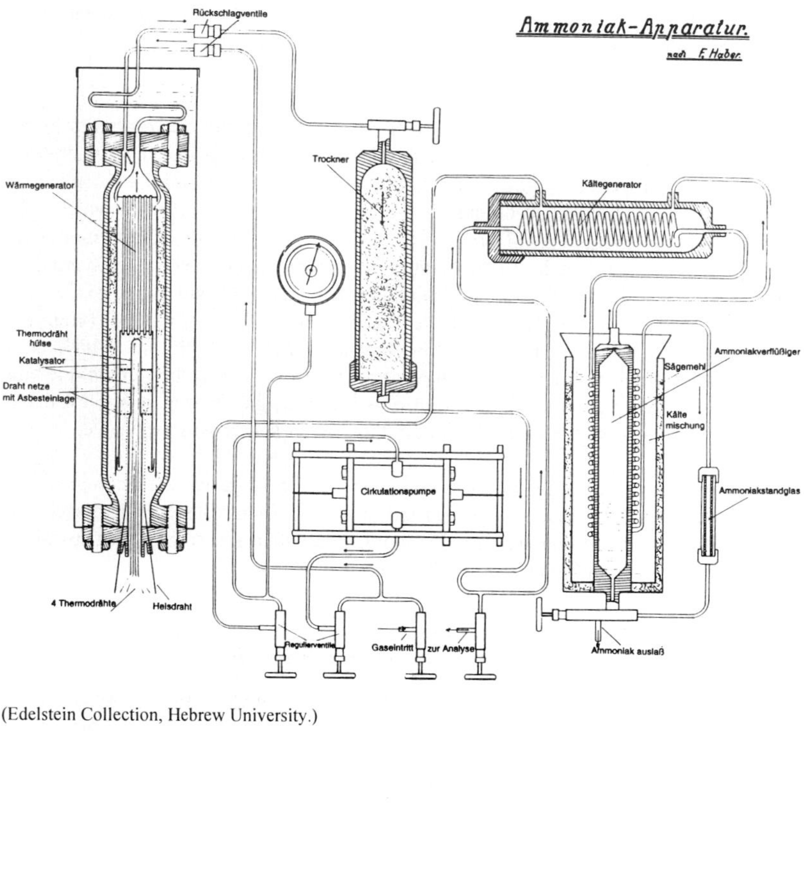


(Edelstein Collection, Hebrew University.)

hydrogen on the inner surface. The first test converters, made by Mannesmann, self destructed after 80 hours, though fortunately they were not operated with the precious metal osmium as catalyst, which Bosch noted might have led to loss of almost the entire world supply. The first pilot plant converters were 1.8m high and weighed 0.3 tons.

The main step forward was achieved in 1911, when Bosch constructed a double wall converter. The inner wall was made of low carbon soft steel, and allowed hydrogen to diffuse through it. The interior of the inner wall was packed with catalyst. The outer wall was of ordinary steel. Hydrogen diffusing through the inner wall underwent loss of pressure and then came into contact with the outer wall, which though hot from external heating did not become brittle. Internal heating of the converter by combustion of gas was found to be safe, though this was later replaced by internal electrical heating. With these arrangements, the stress on the outer wall was reduced considerably, particularly when a little hydrogen was allowed to escape into the atmosphere through small holes (Bosch holes).

There were several other innovations, including grooves on the outer surface of the inner wall in which hydrogen accumulated and was then swept away by a current of nitrogen. This contributed towards a reduction of pressure on both sides of the inner wall, further minimizing any reaction between hydrogen and carbon.[10]

**Hydrogen, nitrogen, and compressors**

During the early pilot plant stage, Bosch used hydrogen from the electrolysis of brine. However, in 1910, hydrogen became available from the reaction between steam and red hot coke, which afforded hydrogen and carbon monoxide. Liquid nitrogen condensed out the carbon monoxide, according to the Linde-Frank-Caro process, which was investigated by W. Wild at BASF from 1912, and developed into the carbon monoxide shift conversion, perfected in 1915. In this, the carbon monoxide was converted to carbon dioxide with steam over an iron-chromium catalyst, and the dioxide was removed by scrubbing. It was called the Bosch-Wild or Bamag process (after the engineering firm Berlin-Anhaltische Maschinenbau A.G., of Berlin, which was responsible for much of the special equipment). The mixture of carbon monoxide and hydrogen was known as water gas, and also as synthesis gas. Nitrogen was available from the Linde process for liquefaction of air, though this source was expensive.

The most important part of the physical processing of the gases was their compression. To achieve this under previously untried industrial conditions Bosch had to seek out powerful leak-proof gas compressors, far more robust than those employed in mines and the low temperature Linde liquefaction process.[11]

---

[10] Carl Bosch, *Nobel Lectures: Chemistry, 1922-1941* (Amsterdam: Elsevier, 1966), pp. 326-340.

[11] Air liquefaction processes, compressors, gas cylinders and allied equipment available in the early 1900s are described and illustrated in Georges Claude, *Air Liquide. Oxygène. Azote* (Paris: H. Dunod and E. Pinat,

**The BASF decision**

The success of Bosch's work by 1912 convinced Brunck that BASF should drop all other nitrogen fixation work, mainly the electrochemical combination of nitrogen with oxygen, and concentrate on bringing the Haber process to perfection. Pressure-tight joints and new valves and fittings were designed and constructed. A multitube converter was made up, not unlike the arrangement of tubes used in the contact process (sulphuric acid) apparatus (and which in turn had been copied from steam locomotive technology). This was an important feature of the converter, since it maximized heat exchange. Moreover, the material balance, concerning gas flows and compositions of streams, and the energy balance, related to heat and compression requirements, were important factors in the design and construction of full size plant. The recycling of unreacted gases introduced novel approaches to optimization and control loops.[12] Monitoring of the various physical processes and the chemical process that took place in the reactor required instruments for measuring gas flow at high pressures, gas density, and product mixture composition. Fast-acting magnetic shut-off valves were also required. It was almost new technology from start to finish.

To varying extents, several technological components, not all of them related to mainstream chemical manufacture, converged to create the prerequisite for extremely high pressure chemical synthesis involving the reactive hydrogen. These extended back to improvements in iron for cannons and seige guns after the Crimean war, and led to mild steels (Henry Bessemer 1856, and Sidney Gilchrist Thomas, 1879), the introduction of compressors in mining and tunnelling from the 1860s, the design of locomotive boilers, and the availability of liquefaction and refrigeration units, gas engines and powerful electric motors. Steels from Alfred Krupp, an early user of the Bessemer process (and whose son Friedrich introduced chrome nickel steel), and Rudolf Diesel's engine (1892), which operated at high temperatures and pressures, also prepared the way. Of more than passing interest is the fact that in 1876 Carl von Linde used the physical properties of ammonia to employ it as a refrigerant.

To these should be added the use of steam boilers and pressurized chemical process equipment operated at around 50 atmospheres that had become widespread from the late 1860s, initially in the manufacture of dye intermediates, and in the early synthesis of alizarin (1870), and blowers used in the Solvay process.[13] Last but not least, was progress in coal technology, including the growth of producer gas generators to satisfy

---

1909).

[12] These are discussed without excessive technical detail in M.D. Wynne, *Chemical Processing in Industry* (London: Royal Institute of Chemistry, Monographs for Teachers, no. 16, 1970), pp. 1-8.

[13] Willem J. Hornix, "From process to plant: Innovation in the early artificial dye industry," *British Journal for the History of Science*, **25** (1990), 65-90, esp. 75-81.

the demand for gas engines (that were widely used to power compressors), and water gas production.

At the organizational level, BASF was the systems leader. It was replete with well equipped workshops, had pioneered academic-industrial collaborations, and inaugurated industrial research laboratories dedicated to the realization of new innovations. BASF engaged in risky ventures, exemplified by the manufacture of alizarin and indigo, and acquired a well deserved reputation for seeing such ventures through to their satisfactory conclusions. Furthermore, through the indigo work, BASF was introduced to both industrial electrochemistry, for alkali production, and novel gas reactions in the presence of catalysts for preparing sulphuric acid (contact process). Electrochemical production of alkali also gave hydrogen as one of the products. Unknown, however, at least until Bosch began his investigations, was the problem of steel fatigue resulting from combination of reactive hydrogen with carbon in the steel at four times the pressure of early alizarin reactors (in which hydrogen was produced), and eighteen times the working pressure of a locomotive boiler.

Despite the technical progress made possible by these various advances, and the investigations in the BASF laboratories and workshops, Bosch almost confronted a major setback when Brunck died in 1911. The other board members were less comfortable with the high pressure process, particularly with the great strain on the company's financial resources. Fortunately for Bosch, work on the new Oppau ammonia factory, north of Ludwigshafen, was advancing rapidly. It opened in September 1913. The 8m high, 8.5 ton converter proved its worth, and began turning out ammonia. An alternative and much cheaper source of nitrogen had became available by the action of air on hot coke. The product mixture was carbon monoxide and nitrogen (producer gas). Later, a mixture of nitrogen and hydrogen in the proportions required by the chemical equation (1:3) was made from producer gas, water gas and steam. By 1915, ammonia converters of 12m height and 75 tons in weight were in operation. In the spring of that year synthetic ammonia was converted into nitric acid for munitions production in Germany.[14] A second, larger, plant was opened in 1917 at Leuna, near Merseburg, and close to deposits of lignite.

**Post-war impact**

The success of the Haber-Bosch process from 1915 in sustaining the output of German nitro compounds for explosives certainly impressed the Allies, who made little headway in attempts to imitate the process before 1918. After the war, high pressure synthetic ammonia and dyestuffs plants both became strategic sectors of economies throughout the world. Whereas dyestuffs technology had already matured and was readily mastered, the high pressure process continued to elude investigators outside Germany,

---

14 Anthony S. Travis, "The Haber-Bosch process: Exemplar of 20th century chemical industry," *Chemistry & Industry*, 2 August 1993, pp. 581-585.

and even those who made official visits to German factories.[15]

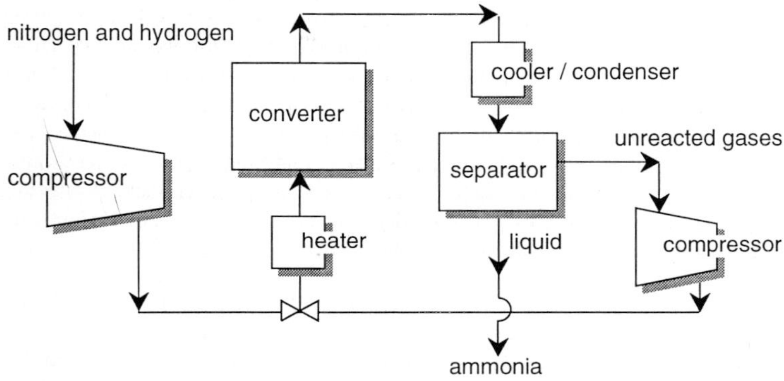


Flow chart for Haber-Bosch process.

Certainly, the great financial investment required did not deter imitators, nor did the modes for acquiring the necessary technical information. Thus Brunner, Mond & Co. in England succeeded in opening an ammonia factory at Billingham in 1923 only after it purchased the secrets from two Alsatian engineers who had worked at German factories during the war.[16]

Luigi Casale and Giacomo Fauser in Italy and Georges Claude in France independently of each other invented high pressure ammonia processes. The Italian and French processes differed from the Haber-Bosch process in that they employed higher pressures and different catalysts. Casale was a former assistant of Walter Nernst, whose process, which operated at 650-750 atmospheres, and incorporated a circulating system similar to that used in the Haber-Bosch process, was invented in 1924. It was rightly seen as a threat by BASF (and later I.G. Farben). The Casale process as operated at Union Chimique Belge, in Ostend, used hydrogen from coke oven gas.

---

[15] See Lothar Meinzer's contribution to this volume.

[16] William J. Reader, *Imperial Chemical Industries: A History*. 2 vols. (London: Oxford University Press, 1970-75), vol. 1: *The Forerunners, 1870-1926*, pp. 347-370. The British Nitrogen Products Committee, set up in 1916, was in many ways a forerunner of the postwar research associations. Research was carried out at University College London, under the direction of H.C. Greenwood, who had worked in Haber's laboratory. However, little of industrial value was achieved beyond a number of gas studies.

Fauser, a consulting engineer, had at first worked independently, using a 250mm cannon as reactor in the garden of his home, and, apparently, without any of the information about the Haber-Bosch process that was gleaned by other investigators.[17] From 1925, his studies were backed by Montecatini, and a viable ammonia process working at 300 atmospheres was established. Montecatini, at least at its Merano factory, in the north of the country and close to cheap hydroelectric power, relied on hydrogen obtained by electrolysis of water. The gas was quite pure, and traces of oxygen were removed by reaction with the hydrogen over a platinized asbestos catalyst. Later, a copper deoxidizer was employed. The Fauser process was made available outside Italy by the Fauser-Montecatini licensing and consulting firm. It was first adopted, during 1928-29, in Belgium (Coppe) and Holland (Sluiskil), and soon after in British Columbia (Trail), Poland, Czechoslovakia, Japan, France, and, again, in Holland (Lutterade).[18] Poland's total nitrogen capacity was considerable; apart from cyanamide plants, there were three synthetic ammonia factories in Upper Silesia.[19]

Georges Claude, whose work on liquefaction had improved the fortunes of Société L'Air Liquide, investigated high pressure ammonia synthesis from 1917 using pressures of around 1,000 atmospheres. This was later tested at the Monterau plant of L'Air Liquide, and its success enabled L'Air Liquide to participate in the formation of Société Belge de l'Azote in 1923.[20] Hydrogen for the Claude process normally came from coke oven gas. The ammonia process was worked such that the gases were passed through a number of converters placed in series, without recirculation. Because of the high pressure the yield was high, and only 10 per cent of unreacted nitrogen and hydrogen remained after passage through four convertors. By 1929, five French factories employed the indigenous Claude process and nine the imported Casale process. With technology made available from BASF, a state owned ammonia factory employing Haber-Bosch technology was opened at Toulouse.[21] In 1929, also, the Haber-Bosch process was introduced into Norway by a 1927 arrangement between I.G. Farben and

---

17 Basil G. Achilladelis, "Process Innovation in the Chemical Industry," University of Sussex Ph.D. thesis, 1973, pp. 148-151.

18 For the Trail plant see "Canadian nitrogen fixation," *Chemistry and Industry*, **48** (1929), 769. The Compagnie Néerlandaise de l'Azote, formed in Brussels during 1929, opened the Sluiskil factory. The Dutch State Mines operated the Lutterade factory.

19 The Moscice State Nitrogen Compounds Factory, Tarnow, Poland, which opened on 18 January 1930, was named after the then president of the Republic of Poland and expert on nitrogen fixation, Ignacy Moscicki. It employed Fauser technology. For the diffusion of synthetic ammonia technology see L.F. Haber, *The Chemical Industry 1900-1930*, op. cit. (1), pp. 325-328.

20 Fred Aftalion, *A History of the International Chemical Industry*, translated by O.T. Benfey (Philadelphia: University of Pennsylvania Press, 1991), pp. 92-93, 133.

21 For the background to the adoption of the Haber-Bosch process at the French state factory see Maurice Deschiens, "The position of the nitrogen industry in France," *Chemistry and Industry*, **43** (1924), 507-508.

Norsk Hydro. At this time six Italian factories used Fauser, one used Casale, and two used Claude processes.[22]

Fauser, Casale and Claude processes were also used in Japan, where there were close links with I.G. Farben, including licensing arrangements over the Haber-Bosch process.[23] In 1925, Soviet high pressure specialist Vladimir N. Ipatieff visited German and Italian factories to investigate possibilities for licensing nitrogen fixation processes that could be employed in the Soviet Union.[24] In the following year he was appointed consultant to German industrial firms engaged in ammonia synthesis.

A consortium of German mining companies, unable to come to an agreement with BASF over the Haber-Bosch process, funded the development of what became known as the Mont Cenis process for ammonia, which operated at only 100 atmospheres (1926-27). The patent holder was Gasverarbeitungs-Gesellschaft m.b.H. (GAVEG) of Herne-Sodingen, Westphalia. Subsequent patent litigation with I.G. Farben revealed the nature of the catalysts. In 1927, Royal Dutch/Shell, anxious to diversify into chemical production and aware of the potential market for fertilisers in the still highly agricultural Netherlands, investigated both Mont Cenis and Casale processes, and late in the year opted for the former. Its IJmuiden plant began to operate in September 1929. Despite many initial difficulties, involving poisoned catalysts, and problems with compressors, heat exchangers, gas losses, and refrigeration units, the improvements made over the following years enabled a break with the German patent holders in February 1934, and the decision not to change over to I.G. Farben Haber-Bosch technology. The IJmuiden factory was supplied with hydrogen from coke oven gas. Hydrogen from petroleum gases was considered for ammonia plants outside Holland, though plans for these were shelved.[25]

Research into high pressure ammonia synthesis had started in the United States during

---

22 For developments in ammonia synthesis during the late 1920s, and in particular of compressors see "High pressures in the manufacture of synthetic ammonia. Recent designs in compressing plant," *Chemistry and Industry*, **48** (1929), 591-598.

23 Akira Kudo, "Japanese technology absorption of the Haber-Bosch Method: The case of the Taki Fertilizer Works," in David J. Jeremy, ed., *The Transfer of International Technology: Europe, Japan and the USA in the Twentieth Century* (Aldershot: Edward Elgar, 1992), pp. 33-56; and *idem*, "I.G. Farben in Japan: The transfer of technology and managerial skills," *Business History*, **36** (1994), 159-183.

24 Vladimir N. Ipatieff, *The Life of a Chemist: Memoirs of Vladimir N. Ipatieff*, edited by J. Eudin, et al., and translated by Vladimir Haensel and Mrs. Ralph H. Lusher (Stanford: Stanford University Press, 1946), pp. 409-413. Some useful historical information on the nitrogen fixation industry in the USSR will be found in Mathew J. Sagers and Theodore Shabad, *The Chemical Industry in the USSR: An Economic Geography*. ACS Professional Reference Book (Boulder, Co.: Westview Press, 1990), esp. pp. 115-117.

25 R.J. Forbes and D.R. O'Beirne, *The Technical Development of the Royal Dutch/Shell 1890-1940* (Leiden: E.J. Brill, 1957), pp. 503-507.

1913 at General Chemical Co., which after the war opted for a process similar to that employed by BASF. In 1920, General Chemical in cooperation with Solvay Process Co. formed the Atmospheric Nitrogen Corporation. In the same year General Chemical and Solvay interests were merged, with others, into Allied Chemical & Dye Corporation. Atmospheric Nitrogen was one of several corporations, some representing European interests, whose titles proclaimed their close connections with nitrogen fixation. By contrast, the American designed process of Nitrogen Engineering Corporation was used in Europe as well as in the United States. Thus in 1925 a small plant was erected in Norway (though future expansion there relied on Haber-Bosch technology).[26] The process was also adopted by Etablissements Kuhlmann near Paris. More impressive was the series of orders from Russia and the Far East, especially after Nitrogen Engineering was absorbed by the American Cyanamid Company's subsidiary Chemical Engineering Corporation, which offered design and construction services for high pressure plants and equipment. In 1916, federal involvement had begun when the Department of Agriculture commenced experiments on ammonia synthesis. Three years later the Secretary of War converted one laboratory of the Chemical Warfare Service into a fixed nitrogen laboratory. This was taken over by the Department of Agriculture in 1921. By 1924, the Fixed Nitrogen Research Laboratory, in Washington, DC, had developed an ammonia process, but this was worked only on a small scale at two sites. Allied Chemical and Du Pont, which at first adopted Claude technology but soon moved over to the Casale process, were the main producers of ammonia in the United States.[27] The Ammonia Corporation owned United States rights to the Casale process, and its subsidiary Niagara Ammonia Company operated an ammonia factory at Niagara Falls (1924).[28]

From 1927, there was massive international growth in ammonia output, propelled by the perceived need to prove the worth of the new high pressure technology and overoptimistic forecasts of demand for fertilisers. This turned into a substantial glut by 1929; the situation did not improve for many years, not even with the formation of an international cartel, initiated by German, British and Norwegian interests.[29] In the

---

26 Haber, op. cit. (1), p. 327.

27 David A. Hounshell and John Kenly Smith, Jr., *Science and Corporate Strategy: Du Pont R&D, 1902-1980* (Cambridge: Cambridge University Press, 1988), pp. 183-189, esp. 186. For a less than dispassionate view of the situation in the United States where, as in Europe, companies and inventors jealously fought over the rival merits of ammonia processes, see H. Wigglesworth (representing General Chemical's point of view), "The nitrogen industry in the United States," *Chemistry and Industry*, **46** (1927), 313-315.

28 For a review of nitrogen fixation in the United States see Williams Haynes, *American Chemical Industry*. 6 vols., vol. 4. *The Merger Era, 1923-1929* (Toronto: D. Van Nostrand Co., 1948), especially chapter 6, "Nitric acid from ammonia," pp. 85-95.

29 The French avoided involvement in the cartel, as did the Belgians, at least until the International Nitrogen Conference held in Ostend during the summer of 1930. Other nations were also concerned about the impact of curbing production, in accord with the demands of the nitrogen cartel, since they wished to maintain supplies for purposes of national defense. For the nitrogen cartel see Harm Schröter,

United States, ammonia firms reacted by turning to oxidation of ammonia to afford nitric acid by the Ostwald process as pioneered at BASF during 1915. The demand was so great that in 1928 the first special purpose nitric acid railroad tank car was introduced.[30] Ammonia for refrigeration purposes continued to be a good market, particularly in Japan.

An estimate of world ammonia capacity for 1932-33, showed that the Haber-Bosch process represented 53 per cent, Casale 16 per cent, Fauser 11 per cent, Claude 8 per cent, Mont Cenis 6 per cent, Nitrogen Engineering Corporation (USA) 5 per cent, and Showa Hiryo (Japan) 1 per cent. Together the various operational plants had the ability to produce three million tons of ammonia nitrogen, though by 1935 only 43 per cent of this figure was actually manufactured.[31]

High pressure technology transfer and diversification took many avenues, though most new innovations continued to appear from BASF. First, in 1923, was methanol production at the Leuna ammonia factory, and based on the work of Matthias Pier.[32] BASF had patented a high pressure methanol process in 1914, but no further studies were carried out until after G. Patart in France applied for a similar patent (1921). In this case the same equipment could be used to manufacture ammonia or methanol, according to demand. Synthesis gas, the mixture of hydrogen and carbon monoxide, was used directly, without separation, to prepare methanol. In a similar way, isopropanol was manufactured under high pressures.

---

"Privatwirtschaftliche Marketregulierung und staatliche Interessenpolitik. Das internationale Stickstoffkartell 1929-1939," in H. Schröter and C. Wurm, eds, *Politik, Wirtschaft und internationale Beziehungen. Studien zu ihrem Verhältnis in der Zeit zwischen den beiden Weltkriegen* (Mainz: Verlag Philipp von Zabern, 1991), pp. 117-137. The Deutsch-Englisch-Norwegische Gruppe (DEN), comprised of I.G. Farben, ICI and Norsk Hydro, was created on 1 July 1929, and the agreement was for ten years This was followed with the Convention Internationale de l'Azote (CIA) formed for one year on 1 July 1930 by DEN, France, Belgium, the Netherlands, Italy, Poland and Czechoslovakia. Chilean nitrate interests were not involved, but did contribute to the CIA fund. The CIA succeeded in maintaining prices for its members, but there was a crisis when non-members cut prices. This created problems for all countries except those within DEN. The CIA arrangement was renewed on 1 July 1932, providing even better conditions for DEN. The number of member countries grew until 1939 (Chile joined in 1938). Overall, the DEN came out of this period relatively unscathed, while other members faced great difficulties. I thank Harm Schröter for providing this summary. See Reader, *Imperial Chemical Industries*, op. cit. (16), vol. 2, *The First Quarter-Century, 1926-1952*, pp. 98-115, 145-159, and Haynes, *American Chemical Industry*, op. cit. (28), vol. 5. *Decade of New Products, 1930-1939* (1954), chapter 8, "Nitrogen under political intervention," pp. 101-118.

30 Haynes, op. cit. (28), pp. 91-94.

31 Jocelyn F. Thorpe and M.A. Whiteley, *Thorpe's Dictionary of Applied Chemistry* (London: Longmans, Green & Co., 4th edition, 1937), vol. 1, p. 332. By 1935, world production capacity stood at around five million tons of ammonia nitrogen.

32 A. Von Nagel, *Methanol, Treibstoffe* (Ludwigshafen: BASF A.G., 1970), pp. 7-11.

While ammonia production was a sound investment, high pressure coal to oil conversion faced an uncertain future. First developed by Friedrich Bergius in 1913, the coal to oil process was adopted by BASF in the 1920s. The process was brought close to perfection at BASF but was realized on a large scale only once it became part of Hitler's autarky programme. In the meantime, by a complex system of licensing arrangements, the technology was made available to Brunner, Mond & Co. (and later ICI), and Standard Oil of New Jersey.[33]

By the mid 1920s, then, high pressure chemistry had become the fastest growing sector of the post World War I chemical industry. Just as the momentum, ideological as well as organizational, created by the success of electrochemical processes had stimulated Crookes and others to investigate nitrogen fixation, both as nitrogen monoxide and cyanamide, and using electricity, the momentum created by the high pressure ammonia process was the incentive for extending academic and industrial research into gas reactions under high pressures.[34] Notable industrial outcomes were the ICI process for polyethylene (polythene), and Reppe's remarkable acetylene chemistry and related reactions on olefines in Germany.[35] These were followed by more widespread applications of high pressure processes to the manufacture of organic chemicals.

**Conclusion**

Though their contributions were both different and separate, Fritz Haber and Carl Bosch are jointly remembered through the high pressure ammonia process that carries their names. For them, the 'myth' of the heroic inventor tallies with much of what actually took place. This is notwithstanding the fact that high pressure industrial chemistry derived from a nexus of changing needs, new corporate and public ideologies, new technologies, and new directions in academic chemistry. A measure of the level of success of high pressure ammonia synthesis at BASF, and also of the tremendous prestige accorded to Haber and Bosch, was the total failure to imitate the process elsewhere before 1918, and the status of postwar commissions sent to examine the BASF ammonia plant. The British failed on a technical level when they sought to achieve the ambitious goal of catching up with BASF from scratch. Inspections by

---

[33] In 1929, I.G. Farben produced 70,000 tons of gasoline by hydrogenation, of which 40,000 tons were derived from coal. At Birmingham University, the Mining Research Laboratory was undertaking research into catalysts for hydrogenation of coal, supported with a grant from the British Colliery Owners' Research Association. Anthony N. Stranges has written extensively on the German-inspired production of synthetic fuels from coal, most recently "The US Bureau of Mines' synthetic fuel programme, 1920-1950s: German connections and American advances," *Annals of Science*, **54** (1997), 29-68. For ICI's involvement see Reader, *Imperial Chemical Industries*, op. cit. (16), vol. 2: *The First Quarter-Century, 1926-1952*, pp. 162-182.

[34] Thomas P. Hughes, "Technological momentum in history: Hydrogenation in Germany 1898-1933," *Past and Present*, **44** (1969), 106-132.

[35] See Morris's contribution to this volume.

British and French chemical experts, apart from hindrances placed in the way by BASF, yielded little of value; the technology was just too innovative.

Placed in its proper context, Haber's laboratory work, stimulated by academic polemics, demonstrated the possibilities for high pressure ammonia synthesis on an industrial scale. Unlike most other chemists whose activities straddled the worlds of academic and applied organic chemistry Haber opted for the uncertainties of physical chemistry. Not only did he show by clever manipulation of thermodynamics and kinetics that the reaction could be made to take place in yields that others considered impossible, but he also developed the process to the extent that the reaction rate was attractive to industrial chemists. Haber also knew that the only way to convince them would be to devise a continuous apparatus in which the valuable unreacted gases were recirculated to the converter. In July 1909 this apparatus, demonstrated at Karlsruhe, served its intended purpose. Haber's reward was a profit sharing arrangement with BASF, directorship of the new Kaiser Wilhelm-Institut für Physikalische Chemie und Elektrochemie in Berlin (1912), and the Nobel Prize (1918).

Haber, however, played no part in the industrial scale up. That took a team of BASF chemists, metallurgists, and proto-chemical engineers, led by Bosch, delving into a wide range of problems concerning materials, machines, catalysts, instruments, gas supply and purification, and drawing extensively on prior developments in chemical and other technologies, four and a half years to realize. The pursuit of a single invention had, as never before, relied on so many skills and so many individuals.[36] It was driven to a satisfactory conclusion under one unyielding and tenacious leader, Carl Bosch, whose brute force process bound nitrogen with hydrogen on a commercial scale. Bosch's role was recognized by senior management, and his reward was rapid promotion: He became chairman of the board of BASF in 1919, and of I.G. Farben from 1925-35. His role in high pressure synthesis was also acknowledged with the Nobel Prize, received jointly with Friedrich Bergius (1931). This also represented international recognition of the status of high pressure industrial chemistry around 1930.

In summary, high pressure ammonia synthesis was nothing less than another major step towards the industrialization of academic chemistry on the lines pioneered by BASF during 1869-70 with the scale up of the alizarin synthesis, and during the 1890s with the production of synthetic indigo.[37] This created a tradition for highly risky financial

---

[36] See the chapter by Timothy Lenoir "Praktische Vernunft und die soziale Konstruktion von Erkenntnis," in his *Politik im Tempel der Wissenschaft. Forschung und Machtausübung im deutschen Kaisereich* (Frankfurt: Campus Verlag, 1992, edition Pandora, 2), pp. 146-171, and *idem*, "Practical reason and the construction of knowledge: The lifeworld of Haber-Bosch," in Ernan McMullin, ed., *The Social Dimensions of Science* (Notre Dame, Ind.: 1992), pp. 158-197.

[37] Anthony S. Travis, *The Rainbow Makers: The Origins of the Synthetic Dyestuffs Industry in Western Europe* (Bethlehem, Pa.: Lehigh University Press, 1993).

backing of new and untried technical innovations that, once diverted from dyestuffs to high pressure manufacture, ensured BASF's dominant position between the two world wars, and provided the role model for others to emulate.[38] High pressure chemistry also represented the move forward into the 20th century for the chemical industry. The combination of high pressures, high temperatures and catalysts presented challenges that were completely different to those encountered in the manufacture of dyes and traditional heavy organic chemicals. These challenges could only be met with the construction of novel pilot plants, studies on the control and regulation of gas and gas-solid reactions, and the design and testing of full size plant based on the latest engineering practice, including massive converters and compressors, and networks of pipelines.

It was mainly the BASF reticence to license the ammonia process, or release details, that led to the development of significant alternative high pressure ammonia processes, particularly in France, Italy and the United States, and competition from companies anxious to export that technology.

Moreover, the spectacular success achieved with high pressure technology decisively shaped the international view of BASF throughout the interwar period. This became particularly true with attitudes towards coal to oil conversion, in which cash starved I.G. Farben had little trouble in enticing foreign firms into partnership arrangements that proved beneficial only to the German partner. It also inspired others to make risky breaks with the past, including Royal Dutch/Shell whose J.B. August Kessler, anxious to diversify into petrochemicals, advised colleagues in 1927:

> I feel that we should look upon these new departures of ours in an entirely different light from the way we look upon building, say, another cracking unit or another lubricating oil bench. If the calculations show that, for the time being, a new chemical installation for making nitrogen (the binding of nitrogen from the air to hydrogen from natural gas or cracked gases) will not pay, I do not think that should be a reason not to build it. *We should have confidence, energy and courage enough to develop this new chemical part of our business, even if it does not yield profits to start with. The Badische have done the same thing with very great success.* [Emphasis added.][39]

Kessler's bold statement was, however, hardly prescient: High pressure chemistry had

[38] Karl Schoenemann, "The separate development of chemical engineering in Germany," in W.F. Furter, ed., *History of Chemical Engineering* (Washington, DC: American Chemical Society, 1980), pp. 249-271.

[39] Cited from Forbes and O'Beirne, op. cit. (25), p. 456. Royal Dutch/Shell, in common with ICI and Standard Oil of New Jersey, also became closely involved with I.G. Farben in negotiations over high pressure processes for coal to oil conversion. This and other aspects of high pressure industrial chemistry during the interwar period are reviewed in Anthony S. Travis, *The High Pressure Chemists* (Wembley: Brent Schools & Industry Project, 1984).

already become the undisputed leitmotiv of the interwar chemical industry. To neglect it was costly. To adopt it was also costly, but less so in the long term.

**Acknowledgements**
Ernst Homburg, Lothar Meinzer, Peter J.T. Morris, Carsten Reinhardt and Harm G. Schröter are thanked for providing information that has been invaluable to this contribution, and also for critical comments.

# Part 2

## The Impact and Burden of World War I

# CHEMISTRY FOR KING AND KAISER: REVISITING CHEMICAL ENTERPRISE AND THE EUROPEAN WAR

ROY M. MacLEOD
*Department of History*
*University of Sydney*
*NSW Australia 2006*

## Introduction*

Today, it is widely acknowledged that war is primarily a question of technology. From this, as J.F.C. Fuller once said, it is but a short step to the conclusion that "weapons, if only the right ones can be found, constitute ninety-nine percent of victory."[1] While the face of war has changed out of recognition in the 20th century, this appreciation retains its full force. And in the development of new weapons, the Great War retains its role as a pivotal *carrefour* of history during which a new infrastructure of war was invented, and science-based technologies – some new, others familiar – were applied.[2] A decade into the century of General Relativity and quantum mechanics, the war celebrated the destructive power of 'classical' physics. At the same time that new narratives unfolded in aviation, machine guns, armour, and submarine warfare, new international industrial alliances were formed: among the Central Powers, integrating aspects of German and Austrian manufacturing; among the Allies, linking Dominion resources, French designs, British experiments, and American practices of industrialization.[3] In applied chemistry and chemical industry, it became a war of largely familiar knowledge but with new applications, with catastrophic consequences for Europe, and important lessons for

---

* This essay forms part of a larger research project that has been assisted by a grant from the Australian Research Council, and by the award of an Edelstein International Fellowship. To both I wish to express my appreciation.

1 Cited in Martin van Creveld, *Technology and War, From 2000 BC to the Present* (New York: Free Press, 1989), p. 225.

2 See Guy Hartcup, *The War of Invention: Scientific Developments, 1914-1918* (London: Brassey's, 1988).

3 Roy MacLeod, "The Scientists Go To War," paper delivered to the Science and War conference, sponsored by the FRN (Sweden) and the International Council for Science Policy Studies, Malmo, Sweden, June 1988.



*A.S. Travis et al. (eds.),*
*Determinants in the Evolution of the European Chemical Industry, 1900–1939*, 25–49.

science.

The memory of chemical weapons remains among the most enduring legacies of the literature of the war. Sargent's painting of gas casualties on the road to Amiens strikes chords of outrage today. However, by the best estimates we have, poison gas cost fewer than one per cent of fatal battle casualties.[4] A better object of fury, perhaps, would be high explosives, which, notwithstanding the savagery of the machine gun, caused most battle deaths on all fronts. But whether gas or shell, it became, in the language of the day, a war of chemistry. It was not only a war of chemists, of course – empire-building physicists called it the 'acoustics' war, Lloyd George preferred to call it an "engineers' war" – and others have since likened it to a rehearsal for later "wars of the accountants."[5] However, the chemical sobriquet bears weight, for at least three reasons.

First, because, as a global conflict, the war was the first in which all the industrialized countries of the world were engaged, and one which could not have continued as long as it did without chemical industry; second, because the war highlighted the reciprocal engagement of university chemists, government chemistry, professional chemical institutions, and chemical manufacturing in a new nexus, one which prefigured the "academic-military-industrial" estate of the Cold War period; and finally, because the war foreshadowed, and in some cases precipitated, the emergence of modern, international chemical industry, in the form we know it today. Across the world, wartime products became peacetime commodities – producing celluloid for the burgeoning film industry, and aliphatic solvents (notably butyl acetates) for the lacquers that coated the motor cars of the masses.[6]

Twenty-five years ago, Lutz Haber lamented that "an account specifically devoted to the changing relations of academic and industrial chemistry in the leading countries has

---

4 Casualty statistics are notoriously difficult and unreliable, but the best estimates suggest that of the 2.9 million British casualties, and 800,000 fatalities, artillery accounted for 58 per cent, and machine gun and rifle fire, 39 per cent. Altogether, the belligerent armies recorded 1.5 million gas casualties. British gas casualties rose from 7.2 per cent of all wounded in 1917 to 15 per cent in 1918, but gas fatalities fell from 3.4 per cent to 2.9 per cent of all battlefield deaths. French gas casualties numbered 130,000, of which 17.5 per cent were fatal. The Americans had 73,000 gas casualties, and the Germans, 107,000, of which 5 per cent were fatal. Only 17,000 Allied deaths, and up to 90,000 deaths in all belligerent armies, were attributed to gas, but this figure undoubtedly underestimates large numbers in the Russian armies, in the Balkans and Mesopotamia, and in the West who died subsequently from respiratory illnesses and other complications experienced after war records were compiled. See J.M. Winter, *The Experience of World War I* (New York: Cambridge University Press, 1989), pp. 138, 140; Curt Wachtel, *Chemical Warfare* (Brooklyn: Chemical Publishing Co., 1941), p. 44.

5 D.H. Aldcroft and H.W. Richardson, *The British Economy, 1870-1939* (London, 1969), p. 151; Martin van Creveld, *Supplying War: Logistics from Wallensten to Patton* (Cambridge: Cambridge University Press, 1977), chapter 7.

6 Cf. Seymour H. Mauskopf ed., *Chemical Sciences in the Modern World* (Philadelphia: University of Pennsylvania Press, 1993).

still to be written."[7] For the war period, this remains largely true today, and prompts us to ask, why? From an industrial point of view, Haber argued, dyestuffs manufacture and the fixation of atmospheric nitrogen were the "most spectacular manifestations of applied chemistry during the war." From this perspective, the rate of technological innovation in chemical industry was less important to its success than industrial organization and scale. However, the history of wartime chemical weapons and explosives manufacture can be read differently, and perhaps to greater advantage, if one considers them from a different standpoint – for example, in relation to the character and chronology of mobilized industrial organization. In this respect, it becomes useful to ask: what did contemporaries consider significant about their experience of wartime scientific developments; what is known, and what remains to be understood, about the sequential engagements between government practices, military requirements, industrial capabilities, and applied science; and what new evidence and interpretations await historians of modern chemistry?

This paper endeavours to suggest the merits of using a comparative, chronological framework, within which we may begin to look more closely at the narrative of wartime chemistry, at its military, industrial and academic dimensions, and at its consequences for managerial organization and technological innovation. *In extenso,* this represents a major undertaking, towards which this can be no more than an early installment.[8] For discussion, I would like to locate the "chemical war" – incorporating, in McNeill's words, its "managerial metamorphosis"[9] – in the chronology that followed the changing culture of chemical industry between the autumn of 1914 and the winter of 1918. The use of chemical weapons was of great importance in Russia, the Balkans, and on the Italian-Austrian frontier, but those wartime sectors must form a separate study. Similarly, although the British Dominions played a vital part in Allied industrial and economic strategy,[10] the following commentary will confine itself to the situation as it can be broadly recalled in the Atlantic west.

For our purposes, the "chemical war" in France and Flanders unfolded in four phases. The first – the opening gambit – was a period of confusion, hesitation, and improvisation, as the Schlieffen plan for quick German victory was stalled and stopped. A second phase opened in the spring of 1915, and saw two intermediate stages: from mid-July 1915 to the beginning of the battle of the Somme in mid-1916, in which

---

7 L.F. Haber, *The Chemical Industry, 1900-1930: International Growth and Technological Change* (Oxford: Clarendon Press, 1971), p. 408.

8 For a useful comparative approach, see Stuart Wallace, *War and the Image of Germany: British Academics, 1914-1918* (Edinburgh: John Donald, 1988).

9 William H. McNeill, *The Pursuit of Power* (Oxford: Blackwell, 1982), p. 317.

10 See Roy MacLeod, "The 'Arsenal' in the Strand: Australian chemists and the British munitions effort, 1916-1919," *Annals of Science*, **46** (1989), 45-67.

chemical professions and weapons were fully mobilized; and a second, from mid-1916 to November 1917, which saw the full deployment of chemical industry among the Allies, and in Germany the launching of the Hindenberg plan, imitating Lloyd George's 1915 campaign for the expansion of British munitions production.

The third phase – the endgame – began towards the end of 1917 and continued through the Armistice a year later. At that point, once mining and submarine warfare had ceased, and American arms and industry had begun to shift the balance in France, the stage was set for a further elaboration of scientific warfare – including chemical weapons coupled with airpower – which could have led to a chemical *Vernichtungschlacht* in 1919, or else to a continued war of attrition lasting well into 1920, accompanied by huge civilian losses, the saturation bombardment of England and France, and the aerial devastation of Germany.[11] What would have been the outcome if Germany had, in the last great offensive of March 1918, used the gases her chemists had ready? And what would have happened if the huge quantities of American Lewisite, perhaps the world's most toxic gas, sent from New York, had reached the Front before the fighting ended?

Everyone at first expected it to be all over by Christmas 1914, of course. But the unexpected halt of the Germans at the Marne, and subsequent stalemate on the Russian, Alpine and Western Fronts, brought university academics, professional societies, and manufacturers, as well as workers in agriculture, industry, health, communication and transportation, to respond to King and Kaiser with a crusading sense of national purpose which the major powers of Europe had not seen for a hundred years.

## I. The first phase (August 1914 – mid-1915)

The guns of August signalled an immediate end to the familiar operations of the international market; the disruption of normal relations terminated the free flow of scientific ideas, and poisoned the well of collegial trust.[12] In Belgium, France, Britain, Austria-Hungary and Germany, a generation of science students disappeared, until men not already in the fighting were deployed for war work at home.[13] Among institutions, established hegemonies were rearranged; disciplines grew new appendages, or lapsed into suspension, and throughout Europe, the high Humboldtian vision of international

---

11 On the likelihood of continued war in 1919 and 1920, led by American enterprise, see Rexmond C. Cochrane, *The National Academy of Sciences: The First Hundred Years, 1863-1963* (Washington, DC: NAS, 1978), p. 241; L.F. Haber, *The Poisonous Cloud: Chemical Warfare in the First World War* (Oxford: Clarendon Press, 1986), p. 229.

12 Brigitte Schroeder-Gudehus, "Characteristics des relations scientifiques internationales, 1870-1914," *Cahiers d'Histoire Mondiale*, **10** (1966), 161-177.

13 *50 Jahre Kaiser-Wilhelm-Gesellschaft und Max-Planck-Gessellschaft zur Forderung der Wissenschaften, 1911-1961* (Gottingen, 1961), p. 89.

science vanished, except, perhaps, among a thin elite on either side.[14]

In industry, as is well known, Britain found itself without dyestuffs and intermediates needed for chemical munitions, and, in many respects, without the skills needed to establish a new industrial base. Warnings had issued repeatedly from chemical reformers for over half a century, but to little effect.[15] France, which had lost most of its chemical industry in the first days of fighting, was forced to consolidate its intermediates production in the south and west, in the factories of Hoechst at Creil and Badische near Lyon, and to depend upon Switzerland for much of its explosives and dyestuffs. Britain, forced at the outset to order dyestuffs from neutral Switzerland, used every diplomatic means to encourage the Swiss to manufacture military dyes from British intermediates.[16]

At the end of August 1914, spurred by necessity, Walter Runciman, President of the British Board of Trade, set up a Chemical Products Supply Committee, chaired by Lord Haldane, whose brief was to "consider and advise as to the best means of obtaining for the use of British industries sufficient supplies of chemical products, colours and dyestuffs of kinds hitherto largely imported from countries with which we are presently at war." The same month, an Explosives Supply Committee was created at the War Office, chaired by the aged but distinguished John Fletcher (later Lord) Moulton, among the most capable men in Britain's wartime administration.[17] Chemists were appointed to neither. As *Nature* observed, a generation of industrial indifference towards science could not be remedied overnight.

---

14 Fritz Ringer, *The Decline of the German Mandarins* (Cambridge, Mass.: Harvard University Press, 1969); see also Roy MacLeod, "'*Kriegesgeologen* and practical men': Military geology and modern memory, 1914–1918," *British Journal for the History of Science*, **28** (4), (1995), 427-450.

15 See William Reginald Ormandy, "Britain and Germany in relation to the chemical trade," *Royal Society of Arts*, 2 December 1914, reprinted in *Chemical News*, **111** (1 January 1915), 3-8; "The war and British economic policy: The aniline dye industry," *Chemical News*, **111** (16 February 1915), 97; "The war and British chemical industry," *Nature*, **95** (1 April 1915), 119-20. See also Percy F. Frankland, "The chemical industries of Germany," *Journal of the Society of Chemical Industry*, **24** (7), reprinted in *Chemical News*, **111** (18 May 1915), 255-257; (4 June), 266-268; (11 June), 277-280, 291. The well-known facts of the coal tar dye industry and its 'takeover' by Germany were assembled for a popular readership in *The British Coal Tar Industry: Its Origin, Development and Decline*, edited by W.M. Gardener (London, 1915).

16 As a consequence of the war, Switzerland overtook Germany as the leading exporter of dyestuffs. See Haber, *Chemical Industry*, op. cit. (7), pp. 162-164, 197-198, 308-309.

17 In 1894, Moulton was the leading counsel for Nobels in the famous "cordite case" against the British Government. His expert knowledge of nitroglycerine, TNT and explosives manufacture was unrivalled. The suit failed, but on grounds that were invidious to the government scientists involved. Moulton became a Lord Justice of Appeal in 1906. At the time of his appointment to the Ministry of Munitions, he was nearly seventy. See Robert W. Reid, *Tongues of Conscience: War and the Scientist's Dilemma* (London, 1969), pp. 1-9, 36.

It was not much different in Paris, where the fighting services had no extensive experience of civilian chemists; nor even in Berlin, where the overtures of Fritz Haber, Emil Fischer, and Richard Willstätter, the leading chemists of the Kaiser Wilhelm Gesellschaft (KWG), were politely declined by the Prussian military bureaucracy.[18] The German High Command insisted that the war would be over in a few months – a politically necessary illusion, since to contemplate any other possibility would disclose a sad history of inconsistent efforts at industrial planning and stockpiling. There is little evidence of any pre-war German strategy to produce synthetic substitutes for raw materials that might be denied in time of war.[19] On the contrary, the Schlieffen Plan, which initially assumed the neutrality of Holland for the purpose of preserving access to vital commodities, had been modified in 1911 to eliminate that option; there was no alternative but a speedy victory.

With September emerged an unexpected war of fixed positions, a line of trenches that soon reached from the Channel to Switzerland. The nature of this war caught both sides unprepared. For the British and French, shrapnel shell was first exhausted, then made redundant, by a dramatically increased need for high explosives. This was the beginning of the "nitrate war."[20] Germany's chief chemical problem was the production of nitrates for fertilisers and explosives. There were three alternatives: the importation of Chilean nitrates, a possibility limited by the British blockade, and closed altogether in December after the defeat of von Spee's squadron in the South Atlantic; the electric arc process, in which oxygen and nitrogen were combined to produce nitrous oxide, thence nitric acid; and the manufacture of cyanamide, from atmospheric nitrogen and calcium carbide.

To these three, was added a fourth, the application of the Haber-Bosch process for the high-pressure catalytic combination of atmospheric nitrogen and hydrogen to produce ammonia. This was also the most promising route to the nitric acid needed for munitions production. The work of Fritz Haber had been financed by the Badische (BASF), patented in 1909, and developed at the BASF factory in Ludwigshafen between 1910 and 1911.[21] The first ammonia factory was opened at Oppau in 1913. The

---

18 See Richard Willstätter, *From My Life: The Memoirs of Richard Willstätter* (New York: W.A. Benjamin, 1965), pp. 263-282.

19 Jeffrey Allan Johnson, *The Kaiser's Chemists: Science and Modernization in Imperial Germany* (Chapel Hill: University of North Carolina Press, 1990), pp. 185, 185.

20 For reflective commentary, see J.R. Partington, *The Nitrogen Industry* (New York: Van Nostrand, 1923), and Harry A. Curtis, ed., *Fixed Nitrogen* (New York: Chemical Catalogue Co., 1932); for a range of contemporary perspectives, see "The nitrogen problem in relation to the war," *Scientific American Supplement*, **86** (1918), 98-99; "Why we must get nitrogen to wage war on a great scale," *Current Opinion*, **63** (1917), 61-62; Allerton S. Cushman, "The role of chemistry in the war," *Journal of the Franklin Institute*, **181** (1916), 163-190; Grinnel Jones, "Nitrogen, its fixation, its uses in peace and war," *Quarterly Journal of Economics* (May 1920), 391-431.

21 See "The Haber Process at Oppau," *Journal of the Society of Chemical Industry*, **40** (1921), 99.

new process produced 7,000 tons of nitrogen per year, or 5.7 per cent of Germany's total output, and by July 1914, over 14,000 tons per year, or nearly eleven per cent.[22] The war made extensive what the prewar plant had begun.

To secure an adequate supply of nitrates had been a European agricultural problem long before it became a military one. Since the 1890s, a rising population and its demand for nitrogen-based fertilisers brought about an increased demand for sodium nitrate and ammonian sulphate. War-shortages were a predictable eventuality that Fritz Haber, director of the Kaiser Wilhelm Gesellschaft's Institute for Physical Chemistry, had discussed for years. This institute, established in 1911 with the blessing of the Kaiser, but maintained by industry, provided a platform for Haber's research. Before the war, Haber had been at least twice to the United States, and saw the rising American chemical giant — and not British chemical industry — as Germany's principal scientific and industrial rival of the future. At the same time, he saw a need to modernize German science, to move beyond the achievements in organic and dyestuffs chemistry that had proved so profitable during the last century, and to seize leadership in the fast-developing fields of inorganic and physical chemistry.[23] "Standstill is retreat," said Carl Duisberg, a director of Bayer and a keen chemical imperialist, in 1906, and the coming experience of war made his point.

In late September 1914, the Kaiser Wilhelm Gesellschaft informed the German government that the British blockade would reduce by half Germany's average annual consumption of nitrates. Haber, Fischer, and Walther von Rathenau (head of AEG), were among the few to foresee the huge gaps that would emerge between production and requirements in toluene, nitric acid, sulphuric acid, petroleum and rubber. Fischer urged the Imperial War Ministry to coordinate German coal producers to increase ammonia production, and set his staff to find a substitute for sulphates normally extracted from Spanish pyrites. Eventually, Germany began to produce sulphuric acid using gypsum from countries in central Europe, over which it had territorial control.[24]

The arc and cyanamide processes for the capture of atmospheric nitrogen used familiar technology that was relatively simple and widespread. While less dependent on cheap power, the Haber-Bosch process was capital intensive, and required complex and costly plant. It yielded savings with increasing scale,[25] but required an experienced

---

22 A.S. Travis, *The High Pressure Chemists* (London: Brent Schools and Industry Project, 1984), pp. 51, 53. This introduction, intended for secondary school teachers and students, should be required reading for undergraduates.

23 Johnson, op. cit. (19), pp. 36, 39.

24 Michael Sanderson, *The Universities and British Industry, 1850-1970* (London: Routledge and Kegan Paul, 1972), chapter 8.

25 Haber, *Chemical Industry*, op. cit. (7), pp. 84-90.

combination of engineering and chemical skills, as well as metallurgical and process technologies that only the most advanced industrial societies could afford. At the outbreak of the war, BASF agreed to produce 30,000 tons of synthetic ammonia for the government by the Haber-Bosch process. It was the beginning of a policy of Ersatz et al. manufacture that would eventually overtake the whole of German industry.[26]

In Britain, the technical details of the Haber-Bosch process remained unknown – despite much research at University College London, and elsewhere until the end of the war. The production of cordite, the principal British propellant, required huge quantities of nitric and sulphuric acids; and picric acid explosive (= Lyddite, or Melinite in France), made by the nitration of coal-tar phenol, required a large capacity for aromatic nitrations, which at first was absent. Britain had for decades produced sulphuric acid from Spanish pyrites or Sicilian sulphur, and nitric acid from Chilean saltpetre (sodium nitrate). The first was prohibitively expensive; the second could be maintained, at a cost. But supplies of both depended upon open sea lanes, already threatened by German U-boats. The British government was forced to rely upon a handful of firms – including Read Holliday and Clayton Aniline – for the production of almost all its key intermediates, including nitro compounds of phenol and toluene. Picric acid soon gave way to the more stable TNT, but shortages of glycerine, phenol, toluene and acetone solvent (for cordite) forced the mobilization of every soap and gas works and coal and coke distillery in the country. In the United States, chemical manufacturers watched these developments with increasing concern. The British blockade helped American subsidiaries of German firms build up profits, but the United States government began to ask for ways of reducing America's near-total dependency on German organic chemicals.

By December 1914, the "war of illusion" was over. Chemistry had also declared war. The ninety-three German professors who signed the infamous October manifesto, pledging science to the military and the Fatherland, included four of the six directors of the KWG, and all of its leading chemists. An opposing manifesto, circulated by Einstein, attracted only four signatures, including his own. Emil Fischer at first defended his commitment to internationalism: "science is, and remains, international," he protested.[27] But the distinguished English chemist Sir William Ramsay, who for years had applauded German methods, now denounced German aggression, and went out of his way to denigrate German chemistry as imitative and unoriginal.[28] However,

---

26 Johnson, op. cit. (19), p. 189.

27 Cf. Bernhard vom Brocke, "Wissenchaft und Militarismus," in William M. Calder, III, et al., *Wilamowitz nach 50 Jahren* (Darmstadt, 1985), p. 674; Fredrick Herneck, *Abenteuer der Erkenntnis: Funf Naturforscher aus drei Epochen* (Berlin-DDR, 1973).

28 M.W. Travers, *A Life of Sir William Ramsay* (London: Arnold, 1956); Frank Turner, "Public science in Britain, 1880-1919," *Isis*, **71** (1980), pp. 603-606.

verbal insults were poor weapons against high explosives, in the manufacture of which Germany apparently had the upper hand. Eventually, even Emil Fischer accepted the war as a patriotic duty, conceding that although it would be a hard struggle, German chemical industry would emerge intact, and in command of a continental economic bloc strong enough to compete with the United States.

In late 1914, however, this outcome was not self-evident. From Nobel's Somerset West Explosives Factory near Cape Town, South Africa, Lord Moulton summoned the American chemical engineer K.B. Quinan, and contracted with him to produce vast quantities of TNT in place of picric acid. Quinan began to recruit chemists with engineering experience, and engineers who knew chemistry, to help build and operate two huge national factories, one at Oldbury for TNT and another at Queensferry for guncotton. Within five months, with a combination of Quinan's design and Chance and Hunt's construction expertise, Oldbury had transformed the production of TNT by a new continuous process using petroleum-derived toluene.[29] It was a harbinger of things to come, as Quinan and his colleagues began a rapid extension of factory plans to Sandycroft and Swindon, and over fifteen other locations, including a huge green-field site near Gretna, all of which were to come on line within twelve months.[30]

By the beginning of 1915, it was becoming clear that the war would involve, to an unexpected degree, an unprecedented combination of (1) geopolitical control over raw materials; (2) the mobilization and deployment of skilled manpower; (3) the rationalization of component industries, with linkages between the supply of raw materials, the production of intermediates, and the manufacture of munitions; and (4) clear lines of communication between applied research, chemical industry, government procurement, and military demands. In Britain, as in Germany, the nature of chemical

---

[29] See C.S. Robinson, "Kenneth Bingham Quinan," *The Chemical Engineer*, No. 203 (November 1966), 290-297; Jas. Gray, "Some chemical discoveries which have influenced the development of South Africa," *South African Journal of Science*, **30** (1933), 32-34; G. Taberner, "The explosives and chemical industry in South Africa," *South African Journal of Science*, **33** (1937), 259-69. Quinan was initially 'loaned' by De Beers to the Committee on Supply of High Explosives in December 1914 for nine months, at a salary recovery of £4000 p.a. plus expenses. In December, 1915, he was reappointed for a further six months, and then for the duration, until he was 're-loaned' to the Americans in June 1918.

A self-taught engineer, 'KBQ' was the nephew of W.R. Quinan, an American army officer who was recruited from California to South Africa by De Beers' American general manager, Gardner Williams, in 1898, as part of a plan inspired by Cecil Rhodes to create a British counterpart to the German-backed explosives factory at Modderfontein outside Johannesburg which, in due course, fell to the Boers in the war of 1899-1901. The relationship between American, South African, and Australian explosives, which pivoted on the needs of the mining industry, suggests the imperial bearing of Britain's munitions effort. See also A.P. Cartwright, *The Dynamite Company: The Story of African Explosives and Chemical Industries, Ltd.* (Capetown, 1964), pp. 82, 127, 137. I am indebted to Mrs. L.D. Arnott, Assistant Archivist of Barlow Rand, Ltd., of Johannesburg, for this information.

[30] See R.P. Ayerst, *et al.*, "The role of chemical engineering in providing propellants and explosives for the U.K. armed forces," in William F. Furter, ed., *History of Chemical Engineering* (Washington, DC, 1980), pp. 367-391.

warfare required a new form of 'mixed' economy, in which public and private industry worked side by side. These were the conditions of total war. In the second phase, Germany and Britain revealed different approaches to these requirements, which gave the former an early advantage, but which the latter would soon overcome.

## 2. The second phase (April 1915 – June 1916)

In the spring of 1915, Germany was the first to be driven by the circumstances of trench warfare and the shortages in strategic materials to begin a new chemistry, and a new warfare, based upon synthetic products. Once gas warfare was possible, it was necessary. As Jeffrey Allan Johnson has acutely observed, the new poison gases were to be Ersatz weapons in themselves, substitutes for conventional weapons. The first use of chemical weapons on the Western Front, in April 1915, has been traced to arguments put by Fritz Haber to the German General Staff, which had been persuaded by the evidence of their senses that further shortfalls in explosive shell production would bring their forces to a halt within weeks. The ailing German dyestuffs industry needed time to expand, to test large scale processes for nitric acid manufacture, and to organize its own army of workers in overalls – almost entirely men, it seems, in competition with soldiers in uniform. The generals, obliged to swallow their pride, asked their chemists to clear the enemy trenches as best they could. Chemical weapons began as a trenching tactic, and became a standard tool.[31]

Since at least the middle of the 19th century, all major European powers appear to have experimented with the use of lethal and irritant gases; and the Hague Conventions in 1899 and 1907 were specifically intended to restrict development of precisely the kind that was underway at the outbreak of the war. Nonetheless, during the autumn of 1914, all the belligerents employed non-lethal tear gases, mostly delivered by shell and grenade, although none did so in a systematic or strategic manner. Lutz Haber argues that the lack of interest and absence of urgency on both sides owed more to material than ethical considerations.[32] This is probably correct, as neither side had a monopoly on morality. However, German chemistry was better placed to act. When the German general staff requested an offensive gas, they turned to Fritz Haber, and Haber turned for the duration to gas warfare.

Given sufficient time, Haber would have preferred to launch his gas war with phosgene, a far more devastating weapon, but in early 1915 this was not yet available in sufficient

---

[31] The classic American text is Amos A. Fries, *Chemical Warfare* (New York: McGraw Hill, 1st ed., 1921). For the postwar tactical literature, see Augustin M. Prentice, *Chemicals in War: A Treatise on Chemical Warfare* (New York: McGraw Hill, 1937); Col. Alden Waitt, *Gas Warfare, the Chemical Weapon: Its Use and Protection against It* (New York: Duell, Sloan and Pearce, 1943); Frederick A. Hessell, *Chemistry in Warfare: Its Strategic Importance* (New York: Hastings House, 1942).

[32] Haber, *Poisonous Cloud*, op. cit. (11), p. 27.

quantities, and the German Army lacked masks, shells and heavy howitzers to deliver it. Accordingly, Haber provided the army with chlorine instead, to be delivered not by shell but by the innovative expedient of a gas cloud from cylinders, while Haber dissembled its use as not inconsistent with the Hague Conventions. That this new weapon – given prevailing winds and the inevitability of reprisal attacks – would be as hazardous to those who used it, as to those against whom it was used, was accepted as a strategic gamble. In the event, the operation that unfolded at Ypres on 22 April 1915 took two and a half months to prepare, and was code-named 'Disinfection,' possibly because it was assigned to a reserve Pionierkommando unit, or Disinfektionstruppe.[33] Duisberg gave the Allies six months to retaliate.[34] He was correct. The British counterattacked at Loos in September.

German industry used those six months creatively, in expanding its production of high explosives. Within a month after the gas attack at Ypres, Carl Bosch announced completion of prototypes for improved nitric acid plants.[35] From mid-1915, the Imperial government funded new plant for the cyanamide process at the Bayerische Stickstoffwerke in Bavaria, and at the AG für Stickstoffdünger in Knapsack near Cologne. Haber-Bosch ammonia production was increased at the Badische (BASF) factory at Oppau, near Ludwigshafen. By the end of the year, the Oppau plant was working to full capacity, and, fearing Allied bombing, the Badische built a second, even larger plant for the Haber-Bosch process at Leuna, near Merseburg (Leipzig), again with government funding.[36] This soon began producing 130,000 tons of ammonia a year and would treble its production by 1928.[37] BASF plants produced concentrated nitric acid well suited to munitions production. As Tony Travis has noted, the German chemists were so successful that they soon exceeded Allied production of nitric acid.[38] While correct, this masks a number of questions concerning Allied production technologies, about which much remains to be written.

Throughout the war, Chilean nitrates remained a British monopoly, but, without a mobilized industrial base, Britain could not meet increasing demands. Among the principal problems facing British chemical manufacturers in early 1915 was the supply of intermediates. Britain's two main dyemaking firms – Read Holliday at Huddersfield

---

33 Haber, *Poisonous Cloud*, op. cit. (11), p. 30.

34 Johnson, op. cit. (19), p. 190.

35 Travis, *The High Pressure Chemists*, op. cit. (22), p. 61.

36 Haber, *Chemical Industry*, op. cit. (7)

37 See Curtis ed., op. cit. (20), p. 84.

38 Travis, op. cit. (22), pp. 61, 65.

and Levinstein at Manchester – increased production, but in May 1915, to speed up progress, the former was taken over by the British government and thereafter known as British Dyes, Ltd. In July, reeling from the memorable "shell crisis" of the early spring, Asquith's new coalition government created an unprecedented new organization, the Ministry of Munitions, to assume control of industrial production. This was placed under the direction of Lloyd George and Christopher Addison.[39] The same month, the British government established an equally innovative Committee of the Privy Council for Scientific and Industrial Research (the forerunner of the Department of Scientific and Industrial Research, or DSIR), with Treasury funds to forge links between academic science and industry.[40] The DSIR-model quickly spread throughout the British Empire, and was reflected in the establishment of America's new National Research Council, in 1917.[41]

The Ministry of Munitions took over propellent and explosives manufacture from the War Office, and soon ran over 100 factories, of its own, whilst supervising nearly the whole of British industry.[42] Along with many smaller firms, Levinstein manufactured intermediates and dyestuffs alongside the new government plants. As head of explosives supply, Moulton championed the production of 'amatol,' a mixture of TNT and nitroglycerin developed by (Sir) Robert Robertson's Research Department at Woolwich, which was processed industrially in proportions that saved vast quantities of material with improved effect.[43] Innovation also took place in small firms — Johnson of Hendon, is a good example; it specialized in import substitutions in photographic and other fine chemicals.[44]

In February 1915, official steps were at last taken to conserve technical manpower. Early in the war, a leaflet issued by the British Parliamentary Recruiting Committee calling for volunteers indicated that classes of worker which could be most readily spared for service were "navvies, tunnellers and chemists."[45] Responding to the call,

---

39 Chris Wrigley, "The Ministry of Munitions: An innovatory department," in K. Burk, ed., *War and the State: The Transformation of British Government, 1914-18* (London, 1982), p. 33.

40 Roy MacLeod and E.K. Andrews, "The origins of the DSIR: Reflections on ideas and men, 1915-1916," *Public Administration*, **48** (1970), 23-48; and Ian Varcoe, "Scientists, government and organized research in Great Britain, 1914-16: The early history of the DSIR," *Minerva*, **8** (1970), 192-216.

41 Cochrane, op. cit. (11)

42 Michael Pattison, "Scientists, inventors and the military in Britain, 1915-19: The Munitions Inventions Department," *Social Studies of Science*, **13** (1983), 521-567; cf. also *The History of the Ministry of Munitions* (London: HMSO, 1922), 19 vols.

43 George A.B. Dewar, *The Great Munitions Feat, 1914-1918* (London, 1921).

44 Travis, op. cit. (22), p. 66.

45 Prof. N.T. Wilsmore, "Presidential Address to Section B," AAAS (Hobart, 1921), 22; cf. J.J. Dobbie,

178 Fellows and Associates of the Institute of Chemistry were soon in uniform. The first was killed on 17 August 1914. Eventually, Britain, France and Italy would withdraw chemists from the ranks; Canada did not allow them to enlist. Australia mobilized most of its chemists directly into the war industries.[46] The summer of 1915 saw the first deployment of British gas troops for cylinder attacks in France, culminating in the misjudgement of Loos in November. Meanwhile, France pushed ahead with a smaller gas project, linking offensive and defensive warfare within the same organization, but failing to secure a close integration of science, manufacture, and deployment. Autumn saw increasing chemical casualties at the Front, and continuing confusion at home, as both British and French governments attempted to prepare defensive measures, and rationalize production of gas weapons.

By the end of the year, chemical warfare took a new turn, with the first use of phosgene by the Germans on 19 December, followed by the French at Verdun in the following February.[47] This gas the Germans could make easily in the form of diphosgene, and it was produced at Bayer's Leverkusen works from April 1916, and later in Frankfurt by Hoechst. But it was a gas the British made with great difficulty. Efficient production depended upon the choice of process, and at least two routes were possible – either reacting sulphuric acid with carbon tetrachloride (as at the United Alkali Company), or passing equal volumes of chlorine and carbon monoxide over a charcoal catalyst (as used by Hoechst). Some German chemical firms refused to fill shells with gas, and the government had to build special plants for the purpose. In Britain, gas production was dogged by accident and delay. That year, as Haber puts it, "the allies intention to fight a vigorous and effective gas war were beset by ... many obstacles, none of them caused by the enemy."[48]

While the situation remained bleak throughout 1915, few in Britain knew the organizational difficulties their counterparts in Germany were facing. Just before Christmas, 1915, E.B. Poulton, professor of zoology at Oxford, devoted his Romanes Lecture to the subject of "Science and the Great War." Britain's most egregious failings, he said, were due to its "national neglect of science." His message lit a slow fuse that exploded in a letter to *The Times* in February 1916, and then in a highly publicized conference in London in May. Of the six convenors of the conference, two were chemists, who used the history of Britain's poor industrial performance to rally support for the reform of science education and for the dedication of public money to

---

"Presidential Address," *Proceedings of the Institute of Chemistry*, (1918), (I), 33-34.

46 *Science*, **46** (13 July 1917), 32-33; Roy MacLeod, "The 'Arsenal' in the Strand," op. cit. (10).

47 Haber, *Poisonous Cloud*, op. cit. (11), p. 94.

48 Haber, *Poisonous Cloud*, op. cit. (11), p. 90.

applied research.[49]

W.J. Pope, professor of chemistry at Cambridge, was a determined reformer, not above using the war as an occasion for special pleading. The British government, he said, had applied its "usual methods" in establishing in the Board of Trade an organization in which men in control of the chemical industry knew nothing about chemistry.[50] By mid-1916, however, this situation was changing, as Britain carried through a spectacularly successful mobilization of its scientific workers. In March 1916, the Royal Society set up a Conjoint Board of Scientific Societies to coordinate academic advice in Britain, whilst in France the teaching laboratories at the Sorbonne were transformed into consulting rooms for industry and the French Army's Service des Poudres.[51] In both countries, government agencies determined production, fixed profits, and established standards of industrial efficiency.

In Britain, academic scientists began to go about their business with greater "industrial efficiency" than anyone had expected. Whilst more was heard about German chemistry, few names of chemists were widely aired; whereas in Britain, many chemists were pushed into prominence.[52] Just as the academies of Belgium, France and Germany began to expunge the names of foreign members from their lists, their own scientists worked with vigour and purpose. In Germany, the defence of the Fatherland moved professors into print, asserting that national pride did not compromise humanity.[53] The British countered, using words as weapons. The gulf between the Allies and the Central Powers was ethical, not intellectual, pronounced Sir William Osler, as Allied science and industry claimed the moral high ground.[54] The public voices of Pope and Henry Armstrong sent a message that the Allied military use of chemistry was vital if this were, in fact, to be the "war to end all wars."

---

49 Cardwell, op. cit. (58), p. 223-224.

50 Quoted in *Science*, **46** (2 November 1917), 427-428.

51 L. Tissier, "Les Poudres et Explosifs," *Chimie et Industrie, Dix ans d'efforts scientifiques et industriels* (Paris, 1926), vol. 1, pp. 1320-1345.

52 Roy MacLeod, "The chemists go to war: The mobilisation of civilian chemists and the British war effort, 1914-1918," *Annals of Science*, **50** (1993), 455-481. This statement may be qualified in the light of new work I am now doing on German chemical pamphlets issued as propaganda texts in 1915 and 1916.

53 Max van Laue, Max Planck, Johannes Stark, and Richard Willstätter protested their patriotism in civilian dress. See J.L. Heilbron, *The Dilemmas of an Upright Man: Max Planck as Spokesman for German Science* (Berkeley: University of California Press, 1986).

54 Sir William Osler, *Science and War* (Oxford, 1915), reviewed in *Nature*, **96** (16 December 1915), 431.

**3. The third phase (mid-1916 – November 1917)**

On 1 July 1916, the British Army suffered more casualties than on any day before or since in British military history. The sea of mud on the banks of the Somme, and the German emplacements above, would be etched on the memory of a generation. Chemistry had no especial role in this episode of mass carnage, which eventually cost both sides 500,000 men. But the experience pressed both sides to seek scientific expedients that could bring the war to a grinding halt. Some of the proposed solutions were chemical. By mid-1916, a "chemical army" was beginning to take shape on both sides of the Rhine and across the Channel. We do not yet know precisely its size, but in Britain it included approximately three-quarters of the 2,400 Fellows and Members of the Institute of Chemistry, and a large proportion of the 3,700 members of the Chemical Society – or well over 2,500 – together with several hundred chemists with practical works experience.[55] Lutz Haber estimates that 1,460 men took part in British chemical warfare work,[56] but this does not count another 800-1,000 who worked in the explosives factories and a host of other war-related industrial chemical occupations. Many of these were chemists with engineering experience.

Against them, stood a German army of perhaps 9,000 chemists, of whom possibly 2,000 were employed in the munitions industries, and 2,000 in chemical warfare. Carl Duisberg's Bayer plant at Leverkusen employed 203 chemists.[57] To Haber's calculation of 5,700 chemists on both sides, we may thus add close to another 3,000. To this must be added a large number of engineers, whom industry appointed to work alongside its chemists.[58] France contributed a small additional number, perhaps 200 in total. The opposing sides were thus not equally positioned, but neither was the picture one-sided. What made the difference by the end of 1917, was the prospect of "chemical reinforcements" from the United States, which contributed perhaps 1,900 to chemical warfare, but which, including the munitions industries, could well have summoned as many as 2,500.[59]

---

55 Calculations from published registers give 179 Fellows, 267 Associates and 165 Students, or 611 in total. *Proceedings of the Institute of Chemistry*, 1918, (iv), 30. However, at the end of the war, the Institute reported that 800 members (25 per cent of total membership) had been on active service. *Proceedings of the Institute of Chemistry*, 1919 (iv), 30.

56 Haber, *Poisonous Cloud*, op. cit. (11), p. 107.

57 Jean-Claude Guédon, "Conceptual and institutional obstacles to the emergence of unit operations in Europe," in William F. Furter, ed., *History of Chemical Engineering* (Washington, DC, 1980), p. 64.

58 Klaus Buchholz, "*Verfahrenstechnik* (Chemical Engineering) - Its development, present state and structure," *Social Studies of Science*, **9** (1979), (1), 33-62.

59 "Report on chemical instruction in Germany," quoted in Peter Alter, *The Reluctant Patron: Science and the State in Britain, 1850-1920* (Oxford, 1987), p. 225; Lothar Burchardt, "Professionalisierung oder Berufskonstruktion? Das Beispiel des Chemikers im Wilhelminischen Deutschland," *Geschichte und Gesellschaft*, **6**, (1988), 326-348. Donald Cardwell suggests there were 4,000 chemists in Germany's

We know very little about most of these men, and a good deal about a few, notably the academics – in Germany including Haber, Fischer, Nernst and Willstätter, in the USA, Lewis and Conant, and in England, Hartley and his senior officers. Apart from these 'officers,' we have yet to learn about the "other ranks" in the chemical factories and laboratories. By mid-1916, with "chemical armies" of comparable size, British and German government-assisted industries found themselves almost equally matched. Spurred by necessity, university chemists were given contracts to end Britain's prewar dependence on Germany, and produced drugs, anaesthetics, and anti-bacterials at a fraction of their prewar cost.[60] The British government's timely acquisition of Chaim Weizmann's process of starch fermentation, and its diffusion through converted distilleries throughout the kingdom, ended the acetone shortage that until the end of 1915 had crippled British cordite manufacture.[61]

In high explosives, Britain also came abreast of Germany, and in mustard gas at least, British unit output was cost-effective.[62] By late 1916, France and Britain had become independent in critical chemicals previously obtained from Germany, and Britain, at a huge cost to its economy, was supplying Russia and France, and eventually Italy, with explosives and other munitions.[63] France enlarged its cyanamide works to meet agricultural needs, and produced sufficient chlorine gas to mount its own offensives, and to supply British needs.[64] But production was not only a matter of raw materials. In Britain, the Ministry of Munitions and the Ministry of Labour introduced arbitration and safety practices to keep factory staff on schedule. But in Germany, strikes, food scarcities and wage disputes forced shell production into arrears from 1916 to the end of

---

chemical industry. See D.S.L. Cardwell, *The Organisation of Science in England* (London: Heinemann, 1972), pp. 215, 217. Contemporary American estimates of chemistry populations were much higher: e.g., 35,000 in Germany, 5,000 in Britain, 2,500 in France and 15,000 in the United States. See William A. Noyes, "America's opportunity in chemistry," *Scientific Monthly*, **24** (1927), 205.

60 Sanderson, op. cit. (24), chapter 8.

61 See Barnet Litvinoff, *The Essential Chaim Weizmann* (London: Weidenfeld and Nicolson, 1982), p. 244; Jehuda Reinharz, *Chaim Weizmann, The Making of a Zionist Leader* (New York: Oxford University Press, 1985), chapter XV; and Robert Bud, *The Uses of Life: A History of Biotechnology* (Cambridge: Cambridge University Press, 1993), pp. 37-45.

62 Henry Louis, "Presidential address to the Society of Chemical Industry," *Journal of the Society of Chemical Industry*, **38** (31 July 1919), 201T-206T.

63 Roy MacLeod, "'Full of honour and gain to science': Munitions production, technical intelligence and the wartime career of Sir Douglas Mawson, FRS," *Historical Records of Australian Science*, 7 (2), (1988), 189-203.

64 Haber, *Chemical Industry*, op. cit. (7), p. 210; see E. Vinet, "La Guerre des gas et les travaux des services chimiques français," *Chim. et Ind.*, **2** (Nos. 11-12), (1919), 1377-1415.

the war.[65] In July, in the absence of a single over-arching government-industry model such as Britain had invented, the leading German chemical companies – including BASF, Bayer, Hoechst, Kalle, Cassella, Agfa, Weiler-ter Meer and Griesheim – formed their own organizational collective, that eventually became the I.G. Farben group.[66] A comparison of the British and German industrial combines, public and private, has yet to be attempted.

Germany's decision to begin unrestricted submarine warfare on 1 February 1917 signalled the beginning of the end of its attempts to forestall the supply of Allied forces. On the eve of the Russian revolution – and of Britain's impending bankruptcy – there were already signs that the outcome would be decided by economic, rather than exclusively military, considerations. For the next year, however, victory in the 'nitrate' war remained the key ingredient in the war of the trenches. By the end of 1917, following bloody campaigns on the Somme and at Verdun, a war of attrition was the only alternative actively considered by the general staffs on either side. The Germans suffered their worst gas casualties in 1916 from French phosgene released in shells at Verdun, and in 1917 from the new Livens projectors then deployed by the British.[67] By itself, a war of gas was not winnable, but an industrial war combining gas and explosives had some way to go, at least before larger economic factors could render the whole effort null and void. Following the withdrawal of Russia from the Allies, the Western Front become a war of chemical imitation, with new gases from Haber's Institute generally leading, and the Allies usually responding, in kind. In a host of ways, Germany's attempts to introduce synthetic chemicals, notably oil and rubber, proved too ambitious to be completed in time for this war, but remained promising opportunities for wars to come.

On both sides, relations between chemists and the military remained problematic. "German generals, scientists and technologists lived under the same roof," Fritz Haber recalled after the war, "they greeted each other on the staircase." However, the facts suggest otherwise. Courtesies aside, mutual understanding improved little and, for most of the war, chemists were held "in little esteem by professional soldiers."[68] Haber

---

65 Wachtel, op. cit. (4), p. 58.

66 There has been much written on I.G. Farben since Lutz Haber's account, which relied upon F. ter Meer, *Die I.G. Farben* (Dusseldorf: Econ, 1953); see, e.g., R.G. Stokes, *Divide and Prosper: I.G. Farben under Allied Authority, 1945-51* (Berkeley: University of California, 1988); Gottfried Plumpe, *Die I.G. Farbenindustrie AG: Wirtschaft, Technik und Politik, 1904-1945* (Berlin: Duncker and Humblot, 1990); and Peter Hayes, *Industry and Ideology: I.G. Farben in the Nazi Era* (Cambridge: Cambridge University Press, 1987).

67 Cited in Hartcup, op. cit. (2), p. 114.

68 Haber, *Poisonous Cloud*, op. cit. (11), p. 273; see also J.E. Coates, "Haber Memorial Lecture," *Memorial Lectures Delivered Before the Chemical Society, 1933-1942* (London, 1951), vol. iv, 127-57.

remained a captain, while his opposite numbers in Britain and the United States became generals. However, the advisory relationship was little better for British or French chemists, and in the American forces military traditionals were uneasy when exposed to scientists in uniform. In the United States, there was even opposition to the creation of a separate 'scientific' chemical warfare service, which came into existence in June 1918, but which was demobilized after the war.[69] Chemists, it was said, suffered from the negative image attaching to their work; much better relations were enjoyed with the military by academic physicists engaged in sound-ranging, and by mathematicians in anti-aircraft research.[70]

**4. The fourth phase (winter 1917 – November 1918)**

It is often said that 1917 was the turning point for chemical warfare.[71] Events had reached a pass where the supply and distribution of materials counted for more than tactics, and shortages, technical failures, and incompetence were the limiting factors. These are generally seen as military failures, rather than industrial failures, as few historians have as yet looked closely at industrial case studies. But they have their roots in industrial production. Chemistry helped the war continue. By 1918, Germany was nearly self-sufficient in nitrogen.[72] Germany was also producing new war gases in increasing amounts, manufactured by factories in the east and in Finland. The nature of the gas war intensified with the introduction of mustard gas in July 1917, and arsenical compounds (Blue Cross, in German usage) whose function was to incapacitate, rather than kill. Sometimes they did both, using mustard gas to force soldiers to take off their masks, and then killing them with phosgene.[73]

The rate of innovation in war gases was higher in Germany than among the Allies until the autumn of 1918.[74] But Germany failed to increase production of gas to one half of

---

[69] G.F. Whittemore, "World War I, poison gas research and the ideals of American chemists," *Social Studies of Science*, **5** (2), (1975), 135-63.

[70] See Roy MacLeod and E. Kay Andrews, "Scientific advice in the war at sea, 1915-1917: The Board of Invention and Research," *Journal of Contemporary History*, **6** (2), (1971), 3-40; W. Hackmann, *Seek and Strike: Sonar, Anti-submarine Warfare and the Royal Navy, 1914-54* (London: HMSO, 1984); J.R. Innes, *Flash Spotters and Sound Rangers* (London: Allen and Unwin, 1935); D.J. Kevles, *The Physicists: The History of a Scientific Community in Modern America* (New York: Knopf, 1978).

[71] Cf. Haber, *Poisonous Cloud*, op. cit. (11), p. 206; see also Rudolf Hanslian, ed., *Der Chemische Krieg* (Berlin, 1925, 3rd ed., 1937).

[72] Haber, *Chemical Industry*, op. cit. (7), p. 203.

[73] Cf. Mario Sartori, *The War Gases: Chemistry and Analysis* (New York: Van Nostrand, 2nd ed., 1940).

[74] Haber, *Poisonous Cloud*, op. cit. (11), p. 223.

its total ammunition output. Some idea of relative industrial production is suggested by the calculation that, by the end of the war, Germany and Austria had delivered 66,400 tons of toxic gases; the Allies, notably Britain, 57,800 tons; and the US, only 1,100 tons.[75] However, by early 1918, the Allies, aided by America, were beginning to out-produce Germany in almost all war materials, and the gap widened in the coming months. Moreover, improved British (and later American) defensive masks made gas attacks more bearable.[76] Indeed, what was initially welcomed as an offensive weapon, became more useful in defence.[77]

America's early record as an arsenal of democracy was better than its record as an explosives manufacturer. The American government, which authorized funds to develop methods of nitrogen fixation as early as June 1916, and which lavished expenditure on the two alternatives, failed to bring its first designated plant (at Sheffield, Alabama) into operation before the war ended. Not until 1917 did the government contract with American Cyanamid to build a new nitrogen fixation plant, in Tennessee; this was completed within a year, but scarcely a fortnight before the Armistice, and it, too, never saw 'action.'[78] Gas weapons production was non-existent until a new dedicated arsenal was built at Edgewood, not far from Washington, DC (and Camp American University, where the Chemical Warfare Service based its testing). When American forces arrived in France in 1918, more than one quarter of all German shells fired were filled with gas, and gas defense was given the highest priority. University chemistry departments were mobilized, and gave hundreds of research students their first taste of applied research. But little was done in time to develop a programme of American offensive gases.[79] That was for the future. In the meantime, according to a contemporary observer, the brief eighteen months of America's war were sufficiently dramatic, in adding coal tar and cellulose to American industrial chemistry, and forcing explosives firms to undertake "numberless technical improvements." The war "taught laboratory staffs a great deal of chemistry," and, by attempting to perform delicate reactions on a large scale, stimulated studies of automatic chemical controls. Even more significant, America's short war convinced the managements of Du Pont, Hercules and Atlas of the long term possibilities of an immensely profitable future.[80]

---

75 Wachtel, op. cit. (4), p. 3.

76 Ibid., pp. 33-34; see also A.M. Prentiss, *Chemicals in War* (New York, 1937).

77 Haber, *Poisonous Cloud*, op. cit. (11), p.117, 266.

78 Its facilities, including its power plant and dams, later became part of the Tennessee Valley Authority, and found their use in World War II. For the history of American munitions mobilization, see Benedict Crowell, *America's Munitions, 1917-1918* (Washington, DC, 1920), 2 vols.

79 Williams Haynes, *The Chemical Front* (New York: A. Knopf, 1943).

80 See Williams Haynes, *American Chemical Industry*, vol. 3, *The World War I Period: 1912-1922* (New

The final German offensive in March 1918, and its collapse in the autumn, put Germany's sophistication in chemical warfare to its final and most exhaustive test. Opinion as to its success remains divided. Massive barrages of high explosive and altered artillery tactics did finally lift the war out of the trenches,[81] but huge volumes of gas – estimates suggest 25 per cent of all shell fire – and increasing care in its use, failed to overcome Allied defensive measures.[82] While the German use of mustard gas caused high casualties among the recently-arrived Americans, its reciprocal, unsystematic use by the British and French shattered German morale. What Ludendorff described as the "black day of the German army," 8 August 1918, came in consequence of a highly technological advance, managed by Australian and Canadian engineers. But eventually Germany's ability to wage war was decided three months later not by science at the front, but by starvation at home.[83] To simplify the situation that emerged by the autumn, German industry could not, in the end, deliver both guns and butter.

### 5. The denouement

Looking at the strategic and tactical uses to which chemical weapons were put, certain tendencies seem clear. We can see that an unexpected static war placed heavy demands on European industry for huge amounts of high explosives and gas, and gave employment to a small number of excellent researchers, and large numbers of chemists, whose wartime influence and professional experience led to a history of postwar empowerment.[84] We also see how the strategic use of chemical weapons was begun by Germany with the intention of delaying Allied operations, and denying Allied advances, until such time as German nitric acid production could be increased. But we also see how the egregious use of high explosives, and the use of gas, often in poorly developed, uncoordinated assaults, reflected a deep-seated military incapacity, on either side, to use such weapons in such a mix, rate, and saturation that would produce efficient results. Defensive technologies soon caught up, reducing competitive lead times over and over again. In chemical warfare, as in trench warfare generally, the circumstances favoured

York: Van Nostrand, 1945), pp. 204-205.

81 Haber, *Poisonous Cloud*, op. cit. (11), p. 227

82 (Brig. Sir) Harold Hartley, "A general comparison of British and German methods of gas warfare," *Journal of the Royal Artillery*, **46** (11), (1920).

83 Cf. Avner Offer, *The First World War: An Agrarian Interpretation* (Oxford: Clarendon Press, 1989).

84 David J. Rhees, "The chemists' war," *Bulletin for the History of Chemistry*, Nos. 13-14 (Winter/Spring 1992-93), 40-46; Daniel P. Jones, "Chemical warfare research during World War I: A model of cooperative research," in John Parascandola and James Whorton, eds, *Chemistry and Modern Society: Historical Essays in Honor of Aaron J. Ihde* (Washington, DC, 1983), pp. 165-185.

defence.[85]

Under these circumstances, wartime successes of industry were constrained by context. It is futile to conjecture what might have happened had the military situation evolved differently. However, it is possible to consider how the Allied and German chemical munitions industries, public and private, and regardless of the final outcome, measured up to their tasks. Immediate postwar inspections, reported in intelligence documents not released until the 1960s, revealed the extent of Allied interest in German manufacturing operations, shell-filling equipment, and progress made in synthesizing compounds.[86] Brig. Harold Hartley, Controller of the British Chemical Warfare Department, visited Berlin, Munich, Breloh and Stuttgart, and interviewed Haber and his colleagues at the KWG. Hartley also visited individual firms throughout Germany.[87]

British intelligence officers who visited factories in the occupied zone of Germany in February 1919, were struck by the ironies of victory. Such factories had to be "kept going in some reasonable approximation of normal output ... lest the shutting down of the factories [encourage] the growth of unemployment and anarchy."[88] "The German factories remain intact," W.J. Pope reported, in a warning to the British Association in 1919, "and can revert immediately to their former task of providing our Empire with dyes, pharmaceutical and photographic chemicals."[89] Some German factories were dismantled, but others were ready to direct their nitrogen production to agriculture, just when the huge British investment in high explosives plant had become superfluous. Germany had been forced to a military and political stalemate, but apparently not to industrial defeat.[90]

That the German industrial combines were well placed for subsequent civil conversion is well known to economic historians – but individual factories and firms, the ways in which they met procurement targets during the war, and the organizational factors that

---

85 Hanslian, op. cit. (71)

86 Haber, *Poisonous Cloud,* op. cit. (11), p. 7.

87 But the transcripts of these conversations still await analysis, as do the archives of individual firms, such as Bayer in Leverkusen and BASF at Leuna (whose archives, apparently now in Merseberg, await closer study).

88 MUN 7/539. "Technical Mission to Germany, 1919. Memorandum regarding the industrial position at Cologne, 4 March 1919"; *Report of the British Mission appointed to visit Enemy Chemical Factories in the Occupied Zone engaged in the Production of Munitions of War in February, 1919*, 1921 [Cmd. 1137], p. 10.

89 Sir William J. Pope, "Chemistry and the war," *Chemical News*, **119** (17 October 1919), 179-80.

90 For postwar chemical competition, see W.J. Reader, *Imperial Chemical Industries: A History* (London: Oxford University Press, 1970-75), 2 vols.

speeded their rehabilitation, have remained largely unstudied. Instead, there grew a substantial contemporary literature associating German industry with German war guilt, while neglecting the role played by chemical industry in Allied countries. The "Riddle on the Rhine" – to borrow the tantalizing phrase of Victor Lefebure, the British chemical liaison officer in France – had counterpart skeletons in corporate closets on the Mersey and the Delaware.[91] The British and American munitions industries would repay closer study. French archives at Vincennes for the period remain uncatalogued, and American archives in Washington, DC, have only recently been tapped for comparative information.[92]

In Britain, papers concerning the major private firms that worked for the Ministry of Munitions survive within the public domain, but are scattered and difficult to correlate. To date, no one has attempted an integrative history of the Department of Explosives Supply, to match Lutz Haber's magisterial work on the Ministry's Trench Warfare Research and Supply Departments, and their affiliated agencies in the War Office. The Ministry's internal history, written immediately at the end of the war – and still unpublished – reflects more the administrative triumphs its officials achieved than the technical battles that its factories waged. We know well the outlines of Britain's "great munitions feat."[93] But as Europe now approaches the 80th anniversary of the Armistice, it is surely time for a full, comparative history of that feat to be written.

**Conclusion:**

If one were left standing in 1919, what would one have seen as the chemical consequences of the war?

First, in Britain, France, Germany and the United States, the war forced an engagement between the interests of government, industry and academic science that would, in many respects, prove a dress rehearsal for the more durable relationship that developed during and after World War II. The experience of haphazard organizational muddle in 1914 and 1915 – for reasons of culture and politics, much discussed in Britain and the United States, less well advertised in France, and far less well known in Germany – forced attention to the need for cooperation between public and private interests in the encouragement of invention and innovation.[94] In Britain, the organization of research,

[91] Victor Lefebure, *The Riddle of the Rhine: Chemical Strategy in Peace and War* (London: Collins, 1921), chapter 1, 'Esplanatory.'

[92] Roy MacLeod, "Of Chemists, Kaisers and Kings, 1915-1919: Reflections on the 'Chemical War' in the shaping of modern memory," lecture to the Chemical Heritage Foundation, 10 April 1996.

[93] Dewar, op. cit. (43)

[94] For Germany, see General M. Schwarte, ed., *Die Technik im Weltkriege: Unter Mitterwerkung von 45 technischen fachwissenschaftlichen Mitarbeiten* (Berlin: E.S. Mittler, 1920); for Britain, B.H. Liddell Hart, *The Real War* (London: Faber, 1930); for France, C. Moureu, *La Chimie et la Guerre* (Paris, 1920); and for

notably under the DSIR and the Ministry of Munitions – and, after the war, by the Admiralty, the Ministry of Aviation and later the Department of Supply – prompted downstream investment in applied science and technology.[95]

Second, the war saw a consolidation of power in the hands of a few large firms with secure overseas horizons; governments took increasing account of civilian chemical manufacturing as a strategic resource, to be encouraged and protected by economic measures; and as a military necessity, to be safeguarded where necessary in integrated national networks, rather than pre-war arsenals. In Britain, economic keys forged by wartime conditions briefly unlocked the manacles that had held chemical industry captive to protective cartels and tariffs, and a tradition of dependence upon imperial markets. What industry would make of these new opportunities remained to be seen. A reorganization of chemical industry followed the recognition that German competition had, far from weakening, re-emerged to challenge its wartime rivals in peacetime production, distribution and sales.[96]

Third, in Britain and America, the application of science-based mass techniques to chemical products became part of everyday life.[97] At the same time, the war taught lessons in methods by which science could shape mass production, completing a "feedback loop" between laboratory experience and military requirements. The chemical industry was not alone to share this fate, but as the war introduced concepts of productivity, workers' health and, eventually, environmental responsibility, so the industry responded. [98]

Looking back on the wartime years, these sweeping changes, while significant, were more a subject of fascination than of analysis. Today, it is useful to consider the subject comparatively, and to write the history of wartime chemistry as a continuous narrative, rather than merely as the sum of firms and products. The development of nitrogen chemistry and chemical weapons forms an interwoven story of organizational experiments – the Service des Inventions of France, the Ministry of Munitions of

---

the United States, R.M. Yerkes, ed., *The New World of Science: Its Development during the War* (New York: Scribners, 1920).

95 MacLeod, 'DSIR,' op. cit. (40)

96 Cf. W.J. Reader, *Imperial Chemical Industries: A History*, op. cit. (90).

97 Daniel P. Jones, "From military to civilian technology: The introduction of tear gas for civil riot control," *Technology and Culture*, **19** (1978), 151-168;

98 Benedict Crowell, *America's Munitions, 1917-1918* (Washington, DC: Government Printing Office, 1920); E.M.H. Lloyd, *Experiments in State Control: At the War Office and the Ministry of Food* (Oxford: Clarendon Press, 1924).

Britain, the War Department's Ordnance and Engineering Corps in the United States, and the War Raw Materials Department of Germany. Organizational cultures owe something to international borrowings, but in certain respects remain distinctive. To the extent that we can read between the lines of tables of organization, we may learn from the "historical anthropology" of these organizations, their patterns of decision-making, their styles and preferences, and the factors that influence different forms of government-industry-academic linkages.

It is received wisdom that the Great War was a catalyst of most, if not all, the *mentalités* of modernism. In the memorable line of Sellar and Yeatman, the war was "the cause of nowadays."[99] The experience marked a shift away from the idea of international conflict as concerning only the operational requirements of armies and navies, and declared that henceforward, "strategic planning" and 'preparedness' would involve all areas of national life. So also the image of science – and especially chemistry – underwent a major transformation. Arguably, industry has always been at the service of war, and only the mythologies of the liberal state have argued otherwise.

However, it is less well known why men of science abandoned their commitments to internationality and communality – values which, while often honoured more in the breach than the observance, had served usefully in the conduct of European science for over a century. Indisputably, scientists became crusaders, overtaking political and military leaders in their zeal to apply science to the war effort.[100] Nowhere was this feature so pronounced as in chemistry. In the United States, a new willingness on the part of scientists nourished a scientific-military-industrial complex, linked to "special relationships" among the Allies, that has continued to the present day.[101] For Arthur Noyes and other chemists, the war was the "first taste of work that was deemed of demonstrable and immediate importance to the general community, and their first experience in spending money freely."[102] As a consequence, chemistry, interwoven with

---

99 W.C. Sellar and R.J. Yeatman, *1066 And All That: A Memorable History of England* (London: Methuen, 1932)

100 Kay and Roy MacLeod, "The social relations of science and technology, 1914-1939," in Carlo Cipolla, ed., *The Fontana Economic History of Europe: The Twentieth Century* (Glasgow: Fontana/Collins, 1976), vol. 5: *The Twentieth Century,* Part I, pp. 301-335.

101 For an American viewpoint, see D.J. Kevles, "George Ellery Hale, the First World War and the advancement of science in America," *Isis*, **59** (1968), 427-437; H. Wright, *Explorer of the Universe,: A Biography of George E. Hale* (New York: Dutton, 1966); for a German perspective, see H. Kessler, *Walther Rathenau: His Life and Work* (London, 1929). For a later history of the concept, see Carroll Pursell, *The Military-Industrial Complex* (New York: Holt and Reinhart, 1972).

102 See Arthur A. Noyes, "The supply of nitrogen products for the manufacture of explosives," in Robert M. Yerkes, ed., *The New World of Science. Its Development During the War* (Freeport, New York: Books for Libraries Press, 1969), pp. 123-133.

industry, won both praise and blame. It is not surprising that vast efforts were devoted in the 1920s to the popularization of chemistry and the 'rehabilitation' of its image.[103]

There is a view commonly held that the chemical war, like other aspects of the scientific war of 1914-18, was more characterized by rapidity and size, than by innovation or inventiveness.[104] Without denying speed and scale, this easy formula masks the complexity of the task that confronted industry. To neglect this is to downplay the significance of inventiveness and innovation of many kinds, and the skill-base of the people involved. Scaling up from untested designs took time and involved factors that tested ideas, procedures, and processes. Each has its importance in the history of applied science. If such issues are not viewed as sufficiently 'fundamental' to be included in the history of science, this suggests a professional shortsightedness we must resist.

Lutz Haber could not decide whether, in failing to record adequate histories of their wartime achievements, chemical manufacturers were either highly "reticent, or maybe just inarticulate."[105] That responsibility has passed from the practitioner to the historian, who – with the ageing of the "principal actors" and the imminent destruction of industrial archives – may be running out of time. As we redefine the historiography of applied science, we look again at the life-writing, at the emergence of new disciplines, and at individual firms. Biographically, we begin with Rathenau and Duisberg, Albert Thomas and the DuPonts. In professional associations, we look to new disciplines, including chemical engineering, which were *avant la lettre* of professional certification. Even so, there remain gaps in our knowledge of what took place in the wartime factories, how processes were operated, what choices in research and design were made, and by whom; and what role was played by new vocabularies of standards and quantification. Today, more than ever, we have reason to echo Haber's lament that "an account specifically devoted to the changing relations of academic and industrial chemistry in the leading countries has still to be written" – and to do something about it.[106]

---

[103] David J. Rhees, "A new voice for science: Science service under Edwin E. Slosson, 1921-29," unpublished MA thesis, University of North Carolina, 1979; Rod Home, "Science on service, 1939-1945," in R. Home, ed., *Australian Science in the Making* (Cambridge: Cambridge University Press, 1988), pp. 220-251.

[104] Haber, *Chemical Industry*, op. cit. (7), p. 208, suggests this, as does Alex Roland, "Science and war," *Osiris*, 2nd Series, **1**, (1985), 247-272 (263).

[105] Haber, *Chemical Industry*, op. cit. (7), p. 408.

[106] For an auspicious start, cf. Jeffrey Allan Johnson, "The academic-industrial symbiosis in German chemical research, 1905-1939," working paper 5.62. Center for German and European Studies, University of California-Berkeley, May 1997.

# "PRODUCTIVE COLLATERAL" OR "ECONOMIC SENSE?": BASF UNDER FRENCH OCCUPATION, 1919 - 1923

LOTHAR MEINZER
*BASF Archives*
*D-67056 Ludwigshafen*
*Germany*

## 1. BASF before and after World War I

At the turn of the century, BASF of Ludwigshafen could look back on 35 years of almost uninterrupted growth. According to the catalogue of the 1900 Paris International Exhibition, the company's 7,000-strong workforce made it "indisputably the world's largest chemical works."[1] Its business was the discovery, production and marketing of synthetic dyes. Its customer was the textile industry, a mainstay of the industrial scene in virtually every country on earth. The global market in colorants was systematically developed by BASF and the other German dye producers. By the outbreak of World War I, the German chemical industry had achieved a worldwide monopoly in the dye sector; in the year 1913 it was responsible for some 90 per cent of world production of dyes.[2] That year, more than 80 per cent of BASF's sales revenue came from dyes. With a view to breaking away from this near-total dependence on dyes, BASF had since the turn of the century carried out intensive research in a new area of chemistry, the fixation of atmospheric nitrogen. BASF's target was the nitrogen fertiliser market, at that time potentially one of the biggest sectors for the chemical industry with a volume almost ten times that of the dye market. The breakthrough came with the Haber-Bosch process, the

---

1 *Paris Exhibition of 1900* (Berlin, 1900), collective exhibition by the German chemical industry, p. 68.

2 On the development of the dye industry, see (with more extensive references) L.F. Haber, *The Chemical Industry during the Nineteenth Century: A Study of the Economic Aspects of Applied Chemistry in Europe and North America* (Oxford, 1958); *idem*, *The Chemical Industry 1900-1930: International Growth and Technological Change* (Oxford, 1971); Gottfried Plumpe, *Die I.G. Farbenindustrie AG, Wirtschaft, Technik, Politik 1904-1945*, (Berlin, 1990); Anthony S. Travis, *The Rainbow Makers: The Origins of the Synthetic Dyestuffs Industry in Western Europe* (Bethlehem, Pa., 1993).
On the development of BASF, see the unpublished manuscripts (in German) of Walter Voigtländer-Tetzner, especially "History of BASF 1865-1914," "Chronicle of BASF 1865-1940," and "Commercial Development of BASF," all in the BASF company archive.



*A.S. Travis et al. (eds.),*
***Determinants in the Evolution of the European Chemical Industry, 1900–1939***, 51–63.

catalytic reaction of nitrogen and hydrogen to produce ammonia. Fritz Haber, sponsored by BASF, had perfected the continuous high pressure process in 1909, and Carl Bosch and his team at Ludwigshafen undertook pilot plant studies during the following years. In 1913, BASF opened the world's first synthetic ammonia plant at a new works near Oppau, close to Ludwigshafen. Despite the high capital cost of the new facilities, BASF achieved sales of about 120 million marks in 1913, with a clear profit of some 15 million marks. The outcome was a pretty good profit to sales ratio after taxes of 12.5 per cent. The success of the Haber-Bosch process was to prove even more dazzling after 1914.

World War I marked a radical turning point in BASF's progress. The company's trademarks and patents in the nations at war with Germany were seized. Organic chemical production facilities in Britain, France and Russia were sequestered by those states and used to support their own chemical industries. It became apparent at an early stage that whatever form the eventual peace treaty might take, the structure of the world market in dyes would be very different to what it had been in 1914. In Germany, confronted by the naval blockade and the failure of its 'blitzkrieg' strategy, the chemical industry became the key to maintaining the supply of raw materials. For example, nitric acid needed for the munitions industry had previously been obtained almost exclusively from imported Chile saltpetre. Now things had changed: BASF set out to establish a new route to nitric acid. By the spring of 1915, the company had developed a process for converting ammonia to nitric acid, and production plants were built to exploit this new process, first at Oppau, then in 1916-17 at a new works near Leuna in central Germany. Production of dyes, however, was almost completely discontinued, while fertiliser manufacture from synthetic ammonia, still in its early stages of operation, was not expanded greatly. In 1917, BASF's sales totalled 258 million marks, of which fertilisers accounted for 3.2 per cent, dyes for 10.9 per cent and chemicals – including "industrial nitrogen," in other words, intermediates for the munitions industry – 85.9 per cent. The export ratio that year was no higher than 6.3 per cent.

BASF's development in the early post-war years was influenced by widely differing factors. Thus the period 1919-24 saw the first widespread, and sometimes violent, social conflicts and industrial unrest.[3] The world dye market was, predictably, afflicted by massive surplus capacities, partly created by German reparations payments in the form of dyes. The latter convinced dye users of the ongoing superiority of German products, which enabled BASF to maintain footholds in many foreign markets – though at a much lower production level than in 1913. The high quality and low prices, assisted by inflation, enabled the company to fight off its new international competitors and even develop new export markets, especially in the Far East.

---

[3] See in detail Dieter Schiffmann, *Von der Revolution zum Neunstundentag: Arbeit und Konflikt bei BASF 1918-1924* (Frankfurt a. M., 1983).

Synthetic ammonia also played an important part in postwar developments. The main advantage of the nitrogen operations was that they could be converted back to peacetime production at relatively low cost; another advantage was that wartime government loans for the construction of the nitrogen capacities could now be repaid in paper marks.[4] By contrast, shutdowns for months at a time because of coal shortages, the after-effects of a devastating explosion in September 1921, and the consequences of the French occupation, hardly placed BASF in a position to take full advantage of what, until 1922, had been a quite good economic climate. In 1923, the economy collapsed completely under pressures of the conflict in the Ruhr and hyperinflation. For the first time since its foundation, BASF reported no profit and was unable to pay a dividend. In 1924, the first 'normal' year since the war and BASF's last year of independence before it was absorbed into I.G. Farbenindustrie AG, the company recorded sales of some RM 430 million and a net profit of RM 15 million. The sales structure had, of course, changed totally since the pre-war days: almost 58 per cent of sales came from the fertiliser business alone, the products of which were mainly sold within Germany. Dyes accounted for only 23 per cent, and the export ratio was just under 38 per cent.

Table 1. BASF: Sales and Profits, 1910 - 1924*

| Year | Sales | | Profit | |
|---|---|---|---|---|
| | nominal (million marks) | real terms (million gold marks/RM) | nominal (million marks) | real terms (million gold marks/RM) |
| 1910 | 97.5 | | 10.8 | |
| 1911 | 105.2 | | 12.4 | |
| 1912 | 114.2 | | 15.2 | |
| 1913 | 120.7 | | 15.2 | |
| 1914 | 112.9 | | 13.6 | |
| 1915 | 122.4 | 102.2 | 17.2 | 14.4 |
| 1916 | 179.2 | 132.0 | 23.5 | 17.4 |
| 1917 | 258.5 | 191.8 | 30.0 | 22.3 |
| 1918 | 330.6 | 165.3 | 10.8 | 5.4 |
| 1919 | 333.5 | 29.9 | 27.0 | 2.4 |
| 1920 | 2,700.3 | 154.6 | 65.2 | 3.7 |
| 1921 | 4,452.0 | 101.6 | 165.3 | 3.8 |
| 1922 | | 265.8 | 3,901.4 | 2.2 |
| 1923 | | 249.6 | | |
| 1924 | 429.3 | | 14.9 | |

*) Figures from BASF Archives, B4/1175, Reparations. Figures in gold marks calculated on the basis of the dollar rate at 31 December for each financial year.

4 Peter Hayes, *Industry and Ideology: I.G. Farben in the Nazi Era* (Cambridge, Mass., 1987), pp. 12 et seq.

**2. French occupation and the problem of espionage**

On 6 December 1918 the first French officers entered the BASF works at Ludwigshafen and Oppau. They set up a control post within the works, and demanded surveys of inventories, raw materials, production and coal requirements; they also confiscated dyes to the value of 1.5 million francs. Apart from these indignities, the initial impression at BASF was that it might be possible to survive the occupation without major problems.[5] There were, of course, the usual quarrels, and complaints about the molesting of "female persons" in the Hemshof workers colony, just as there were about the 'girls' the soldiers brought onto the factory site. It was, apparently, impossible to persuade the French to observe the ban on smoking in the works. Tankers were requisitioned and bicycles went missing from time to time. The soldiers' "sporting activities" in a field of clover at the Agricultural Research Station caused damage to agricultural investigations, and an employee complained that he had been forced to hand over the postage stamps from business correspondence to an officer with an interest in philately.[6]

It soon became apparent that there were more serious problems afoot. The French officers assigned to Ludwigshafen in ever increasing numbers turned out to be chemists and engineers employed by French private enterprises – uniformed spies instructed to hunt out trade secrets. The aniline department alone recorded a total of 57 visits by these inspecting officers between February and May 1919. They were observed, for example: "14.2 ... looking at the design of the apparatus in Building 22 and making notes ... 28.2 ... 3 Frenchmen in Building 322, taking measurements and checking temperatures ... 23.4 ... H. spent a long time in Building 80, investigating all the pipes and climbing up ladders to do it ... 8.5 ... H. following the course of operations, creeping about everywhere so that he himself has been dyed violet." In a telegram dated 6 February 1919 to the German Armistice Commission, the management urgently demanded a "strong protest against this practice and immediate efforts to protect us against further espionage and interference with our private property."[7] However, more and more new 'commissions' appeared at the factory gates. Britain, the USA, and Italy, too, sent experts to report on production operations at BASF. Eventually, the works management took refuge in a form of passive resistance: chemists were forbidden to provide any information on production processes, and the management also decided "to shut down production plants where we cannot allow the commissioners to observe operations without giving away important trade secrets."[8] When the Versailles Treaty

---

5 BASF Archives, C 11, Minutes of the 51st Meeting of the Central Committee of the Supervisory Board of May 21, 1919.

6 Reports in BASF Archives, A 862/IV, French occupation.

7 Memos and telegram in BASF Archives, A 862/Peace Treaty I.

8 BASF Archives, C 11, Meeting of the Central Committee of the Supervisory Board of May 21, 1919; see ibid., Supervisory Board meeting of March 1, 1919. As to the success or failure of this whole espionage business, one can only speculate. It can however be assumed that the factory inspections in conjunction with BASF's foreign patent applications that were impounded during the war were, at the least, no hindrance to

came into force in January 1920, the activities of the inspecting officers came to an end; the last of them pulled out in March 1920, together with the French troop contingent from BASF. Occasional inspections by the Inter-Allied Military Control Commission continued until 1924, but aroused little excitement.[9]

A French military commission visiting BASF, Ludwigshafen, in 1919. (Courtesy BASF Ludwigshafen.)

### 3. Requisitions and reparations

And yet the Treaty of Versailles itself contained certain provisions which posed a dangerous threat to BASF's survival. For a start, there was the matter of reparations. The chemical industry, alone, apart from the coal mines, was singled out in the treaty and required to pay reparations in kind. It was to hand over 50 per cent of its stocks of dyes, chemicals and pharmaceuticals, and until 1 January 1925 it was regularly to make 25 per cent of its total production available for export at pre-war net prices. In addition, the industry was required to deliver 30,000 tons of ammonia and other chemical

the development of the national chemical industry in question.

9 BASF Archives, A 862/IV; A 862/Peace Treaty I.

BASF Ludwigshafen works (centre) and Oppau (right, in background) in 1924, after a painting by Hans Luck. (Courtesy BASF Ludwigshafen.)

raw materials to France every year for three years.[10] Ultimately, of course, it was the German Reich that paid for these shipments, though generally after a long delay and at a price that fell far short of covering costs.[11] Nevertheless, there was the tremendous advantage that the compulsory shipments maintained the continuing presence of the industry's products in export markets, and this ensured that at least some production capacity continued in use.[12]

Then came the conflict in the Ruhr, which made the reparations problem considerably more serious, especially for BASF which, in this second phase, became part of the "productive collaterals" which France took in the Palatinate as well as in the Ruhr. At first it seemed that BASF would be only indirectly affected by the impounding of payroll transports and, especially, coal shipments. But on 15 May 1923 the

[10] *Versailles Peace Treaty*, Part VIII, Annex V § 8 and Annex VI.

[11] The Dyeworks Centre (an office opened by the chemical industry to handle the reparation burden in Frankfurt), in a letter to the Court of Auditors of the German Empire dated September 21, 1926, calculated the financial losses resulting from reparation shipments as follows:

- interest and currency losses on shipments from 50 per cent of inventories: about 25 million goldmarks.
- interest, currency and price losses on shipments from 25 per cent of daily production: about 31 million goldmarks.

BASF Archives, B 4/1175, Reparations.

[12] Haber, *Chemical Industry 1900-1930*, op. cit. (2), p. 250. See also Hayes, op. cit. (4), p.13.

Ludwigshafen works was occupied by French troops.[13] French officers arrested three deputy directors (all members of the Board of Executive Directors had been able to make good their escape across the Rhine bridge) and explained to them that the purpose of occupying the factory was to confiscate the dyes stored there and use them to offset overdue reparations consignments. In accordance with the rules of passive resistance, the BASF directors refused to cooperate in any way. They were retained as hostages to guarantee that the lights stayed on at the factory and the necessary safety, emergency and fire services remained on site. Next day, some fifty French workers began loading and removing the stocks of dyes, and over the next fourteen days some 500 railway wagons were loaded and hauled off towards Strasbourg.[14] Eventually, after they had impounded all the stocks of wood, the French moved out on May 30, but only now to occupy the Oppau works.[15]

---

[13] BASF Archives, A 862/IV, French Occupation. The Fire Chief of the Ludwigshafen complex described the occupation as follows:

> Tuesday, May 15, 1923. Reported at about 3 am that fairly large contingents of French troops had left Oggersheim to occupy the Aniline works. Prompt departure of the Board members living in Ludwigshafen. All gates to the works occupied by 4.30 am; even the smallest exits are manned by French sentries, including Moroccans. Chain of sentries to Oppau, where incidentally only the main gates of the works are occupied. LU[dwigshafen] works entered, occupation in particular of Building 1 and the big dye store. Night shift prevented from leaving the works. Clerical and other staff generally not permitted to enter, though certain shift leaders, in particular, given permission to visit their plants on the basis of red passes.
>
> Deputy Directors Krauss, Mehner and Scharff, either arrested within the works or fetched from their homes, are being held as hostages at the works. Roads around the factory sealed off in some cases, no trams allowed in the northern sector of the town until 8.00 am. Spahis, scouts are keeping the main roads clear and are said to have used their sabres on those who were too slow to take evasive action; but not seen out of doors after a few more hours. The strength of the force involved was apparently about a regiment of Moroccans, two squadrons of spahis and a number of machine guns and tanks ...
>
> Fire Chief alerts the French colonel soon after his arrival to the fire risk at the plant and the need for the ban on smoking. The Frenchman regards smoking in the streets as non-hazardous.

The report concludes with the words:

> The commission of French officers demands a meal in the canteen, and gets it.

BASF Archives, A 862/III, French Occupation.

[14] Incidentally, the confiscated product resurfaced later under dubious circumstances and BASF was invited to buy it back.

[15] BASF Archives, C 10, Report to the Supervisory Board dated June 20, 1923; A 862/II, French Occupation.

At Oppau, the situation deteriorated considerably. The removal of some 60,000 tons of fertilisers – which was the purpose of this operation – called for much more comprehensive technical facilities than the straightforward loading of drums filled with dyes. Most essential was electrical power to get the excavators and conveyor belts running; and the French demanded that this high tonnage of material be supplied in an ultimatum of June 4. Those directors still held hostage by the French refused this request following discussion with the board. They, together with the board (in absentia), were promptly brought before the military court in Landau. The French also expelled a number of prominent persons, including the chairmen of the employees' and workers' councils, commandeered a number of traction and gasoline engines in the town of Ludwigshafen and began to remove the fertilisers themselves. Initially, about 30 road wagons were loaded each day, but this figure increased to 100 from the end of July with the assistance of a large traction engine specially brought in from Paris.

The military court hearing took place in Landau on August 10. The indictments against the detained hostages were quashed, and they were released shortly afterwards. However, all members of the Board of Executive Directors – jointly and in absentia – were sentenced to eight years' imprisonment for their refusal to supply power, with an extra two years for the two signatories of the letter of refusal. For a brief period it was even feared that the French were planning a military sortie to Heidelberg with a view to placing some members of the board physically under arrest, but after a while the tension eased again since all the fertiliser silos were emptied. However, the French now began confiscating iron, steel, lead, copper, glass, cement and other building materials, and continued to remove them even after the passive resistance had ended. Both works, Ludwigshafen and Oppau, were still shut down.[16]

In October, the BASF works management, which needed to restart production as a matter of urgent economic necessity, especially in Oppau, opened negotiations with the French. In doing so, however, they regarded six points as essential:

1. The materials which have been impounded but not yet removed must be released;
2. The remaining Frenchmen at Oppau must vacate the works;
3. We must obtain a binding assurance from the responsible French authorities that there will be no further impounding of raw materials, auxiliaries or products;
4. The French must refrain from imposing an export levy on production from Oppau, because the nitrogen products cannot bear that levy;
5. We must be free to sell products from Oppau not only in the occupied territory but also in unoccupied areas and abroad;
6. The members of the Board of Executive Directors who have been

[16] BASF Archives, C 10, Reports to the Supervisory Board dated August 30, 1923 and October 25, 1923.

> sentenced to terms of imprisonment must have the sentences quashed and be granted safe conduct for freedom of movement within the occupied territory.[17]

The senior French representative in Ludwigshafen could not take a decision without consultation. He therefore arranged talks between members of the BASF Board of Executive Directors – under conditions of safe conduct – and General Demetz in Speyer. The meeting took place on October 17 and, according to a report by the BASF representatives, proceeded as follows:

> After an assurance of safe conduct, Messrs ... proceeded to Speyer, but were treated very badly there. The General complained bitterly that no one from BASF had condescended to call on him, and he was particularly indignant that the members of our Board had escaped arrest by their prompt departure from Ludwigshafen in May this year. The General refused to embark on negotiations regarding the start of production at Oppau, but first demanded a declaration from us to the effect that we were willing to cooperate with him in good faith in economic matters. We tried in vain to establish what the General meant by this 'cooperation.' All we could get out of him was agreement that cooperation should be confined to economic matters only and restricted to the operation of our factory in the occupied zone. The General added that, having received the declaration of good faith, he would be willing to negotiate on starting up the Oppau factory and that we would then have no trouble in reaching agreement on the various points of detail - which he, without knowing what they were, referred to as trivialities.[18]

Thus BASF was unexpectedly drawn into the whirlpool of separatist politics, which in the Palatinate peaked in these months. On October 22 the company gave the required undertaking , promising "to work together with the occupying authorities in good faith, on condition that nothing is asked of us which would be incompatible with our national honour." As a result, the company was able to recommence operations at both the Oppau and Ludwigshafen works from October 27 with no further debate.[19] The burden of reparations, however, once again increased significantly from November 1923, when BASF, among others, was obliged to participate in the so-called Micum Agreements and so contribute additional shipments of dyes and fertilisers.[20]

---

17 BASF Archives, C 10, Report to the Supervisory Board dated October 16, 1923.

18 BASF Archives, C 10, Report to the Supervisory Board dated October 25, 1923.

19 Ibid., Annex. There are no further references to the theme of separatism in the BASF files.

20 The "Micum Agreements" were direct supply agreements between the "Mission Interalliée de Contrôle des Usines et des Mines" and industry in the occupied territories, especially the mining industry in the Ruhr,

These events raise important economic questions. In particular, how much financial damage did BASF in fact suffer during 1923-24 as a result of confiscations and reparations? Alternatively, was it actually able to profit disproportionately from the compensation payments paid by the Reich, a charge which was particularly leveled at the Ruhr mining industry? Indeed, a commission of inquiry set up by the Reichstag in 1925 came to the conclusion that "substantial overpayments" had been made to the coalmining industry in the Ruhr, "though to exactly what extent can no longer be ascertained today [1927] for lack of detailed documentation."[21] The Social Democrat rapporteur, Hertz, estimated the figure at 87 million RM. By contrast, Hertz, expressing the view of the commission, found "that claims by the mining industry received much more favourable treatment than those from the other victims of the Ruhr occupation."[22]

As far as the chemical industry was concerned, the commission calculated the damage done by confiscation and Micum shipments at 77.9 million RM, as against payments made by the Reich totalling 71.7 million RM.[23] BASF itself, in internal documents, came up with a calculation which showed that losses exceeded seven million RM (see page 61, table 2.)[24] It seems, then, that there can be no question of any overpayment in this sector, particularly since the compensation paid by the government related only to direct damage in terms of materials and products. The loss of production brought about by coal shortages since as early as February 1923, the occupation and shut-down of the works from May to October 1923, the controls and prohibitions on exports, etc. – all these things were of much greater importance.

### 4. France and the Haber-Bosch process

Apart from the matter of reparations, the Treaty of Versailles also contained a far more serious passage which threatened, in particular, the very existence of the chemical industry. The text in question was contained in paragraphs 164 et seq., which restricted or prohibited the production of arms, ammunition, munitions and poison gases and

---

concerning reparations; see Gerald D. Feldman and Heidrun Homburg, *Industrie und Inflation: Studien und Dokumente zur Politik der deutschen Unternehmen 1916-1923* (Hamburg, 1977), pp. 146 et seq. In two conventions (the Koblenz Convention of November 10, 1923 and the Paris Convention of December 14, 1923) the chemical industry undertook to supply dyes, pharmaceuticals and nitrogen products; BASF Archives, B 4/1175, Reparations.

21 *Reichstag*, IIIrd Legislative Period 1924/27, Report by the 23rd Commission (Commission of Enquiry into compensation in the Ruhr), publication number 3615, Berlin 1927, pp. 4 et seq.

22 Ibid., p. 4.

23 Ibid., pp. 108 et seq.

24 BASF Archives, B 4/1175, Reparations.

required disclosure of "the nature and method of production of all explosives, poisons or other chemical products." Precisely what this was supposed to mean was, of course, a matter for interpretation and further negotiation.[25] As far as BASF was concerned, its Haber-Bosch technology – then unmatched anywhere in the world – was at stake. Did the production of ammonia and nitric acid during the war justify the classification of Oppau as a munitions plant? If so, that would involve either the closure and demolition of the works or, at least, the disclosure of the complete production process. In either case, this meant enormous economic losses for BASF which was affected by these regulations more than any other German chemical company.[26]

Table 2. Value of confiscations and reparations during 1923/24 at BASF

| Damage | Value in million RM | Government payments in million RM |
|---|---|---|
| Confiscations | | Payments deducted |
| in 1923 | | in 1923 |
| - wood | 1.8 | 1.0 |
| - coal, iron, other materials, miscellaneous damage | 4.2 | 0.4 |
| - dyes | 13.3 | 8.4 |
| - nitrogen | 10.6 | 7.7 |
| Micum shipments | | |
| 1923-24 | 19.2 | |
| | 49.1 | 17.5 |
| | | Residual government payment in 1924/25 |
| | | 24.7 |
| | | 42.2 |

[25] There was no dispute about the destruction of installations built purely for the purposes of the war, which were dismantled under the supervision of an Allied military commission. The most important of these were the explosives plants built during the war at the Farbenfabriken Bayer. See Plumpe, op. cit. (2), p. 96.

[26] The other chemical companies along the Rhine were, of course, also affected by the French occupation policy. However, BASF was the only one of the "big three" (BASF, Bayer and Hoechst) on the left bank of the river, so that the French had continuous direct access to the works until they withdrew in 1930. Admittedly, Bayer and Hoechst were occupied in 1918-19 and 1923, after the two occasions on which the French established bridgeheads on the right bank. They too were subjected to requisitions and attempted espionage, but only for brief periods, whereas BASF had to deal with the French in the longer term. See, regarding Hoechst, Ernst Bäumler, *Die Rotfabriker. Familien-geschichte eines Weltunternehmens* (Munich, 1988), pp. 257 et seq.

France, in particular, was interested in rapidly building up a national nitrogen industry of its own. Admittedly, central government in Paris had access to the Haber-Bosch patents, which had also been applied for in France; not only that, but the French had had the opportunity for a thorough examination of BASF's ammonia works at Oppau. Even so, it seemed inadvisable to rely on these opportunities alone as a means of appropriating the complex high pressure technology.[27] General Patart, Inspector General at the French War Ministry, therefore offered negotiations with BASF early in 1919. His objective was to obtain the Haber-Bosch process and all the necessary know-how from BASF as the price of refraining from destroying the Oppau works. Carl Bosch, Chairman of the BASF Board of Executive Directors and a member of the German Negotiating Commission at Spa and Versailles, considered that the political and military situation left no other alternative open, and accepted the French offer. On 11 November 1919 – Armistice Day – a contract was signed in Paris between the government of France and BASF.[28] The preamble contained the following passage:

> BASF owns processes for the production of synthetic ammonia, which are used in its works at Oppau and Merseburg. BASF has taken out various patents in France to protect these processes. The French government, which desires to use and carry out these patents and processes in France, has regarded it as expedient, to this end, to secure the voluntary cooperation of BASF in connection with the utilization both of said patents and processes and of the keys to the industrial secrets which are necessary or useful for the purposes of manufacture and operation, and the use of more recent improvements or those which are still at the development stage.

What the contract meant, in real terms, was that BASF handed over to France – for exclusive use in France – the complete Haber-Bosch technology with all improvements to be achieved during the subsequent fifteen years. In return, BASF received a lump sum of 5 million francs plus, for fifteen years, a levy of a maximum of 12 centimes per kilo of nitrogen produced in France. Both contracting parties, incidentally, were heavily

---

[27] Naturally, the other Allies were also not backward in attempting to procure the sought-after Haber-Bosch technology by illegal means. Like the French chemists and technologists, the British and Americans took the opportunity for a thorough investigation of the ammonia plants at Oppau – evidently with no tangible success. It was only after two former BASF employees sold stolen blueprints to the English company Brunner, Mond & Co. (later ICI) that it became possible to use the Haber-Bosch technology in Britain, and later in the United States as well. See W.J. Reader, *Imperial Chemical Industries: A History*, 2 vols., Vol. 1: *The Forerunners, 1870-1926* (London, 1970), pp. 364 et seq. [Vol. 2: *The First Quarter Century* (1975)]. As in the case of the ammonia technology, France adopted a different route with dyestuffs. At the end of 1920, France concluded an agreement with the I.G. Farben forerunner companies which, like the "Office National Industriel de l'Azote" (ONIA) Agreement, provided for technology transfer in exchange for profit sharing and market agreements. However, this contract was unilaterally terminated by France in 1924, to the detriment of I.G. Farben. See Plumpe, op cit. (2), pp. 122 et seq.

[28] Text of contract in BASF Archives, B 4/479. For the negotiations, see Karl Holdermann, *Im Banne der Chemie. Carl Bosch, Leben und Werk* (Düsseldorf, 1953), pp. 168 et seq.; Plumpe, op. cit. (2), p. 221.

criticized in their own countries, especially in 1923 when the contract was, after many delays, discussed by the French parliament. The argument advanced was "Why should we pay for something which we could have got for nothing?" In Germany, BASF was accused of treason, especially in view of the conflict in the Ruhr.[29] Despite all the animosity and many problems on both sides, the contract survived. French nitrogen production began in 1928 at a factory which had been built in Toulouse, and neither the state-owned French "Office National Industriel de l'Azote" (ONIA) nor BASF/I.G. Farben interfered with the contract during the 15-year period from 1928 onwards. ONIA continued to receive BASF know-how connected with ammonia, and concluded a whole series of additional contracts with BASF, the last relating to mixed fertilisers in 1938.[30]

Although in the early 1920s BASF was still mainly influenced by the negative aspects associated with the occupying power – occupation of the works, confiscations and reparations – the positive experience of working with France under the ONIA contract soon became more important. In an article for the *Berliner Börsen-Courier* of 17 February 1931, Carl Bosch took this situation as a basis for arguing in favour of close economic and political ties with France. Political resentments, he said, would have to be overcome on the basis of "economic sense" and "the foundations laid for bringing greater stability to Germany and France, and so to Europe." For BASF/I.G. Farben, at any rate, the ONIA contract became the successful model of a trans-national technology transfer and set the pattern for similar activities during the remainder of the interwar period.

---

29 Cf. BASF Archives, C 10, Report to the Members of the Supervisory Board dated February 19, 1923; ibid., collection of press cuttings.

30 BASF Archives, B 4, ONIA records.

# Part 3

## Science and Industry

# BASIC RESEARCH IN INDUSTRY: TWO CASE STUDIES AT I.G. FARBENINDUSTRIE AG IN THE 1920'S AND 1930'S

CARSTEN REINHARDT
*Universität Regensburg*
*Institut für Philosophie*
*Lehrstuhl für Wissenschaftsgeschichte*
*93040 Regensburg, Germany*

**Introduction**

Historian of technology Wolfgang König has recently defined the often misinterpreted term "science-based industry." König introduced four characteristics for the evaluation of the scientific foundation of an industry:[1]

1. Transfer of knowledge by university graduates entering positions in industry;
2. Direct transfer of research results from university to industry;
3. Fundamental scientific discoveries that later become technological developments in industry; and
4. Technological developments that originated in research performed in industrial research departments.

The chemical industry includes several good examples of each of these. In order to understand better the role that science played in industrial research I here investigate the German chemical industry in the interwar period through a case study which focuses on a particular episode in the development of the industrial research laboratory. I argue that the study of the industrial research laboratory, its organization and the research projects carried out there, is a fruitful method to explore determinants of technological progress in the science-based chemical industry during the 19th and 20th centuries.[2]

---

1 Wolfgang König, "Science-based industry or industry-based science? Electrical engineering in Germany before World War I," *Technology and Culture*, **37** (1996), 70-101, on pp. 71-72.

2 The best example is the study of research at Du Pont. See David A. Hounshell and John K. Smith, Jr., *Science and Corporate Strategy. Du Pont R&D, 1902-1980* (Cambridge, Mass.: Cambridge University Press, 1988). For the German case see John Joseph Beer, "Coal tar dye manufacture and the origins of the modern industrial research laboratory," *Isis*, **49** (1958), 123-131; Georg Meyer-Thurow, "The



*A.S. Travis et al. (eds.),*
***Determinants in the Evolution of the European Chemical Industry, 1900–1939,*** **67–88.**

It is significant that in the 1920s both Du Pont and I.G. Farbenindustrie AG, the two largest chemical companies in the United States and Germany, respectively, introduced basic research in their industrial research departments. Up to this date, research in the chemical industry was usually concerned with the improvement of existing technology, and the invention and development of new products and manufacturing processes. Most commonly this sort of study is referred to as applied research. It is difficult and in some cases even pointless to distinguish between applied, basic, and pure research. These labels were too often introduced for political and tactical reasons by contemporary participants to constitute a definitive foundation for historical research. Their use was especially important for the introduction of the hierarchy model of the science-technology relationship.[3] Nevertheless, what I describe is a change in content and style of industrial research that is historiographically accessible only by the retention of the terms 'applied' and 'basic' research. It is not my intention to revert to the obsolete hierarchy model of technology as applied science; newer historiographic scholarship has demonstrated the autonomy of these two activities and has focused on the relevance of technology as independent knowledge.[4]

One definition of applied science that usefully avoids the hierarchy model was given by Derek de Solla Price in 1984. In the context of the 18th century he wrote about applied science:

> In a strong sense, this term is a misnomer. The process is not an application of basic science to industrial needs. It is an attack by the methods of science on a particular technology. When we study the world of nature, the result is basic science. When we study the artifactual world of techniques, the result is applied science. [5]

---

industrialization of invention: A case study from the German chemical industry," *Isis*, **73** (1982), 363-381; Ernst Homburg, "The emergence of research laboratories in the dyestuffs industry, 1870-1900," *British Journal for the History of Science*, **25** (1992), 91-111; Ulrich Marsch, "Strategies for success: Research organization in German chemical companies and I.G. Farben until 1936," *History and Technology*, **12** (1995), 23-77.

3 See the contribution of Geert Somsen in this volume.

4 Edwin Layton, "Mirror image twins: The communities of science and technology in 19th century America," *Technology and Culture*, **12** (1971), 562-580. See the whole 1976 October issue of *Technology and Culture*, especially Otto Mayr, "The science-technology relationship as a historiographic problem," *Technology and Culture*, **17** (1976), 663-673; the essay of George Wise, "Science and technology," *Osiris*, **1** (1985), 229-246; and, most recently, Ronald Kline, "Construing technology as applied science. Public rhetoric of scientists and engineers in the United States, 1880-1945," *Isis*, **86** (1995), 194-221.

5 Derek de Solla Price, "The science/technology relationship, the craft of experimental science, and policy for the improvement of high technology innovation," *Research Policy*, **13** (1984), 3-20, p. 7.

Apart from the difficulty of distinguishing between the natural and artifactual world, at least in the 20th century, there is another point worth mentioning here: Had I accepted such a definition, then the title of my contribution would be misleading, because nearly all of the research carried out in industry would be labeled "applied science." This very broad definition does not permit a study of the fine structure of research carried out in industry.

The historian of research at General Electric and Bell Telephone, Leonard Reich, has characterized industrial research as follows:

> Industrial laboratories set apart from production facilities, staffed by people trained in science and advanced engineering who work toward deeper understandings of corporate-related science and technology, and who are organized and administered to keep them somewhat insulated from immediate demands yet responsive to long-term company needs. [6]

I would like to draw attention to "work toward deeper understandings of corporate-related science and technology." If we accept this as a fitting definition of industrial research we have to exclude the greater part of research on dyestuffs, pharmaceuticals, plastics, and other materials carried out within and for industry. But the definition of Reich is exactly the definition I would like to use for the term "basic research," which comes close to the definition of Lillian Hoddeson, who refers to "fundamental studies carried out in the context of industry, which may lead to, but do not primarily aim at, application."[7]

Before World War I chemical industrial research was focused on the invention of new products and processes and the improvement of existing technology: it was certainly applied research. Basic research in the sense of the definition of Hoddeson and in part of Reich was a newcomer to the chemical industry in the 1920s. The study of this newcomer seems worthwhile to establish a better understanding of the role of basic research in technology and its relation to applied research. At least it should be possible to show the extent to which leaders of industrial research in the 1920s were willing to contribute not only to technological, but also to scientific progress.

In 1927 the management of Du Pont initiated a basic research programme that led to the discovery of neoprene synthetic rubber and nylon. The programme encompassed colloid chemistry, catalysis, the gathering of physical and chemical data, organic

---

6 Leonard S. Reich, *The Making of American Industrial Research. Science and Business at GE and Bell, 1876-1926* (Cambridge, Mass.: Cambridge University Press, 1985), p. 3.

7 Lillian Hoddeson, "The emergence of basic research in the Bell Telephone system, 1875-1915," *Technology and Culture*, **22** (1981), 512-544, p. 514, note 3.

synthesis, and polymerization.[8] At the same time, the directors of at least two research laboratories at I.G. Farbenindustrie AG started a similar programme.

I.G. Farben was the outcome of a merger of the leading German chemical companies in 1925, including the Badische Anilin- & Soda-Fabrik (BASF), of Ludwigshafen, Rhine, the Farbenfabriken Bayer, Leverkusen, and the Farbwerke Hoechst, Höchst-on-the-Main, and others.[9] It is my intention to analyze the basic research programmes of the Central Research Laboratory at Ludwigshafen and the Ammonia Laboratory at Oppau, near Ludwigshafen, both formerly BASF, in the period between the two world wars. To achieve this it is appropriate to describe the organizational development of the two laboratories and to evaluate the position of basic research in an industrial research laboratory in relation to applied research. I intend to show some of the reasons for the initiation of the basic research programmes at these two laboratories, as well as for the end of the engagement in basic research.

**The research organization at I.G. Farben Ludwigshafen, 1926**

The organization of research at I.G. Farben from 1926 up to 1936 has been well described by Ulrich Marsch.[10] Each of the main plants of I.G. Farben usually had one large central research laboratory, several departmental laboratories, dye testing facilities, technical stations, and analytical laboratories.

In 1926 the situation at the former BASF plant in Ludwigshafen was as follows: Each of the production departments had its own laboratory, with the exception of the aniline and azo dye department.[11] The chemists in the various laboratories undertook research work in the fields of commercial interest for their departments (See table 1).

Table 1: Departmental Laboratories at I.G. Farben's Ludwigshafen Plant, 1926

| **Laboratory** | Alizarin | Acid | Indigo | Analytical |
|---|---|---|---|---|
| **Number of chemists** | 14 | 7 | 13 | 3 |
| **Research fields** | Alizarin and vat dyes | Oxidation catalysis, organic intermediates, titan. | Indigo and related products, solvents and intermediates | Routine analytical work |

8 Hounshell and Smith, op. cit. (2), pp. 223-248.

9 Gottfried Plumpe, *Die I.G. Farbenindustrie AG. Wirtschaft, Technik und Politik 1904-1945* (Berlin: Duncker & Humblot, 1990).

10 Marsch, op. cit. (2).

11 "Bericht über die wissenschaftlichen Laboratorien der Farbenfabriken" (TEA-Beschluss vom 28. Oktober 1926), Ludwigshafen a/Rhein, June 1927. Bayer-Archiv Leverkusen (BAL) 103/17.05.5. The author of this report was Kurt H. Meyer, head of the Central Research Laboratory at BASF, Ludwigshafen.

In contrast, the Central Research Laboratory at Ludwigshafen did not engage in research areas related explicitly to production work. Seven chemists undertook research work on the synthesis of artificial dyestuffs, fifteen on catalytic reactions for the production of intermediates, and ten in the fields of synthetic tanning agents, cellulose, colloid chemistry, dyeing and printing processes, and detergents. Four chemists, including two women, were engaged in service work (analysis, literature, journal reviews).[12] With 36 chemists, the Central Research Laboratory was by far the largest in Ludwigshafen. This research organization had its origins in strategies adopted before World War I.[13]

In 1926 more than 200 research chemists at I.G. Farben were working in the field of dyestuffs synthesis, representing 20 per cent of all research chemists in the whole company. (The 291 chemists and technicians at dye testing stations are not included in this figure.) Therefore, nearly 50 per cent of the research chemists and technicians were involved in research and development work for dyestuffs.[14] However, sales of newly discovered dyestuffs were not sufficient to justify this enormous research effort. K.H. Meyer, director of research at Ludwigshafen, therefore proposed to modify the organization and orientation of research in two directions: First, research and works chemists should be centralized in larger laboratories, in order to allow better communication between chemists working in the same field. Also, the planning of research in specific areas should be centralized all over the company through scientific commissions, provided that such work had not already been initiated.[15] Second, the research topics should be changed from dyestuffs chemistry to projects in the chemistry of synthetic rubber, detergents and cellulose.[16] In a report on the training of chemists and engineers, dated August 1926, K.H. Meyer evaluated current dyestuffs research in somewhat extreme terms:

> Today we must recognize that large parts of our manufacturing departments are laid down. In addition, as a result of the merger [of I.G. Farbenindustrie AG in 1925] a great part of competitive work, formerly done in laboratories

---

12 Ibid.

13 Carsten Reinhardt, "An instrument of corporate strategy: The central research laboratory at BASF, 1868-1890," in Ernst Homburg, Harm Schröter, and Anthony S. Travis, eds, *The Chemical Industry in Europe, 1850-1914; Industrial Growth, Pollution and Professionalization* (Dordrecht: Kluwer, 1998).

14 "Bericht über die wissenschaftlichen Laboratorien der Farbenfabriken" (TEA-Beschluss vom 28. Oktober 1926), Ludwigshafen a/Rhein, June 1927. Summary of Laboratories and Dye testing stations, first half of 1926, p. 11. BAL 103/17.05.5.

15 See Marsch, op. cit. (2), p. 46.

16 "Bericht über die wissenschaftlichen Laboratorien der Farbenfabriken" (TEA-Beschluss vom 28. Oktober 1926), op. cit. (14), pp. 9-10.

to analyze the products of competitors and to circumvent their patents, is unnecessary. Today we see that I.G. employs too many organic chemists. We must take them out of the graveyards of the closed production facilities and confront them with problems in physical chemistry. [17]

Kurt H. Meyer (1883-1952), during 1921-1932 director of research at I.G. Farbenindustrie AG, Ludwigshafen. (Courtesy of BASF Ludwigshafen.)

**The Central Research Laboratory at BASF, 1921-1938**

Until World War I the synthesis of artificial dyestuffs and intermediates was the only research topic of the Central Research Laboratory.[18] Minor exceptions were synthetic

[17] Kurt H. Meyer, "Über die Ausbildung der Chemiker und Ingenieure an Universitäten und Hochschulen als Vorbereitung für industrielle Tätigkeit," unpublished manuscript, August 1926, p. 4. CA BASF W1 Meyer. The citation is in German: "Schon heute sehen wir, daß große Teile unserer Fabriken vollkommen still gelegt sind. Ferner fällt durch die Fusion ein großer Teil überflüssiger Konkurrenzarbeit weg, die früher in Laboratorien zum Zwecke der Analyse fremder Produkte und zum Zwecke der Patentumgehung erforderlich war. Schon heute erkennen wir, daß die I.G. viel zu viel organische Chemiker besitzt. Wir müssen sie aus den Friedhöfen der stillgelegten Betriebe herausnehmen und meistens mit Problemen der physikalischen Chemie beschäftigen." See, for confirmation of the relevance of this opinion, the letters of Carl Bosch and K.H. Meyer to the Ministerium der geistlichen und Unterrichts-Angelegenheiten, and to Richard Willstätter in 1925/1926, quoted in Eric Elliott, "Die I.G. Farbenindustrie. Ein Thema für Wissenschaftshistoriker?," *Mitteilungen der Fachgruppe Geschichte der Chemie*, **5** (1991), 10-15, esp. p. 13.

[18] Reinhardt, op. cit. (13).

tanning agents and some work on catalytic hydrogenation, begun by Otto Schmidt just before the war, and work on detergents by Fritz Günther, originated in 1917.[19]

This situation changed substantially when in April 1921, Kurt H. Meyer was appointed leader of the Central Research Laboratory.[20] Although an organic chemist, Meyer was interested in the emerging field of physical chemistry. In the years from 1921 to 1932 he extended the research topics of his laboratory from dyestuffs chemistry into the field of natural and synthetic fibres, plastics, and related intermediates. But for the purpose of this paper, his attitude towards fundamental questions such as the constitution of natural and synthetic polymers, and the mechanisms of catalytic reactions is more significant. Meyer inaugurated a short period of basic research in the Central Research Laboratory.

The beginnings of basic research at BASF were somewhat hesitant and still related to the needs of dyestuffs chemistry. In collaboration with Hans Fikentscher, Curt Schuster and W. Bülow, K.H. Meyer established the physical and chemical principles of the dyeing process of cellulose acetate and nitrate, silk, and wool.[21] This research ran parallel to the efforts of the laboratory to invent dyestuffs for the new artificial fibres, such as cellulose acetate (rayon).[22]

In the mid twenties several circumstances permitted a revised orientation of both content and style of areas of research at the Central Research Laboratory. In 1925 the Technical Committee (TEA) of I.G. Farben discussed the possibilities for producing artificial fibres.[23] At this time, I.G. Farben was the second largest producer of artificial fibres in Germany.[24] Therefore polymer chemistry became more important for the company at the same time as dyestuffs chemistry lost its former position. However, the science of synthetic, semi-synthetic and natural polymers was not yet established in the same way as structural chemistry was for organic dyestuffs, pharmaceuticals, and intermediates. Colloid chemists regarded substances such as cellulose, silk, and wool as

---

19 O. Hecht, "Otto Schmidt (1874-1943)," *Berichte der Deutschen Chemischen Gesellschaft*, **76 A** (1943), 121-125; Karl Saftien, "Fritz Günther," *Chemische Berichte*, **92** (1959), xxix-xxxv.

20 H. Hopff, "Kurt H. Meyer (1882-1952)," *Chemische Berichte*, **92** (1959), cxxi-cxxxvi. See also the contract of Kurt H. Meyer with BASF, Munich, 29 November 1920. CA BASF W1 Meyer.

21 Summarized in Kurt H. Meyer, "Zur Physik und Chemie der Färbevorgänge," *Die Naturwissenschaften* **15** (1927), 129-134. See the publications of K.H. Meyer and his collaborators in *Melliand Textilberichte* **6** (1925), 737; **7** (1926), 29, 605; **8** (1927), 781; **9** (1928), 573.

22 See Fritz Günther, "Das Hauptlaboratorium der Badischen Anilin- und Soda-Fabrik Ludwigshafen am Rhein, 1868-1938," pp. 39-41. Unpublished manuscript, CA BASF D 101/1.

23 Marsch, op. cit. (2), pp. 53-54.

24 Plumpe, op. cit. (9), pp. 306-308.

relatively small molecules, bound together by special forces. In contrast, scientists such as Hermann Staudinger and others were of the opinion that these substances consisted of long covalently bonded chains, that is, macromolecules or polymers.

In December 1925 Meyer was appointed a member of the board of I.G. Farben.[25] Around the summer of 1926 he decided to establish a research programme in the Central Research Laboratory dedicated to an area which became known as polymer chemistry.[26] The core of the project was the department of Herman F. Mark, one of the leading specialists in x-ray analysis of natural fibres.[27] When hired on 1 January 1927, Mark assembled a team of physicists to study the structures of fibres and light metals. Georg von Susich and Emery Valkó moved with him from the Kaiser-Wilhelm-Institut für Faserstoffchemie in Berlin-Dahlem to Ludwigshafen; Raimund Wierl, Karl Wolf and Josef Hengstenberg joined the group in the following year.[28] Together they occupied a modern laboratory in the former building of the Ludwigshafen alizarin laboratory. Von Susich undertook x-ray analysis on the structure of cellulose and rubber; Wierl used electron diffraction to investigate the structure of relatively simple organic substances; and Wolf and Hengstenberg used analytical instruments to study the structure of polymers and light metals.[29]

The physical laboratory, complete with x-ray apparatus, electron diffraction, viscosity and osmotic measurement equipment, was under the direct guidance of Karl Wolf. When Mark took over it was not the only department of the Central Research Laboratory. There was a synthetic organic laboratory under the leadership of Heinrich Hopff, the personal assistant of K.H. Meyer, and a technical station, called Technion, for the synthesis of amounts of polymer substances sufficient to permit study of their properties. The latter group was under Manfred Dunkel. The research group of Otto Schmidt studied catalytic reactions and synthetic rubber chemistry.[30]

---

25 Biographical information, dated 7 November 1961. CA BASF W1 Meyer.

26 I use the term "polymer chemistry" despite the fact that it came into use only after the mid twenties. The contemporary expression was "fibre research." Chemical Heritage Foundation, Transcript of Interviews conducted by J.J. Bohning and J.L. Sturchio with Herman Mark, 3 February, 17 March and 20 June 1986, pp. 15-16.

27 Herman F. Mark, *From Small Organic Molecules to Large* (Washington DC: ACS, 1993); G.A. Stahl, ed., *Polymer Science Overview. A Tribute to Herman F. Mark* (Washington DC, 1981).

28 Mark, op. cit. (27), pp. 36, 39.

29 Günther, op. cit. (22), p. 76.

30 Chemical Heritage Foundation, Transcript of Interviews conducted by J.J. Bohning and J.L. Sturchio with Herman Mark, 3 February, 17 March and 20 June 1986, p. 19.

The decision of K.H. Meyer to engage in basic research was probably influenced by the example of the pioneering work of Irving Langmuir at General Electric, especially studies on the surface structure of catalysts.[31]

**Polymers**

> The most important ingredient for progress was a clear-cut superiority in one of the leading experimental techniques. We had it in Dahlem in the x-ray field, and we rebuilt it in Ludwigshafen by adding electron diffraction and spectroscopy (optical and IR).[32]

In his autobiography, Mark characterized the preconditions of scientific progress, which he found also in the industrial laboratory in Ludwigshafen. K.H. Meyer and Mark decided to use analytical equipment to investigate the structure of natural and semi-synthetic fibres. In a paper published in 1928 they developed the theory of cellulose chains consisting of primary valency chains. They proposed, in addition, that these chains were aggregated in the form of micelles by special molecular forces, later identified with hydrogen bonding and van der Waals' forces. Meyer and Mark were able to explain the physical properties of cellulose, and soon afterwards of other natural polymers, such as silk and rubber.[33] The theory of Meyer and Mark, best described in a book published in 1930,[34] gave rise to a controversy over priority with Hermann Staudinger, the most famous proponent of the macromolecular view.[35]

Though I do not wish to reiterate the well known history of polymer science, it is worthwhile emphasizing that these two industrial scientists made important contributions to pure science, as they had intended. They investigated fundamental questions concerning the constitution of polymers, without envisaging any immediate use of their work in industry. Moreover, Mark and his collaborators had the opportunity to do scientific work that was not connected with polymers and metals. They published papers on the width of x-ray emission lines, the optical Stark effect, molecular structure

---

[31] Curt Schuster, *Geschichte des Hauptlaboratoriums 1868-1961*, p. 27. CA BASF D 101/2. Curt Schuster was one of the graduate students of Meyer in Munich and joined the Central Research Laboratory at BASF with K.H. Meyer in 1921.

[32] Mark, op. cit. (27), p. 37.

[33] A good historical summary is G.A. Stahl, "A short history of polymer science," in Stahl, ed., op. cit. (27), pp. 25-44, esp. p. 37. The original paper is: K.H. Meyer, H. Mark, "Über den Bau des kristallisierten Anteils der Cellulose I," *Berichte der Deutschen Chemischen Gesellschaft*, **61** (1928), 593-614.

[34] Kurt H. Meyer and Hermann Mark, *Der Aufbau der hochpolymeren organischen Naturstoffe* (Leipzig: Akademische Verlagsgesellschaft, 1930).

[35] Claus Priesner, *H. Staudinger, H. Mark und K.H. Meyer. Thesen zur Größe und Struktur der Makromoleküle* (Weinheim: VCH, 1980).

investigated by electron diffraction, and other seemingly unrelated fields.[36] The electron diffraction work, done by Mark and Raimund Wierl, inspired Linus Pauling to start his famous investigations of molecular structure.[37] The electron diffractometer was originally introduced at Ludwigshafen to study the mercerization of cellulose, and the extension and contraction of rubber, but Wierl and Mark also used the method to establish the bond lengths of relatively small molecules (such as $CCl_4$, $CHCl_3$, $C_2H_6$, and others).[38]

The physicists group of Herman Mark in December 1930. First row, from left: H. Mark, J.A. Prins (University of Groningen), R. Wierl. Second row, from left: J. Hengstenberg, R. Brill, G. von Susich, Pelzer, K. Wolf. (Courtesy of BASF Ludwigshafen.)

**Catalysis**

A second important project of the Herman Mark group was the investigation of catalysis. Catalytic reactions were the underlying principle of many industrial processes, but not well understood at this time. Therefore empirical methods, a sort of trial and error, were the only way to improve upon and find new catalysts.

---

36 G.A. Stahl, "Herman F. Mark: The Geheimrat," in Stahl, ed., op. cit. (27), pp. 61-88, esp. p. 64.

37 Linus Pauling, "Herman F. Mark and the structure of crystals," in Stahl, ed., op. cit. (27), pp. 93-99.

38 Mark, op. cit. (27), p. 50.

There was a certain amount of collaboration of Mark with Otto Schmidt, who led an independent group in the Central Research Laboratory. As far as I has been able to determine, this collaboration focused on the field of synthetic rubbers, Buna-S and Buna-N, a second important topic in the laboratory of Otto Schmidt, and not catalysis.[39] Catalytic reactions were also studied in the Ammonia Laboratory of Alwin Mittasch (mentioned below).

The catalytic group of Herman Mark, and the chemists Hans Dohse, W. Kälberer, and Curt Schuster, started its investigations in the summer of 1927. The group decided to study catalytic phenomena systematically and began with the first stage of heterogeneous catalytic reactions, the adsorption of gases on solid substances. Their intention was to establish:

a) numerical data for the effectiveness of a specific catalyst,
b) the physical properties of the catalyst, and
c) the correlation of these two points.[40]

To achieve these goals they chose a model, the adsorption of certain gases, such as carbon dioxide, ethylene, and argon, on silica gel. Silica gel is not a good catalyst, certainly not for industrial processes, but it is a good adsorbent. This model substance allowed the correlation of the surface of the catalyst, and its activation energy, with the catalytic reactivity. One of the greatest difficulties was the exact measurement of the active surface of the catalyst. The group worked until the beginning of 1929 to solve this problem.[41] Parts of their work were published even before the completion date of the I.G. Farben internal reports.[42]

Between 1925 and 1932, at least 97 scientific publications arose from the Central Laboratory, 45 published by the physical group and 52 by their chemical collaborators.[43] In the same period, many academic scientists were invited to

---

[39] Hecht, op. cit. (19), p. 123. Otto Schmidt, "Systematische Untersuchungen über die Absorption von Gasen an festen Körpern," research report dated 22 January 1927, no. 1719. CA BASF D 02.2/5.

[40] H. Dohse, W. Kälberer and C. Schuster, "Untersuchungen über Adsorption und Katalyse an Kieselsäuregel," research report dated 8 February 1928, p. 1. CA BASF D 02.2/5.

[41] See the research reports of H. Mark and W. Kälberer, "Untersuchungen über Adsorption und Katalyse II," 12 January 1929; W. Kälberer and C. Schuster, "Untersuchungen über Adsorption und Katalyse III," 20 January 1929; H. Dohse, "Bestimmung der aktiven Oberfläche von Aluminiumoxyden und Silikagelen für die Dehydratisierung von Alkoholen (Untersuchung über Adsorption und Katalyse IV)," 1 February 1929. CA BASF D 02.2/5.

[42] W. Kälberer and H. Mark, "Beiträge zur Kenntnis des Adsorptionsvorganges," *Zeitschrift für physikalische Chemie*, **139** (1928), 151-162.

[43] See the lists of scientific publications of the Central Research Laboratory in Curt Schuster, "Die Geschichte des Hauptlaboratoriums 1938-1957," pp. 245-253, CA BASF D 101/3, and Günther, op. cit. (22),

Ludwigshafen to present their research projects.[44] This was part of the strategy of K.H. Meyer to introduce an academic style of research in the Central Research Laboratory.

**Application**

> What was obviously expected from our newly founded laboratory was to use the new basic understandings and insights to arrive at desirable and necessary practical results. Now we would have to expand into synthesis and application.[45]

Certainly the most important task of the Central Research Laboratory was the invention of new products and processes. The polymer group had more freedom than other groups in the laboratory, because in the field of polymers the needs of the production departments were not as dominant as in the field of dyestuffs, detergents, and organic intermediates.[46]

There did, however, exist some departments for the production of artificial fibres at I.G. Farben, and Mark and his group performed important service functions for these departments. Together with chemists of the fibre division they developed a theory of solution spinning, a new high strength cellulose acetate, and the copolymerized synthetic rubbers Buna-S and Buna-N. The most famous new product of Mark and his assistants was polystyrene, then used as a thermoplastic.[47]

As we have seen, the beginnings of Meyer's interest in fibre research were based on the dyeing processes of fibres. One of his collaborators was Hans Fikentscher, who later joined the Mark group. At this time, the mid twenties, the production of cellulose nitrate and acetate was of major importance for the manufacture of films, lacquer, and artificial silk at I.G. Farben. The works chemists had some difficulties explaining the irregularity of cellulose nitrate, which Fikentscher proved to be a consequence of irregularities in viscosity.[48] He established exact measurements of the viscosity of the solutions used and in 1927 proposed an equation which established a correlation between

---

pp. 76-79, CA BASF D 101/1.

44 A list is printed in Schuster, op. cit. (43), pp. 254-257. CA BASF D 101/3.

45 Mark, op. cit. (27), p. 43.

46 Chemical Heritage Foundation, Transcript of Interviews conducted by J.J. Bohning and J.L. Sturchio with Herman Mark, 3 February, 17 March and 20 June 1986, p. 22.

47 Mark, op. cit. (27), pp. 44-51.

48 Schuster, op. cit. (31), pp. 38-39, and H. Fikentscher, "Untersuchung von Nitrozellulose," research report 18 May 1927, CA BASF D 02.2/5.

concentration and viscosity of specific cellulose acetate and nitrate solutions and the size of the particles in solution.[49] The investigation of the viscosity of polymers was critical to the establishment of their molecular weights. In this respect Fikentscher made an important contribution to basic science.

**The end of the period of basic research**

With the onset of the economic crisis of the 1930s the period of basic research at the Central Research Laboratory, Ludwigshafen, came to an end. The management of the Central Laboratory, Meyer, Mark and the group leaders, turned to economy measures. Only individual chemists at the lower levels had enough time to continue their scientific and technical work.[50] The statistical data available show that many of the clerks and labourers in the laboratories were dismissed or removed. The number of chemists and physicists remained constant.

In March 1930 the Central Research Laboratory employment situation, according to official figures, was as in table 2.[51]

Table 2.

| **Group-Leaders** | **Research field** | **Number of chemists** | **Number of clerks and labourers.** |
|---|---|---|---|
| Kunz-Günther | Dyestuffs | 11 | 28 |
| Dye-testing station | | - | 9 |
| Dunkel | Intermediates | 5 | 31 |
| Schuster-Dohse | Catalysis | 2 | 12 |
| Hopff | Synthetic rubbers | 12 | 25 |
| Schmidt | Synthetic rubbers | 8 | 62 |
| Fikentscher | Synthetic rubbers | 8 | 18 |
| Technical Station | - | 1 | 23 |
| Information department and secretary of Meyer | - | 4 | 2 |
| **Total** | - | **51** | **210** |

The group of Mark and his physicists is not included in the sources; also the assignment of the groups of Fikentscher, Hopff, and Schmidt to synthetic rubbers is misleading. Hopff and Schmidt were undertaking research in the field of organic intermediates, while Fikentscher was working mainly in the field of synthetic polymers.

49 See H. Mark, "Über Celluloseesterlösungen," research report 24 December 1927, CA BASF D 02.2/5 and H. Fikentscher, "Die Viskosität der lyophilen Kolloide," research report 29 November 1928, CA BASF D 02.2/5.

50 Letter of Herman Mark to Otto Schmidt, Ludwigshafen, 28 November 1931. CA BASF W1 Schmidt.

51 "Zusammenstellung der in den wissenschaftlichen Laboratorien beschäftigten Chemiker, Angestellten und Arbeiter." TEA-Büro, 21 March 1930. CA BASF D 05/2.

In 1931 the number of assistants (clerks and labourers) in the Central Research Laboratory was reduced to 137, and as a result of further economy measures was later reduced to 100. The number of chemists, then 55, remained stable.[52] Most of the clerks and labourers were transferred to manufacturing departments; the rest were dismissed.[53]

In 1932 the political situation in Germany worsened. The Nazi party won 230 mandates out of the 608 in the German parliament, the *Reichstag,* in July. K.H. Meyer and Herman Mark were of Jewish descent, as were some of their collaborators in the Central Research Laboratory. Meyer left I.G. Farben in January 1932, probably because of disagreements at board level about the future of the company.[54] It does not seem that he resigned because of the political and social circumstances in Germany. In October 1932 he was appointed professor of chemistry at the University of Geneva, Switzerland. Herman Mark left Ludwigshafen in September 1932 for a professorship at the University of Vienna, Austria. In 1938 he emigrated to Canada and later continued his distinguished career in polymer science at Brooklyn Polytechnic, New York. In contrast to Meyer, Mark himself related his departure to Vienna explicitly to antisemitic tendencies in Germany.[55]

At the end of 1932, Arthur Lüttringhaus for a short time received the post of leader of the Central Research Laboratory. He was a vat dyestuff specialist and also the director of the alizarin laboratory in Ludwigshafen. After the retirement of Lüttringhaus in January 1934, Max Kunz and Fritz Günther, who specialized in dyestuffs and detergents respectively, led the laboratory until Walter Reppe took over in 1938.[56] The focal point of research at the Central Research Laboratory changed to more technical problems in the fields of dyestuffs, polymers, and organic intermediates.

The members of the physical group of Herman Mark also changed their sphere of work. Georg von Susich studied dyestuffs using x-ray techniques; Karl Wolf pursued research on the development of emery paper and separation of pulverized substances; Josef Hengstenberg studied analytical methods; and Emery Valkó investigated problems in the use of detergents. Both von Susich and Hengstenberg also investigated the mixed polymerization invented by Hans Fikentscher.[57] Emery Valkó emigrated in 1939 to the

---

52 Letter of Herman Mark to [Fritz] ter Meer. Undated, probably February 1931. CA BASF D 05/2a.

53 Letter of K.H. Meyer to TEA-Office, Ludwigshafen, 30 May 1931. CA BASF D 05/2a.

54 Private communication from Dr. Ute Deichmann, Cologne.

55 Mark, op. cit. (27), pp. 61-63.

56 See the contribution of Peter Morris in this volume.

57 Günther, op. cit. (22), pp. 79-81.

United States.[58]

**The Ammonia Laboratory**

The Ammonia Laboratory had its origins in the investigations of Carl Bosch, Alwin Mittasch and their collaborators studying the high pressure synthesis of ammonia. The group was originally situated in a works laboratory of the acid department at BASF, Ludwigshafen, and was particularly successful in the technical development of the Haber-Bosch synthetic ammonia process from 1909 up to 1913.[59] The Ammonia Laboratory received a more independent status with the foundation of a separate nitrogen department and erection of a plant for the production of synthetic ammonia at Oppau, near Ludwigshafen, in 1912. In April 1918 the staff of the Ammonia Laboratory moved to a new building in Oppau; by 1928 it had become the largest laboratory in the chemical industry of Germany, and one of the largest industrial research laboratories in the world.

Table 3: Number of academic scientists at the Ammonia Laboratory, Oppau, 1914-1938[60]

| | | | | | | | | | | |
|---|---|---|---|---|---|---|---|---|---|---|
| **1914** | **1918** | **1919** | **1920** | **1921** | **1922** | **1923** | **1924** | **1925** | **1926** | **1927** |
| ca. 9 | ca. 10 | 21 | 36 | 40 | 52 | 58 | 54 | 55 | 80 | 120 |
| **1928** | **1929** | **1930** | **1931** | **1932** | **1933** | **1934** | **1935** | **1936** | **1937** | **1938** |
| 165 | 168 | 166 | 157 | 116 | 104 | 98 | 104 | 104 | 111 | 124 |

The research projects were periodically hindered by, successively, World War I, a massive explosion in 1921, which destroyed the larger part of Oppau, and the French occupation in the summer of 1923. The projects included catalytic reactions for production of simple nitrogen and carbon compounds. During World War I the staff of the laboratory developed a process to produce nitric acid by the oxidation of ammonia in the presence of a catalyst that was not based on platinum. The most important project after the war was the technical development of fertilisers, based on ammonia and saltpetre production.[61]

By the end of 1921 an organization had evolved at the Ammonia Laboratory which showed a characteristic division into groups. The composition of the analytical and the administration groups is excluded in the sources.

58 "Emery I. Valkó 16th Olney Medalist," *American Dyestuff Reporter*, (1959), 109-113.

59 See the contribution of Anthony S. Travis on the Haber-Bosch process in this volume and Alwin Mittasch, *Geschichte der Ammoniaksynthese* (Weinheim: VCH, 1951).

60 Alwin Mittasch, *Zur Geschichte des Ammoniaklaboratoriums*, 1933, pp. 6, 25-27, 88-89. Hans Georg Grimm, *Geschichte des Ammoniaklaboratoriums*, vol. 3, 1938, p. 539. CA BASF D 201/2. The numbers of academic scientists are given for July 1914; April 1918; from 1919 up to 1923, for the end of the year; from 1924 up to 1938, for 1 January.

61 Mittasch, op. cit. (60), pp. 5-22, 31 ff.

Table 4: Organization of the Ammonia Laboratory, 1921[62]

| Group | Research field | Number of scientists |
|---|---|---|
| I | Patents and literature; agricultural chemistry; nitrogen (esp. ammonia, nitric acid) | 11 |
| II | Gases | 8 |
| III | Salts; material testing | 3 |
| IV | Catalysis (organic chemistry) | 6 |
| V | Physics | 3 |

The staff of the Ammonia Laboratory increased enormously in the period from 1924 up to 1928 (see table 3). In 1938 the Ammonia Laboratory had roughly 30 groups.[63]

Mittasch himself saw several reasons for this trend, including:[64]

1. The high rate of growth of the synthetic ammonia industry and the corresponding manufacturing departments of I.G. Farben in Oppau and Leuna. These departments called for new chemists to join the production facilities via the Ammonia Laboratory;
2. the addition of new research projects to the repertoire of the Ammonia Laboratory (especially research in plant physiology and biochemistry, additional work on the high pressure hydrogenation of coal, metal and building chemistry); and
3. the availability of modern equipment and many specialists suited to various tasks encouraged other I.G. Farben works to request help.

The most important task of the laboratory was the invention of new processes and compounds, and their technical development. But beside this the Ammonia Laboratory was responsible for pursuing chemical research with respect to specific production processes, patent work, analytical assistance, and pure scientific studies to advance knowledge in specific, technically interesting fields. In some respects, these investigations were not necessary for the technical work. Mittasch mentioned here the salt research of Ernst Jänecke, the catalysis research done by Walter Frankenburger, the physical investigations undertaken by Ernst Hochheim and Otto Eisenhut, and plant physiology. The latter group was founded in 1925 to assist the agriculture station with investigations on the optimum growth conditions for plants. In 1926 two plant physiologists, one enzyme chemist, two bacteriologists and one agricultural chemist began with plant physiology studies. At the end of 1929 the group was separated from

---

62 Ibid., pp. 28-29.

63 Grimm, op. cit. (60) , pp. 542-544.

64 Mittasch, op. cit. (60), pp. 100-102.

the Ammonia Laboratory and became a dedicated 'Biolaboratorium.'[65]

**Physics**

The physicist Ernst Hochheim joined BASF in 1911 as a member of Carl Bosch's staff. In his early years he was responsible for testing construction materials, and general physical questions. He also managed the contacts of I.G. Farben with academic physicists.[66] His group introduced x-ray equipment earlier than did the Ludwigshafen Central Laboratory and investigated on a small scale the scientific principles of the catalytic ammonia reaction.[67]

In 1921, Otto Eisenhut built an x-ray apparatus using the Debye-Scherrer method to study the structure of graphite, alumina, iron, and ammonia catalysts, but in September his equipment was destroyed by the Oppau explosion. In the spring of 1922 he entered the physical control group of the manufacturing department at Oppau. Seven years later he returned to the Ammonia Laboratory; in the same year the group added to its equipment canal rays and electron diffraction methods.[68] Beside analytical work for the production departments, the physical group of the Ammonia Laboratory managed to elucidate the structure of molecules such as iron nitride ($Fe_4N$). The x-ray analysis was at this time done mainly by Rudolf Brill, who also collaborated with Herman Mark's group at the Central Research Laboratory.[69]

The physical group undertook mainly analytical work in the field of technically interesting substances, such as catalysts, fertilisers, synthetic polymers, and metals. It set high standards and improved on the analytical methods. In 1941, Brill left I.G. Farben and joined the Darmstadt Technische Hochschule as professor of inorganic and physical chemistry.[70]

**Catalysis**

Until the mid twenties the catalytic investigations in the Ammonia Laboratory were restricted mainly to applied research. Mittasch and his collaborators used multiple compound catalysts to develop catalytic reactions for the hydrogenation of organic

---

65 Ibid., pp. 392-398, 493.

66 Dr. Hochheim, "Physikalische Arbeiten," in Ibid., pp. 436-451.

67 Ibid., pp. 437-439, 444-445.

68 O. Eisenhut, "Physikalische Arbeiten von Eisenhut und Mitarbeitern," in Mittasch, op. cit. (60), pp. 426-428.

69 R. Brill, "Röntgenographische Arbeiten," in Mittasch, op. cit. (60), pp. 428-435.

70 See Johann Christian Poggendorff, *Biographisch-literarisches Handwörterbuch zur Geschichte der exacten Wissenschaften*, vol. VII a (1956-62), part I, 272.

compounds, the ammonia oxidation, and the catalytic production of hydrogen and methanol.[71] The staff of the laboratory had no means available to investigate the unknown mechanisms of catalytic reactions. New reactions and catalysts were invented in an empirical way.

In 1924, Walter Frankenburger began fundamental investigations on ammonia catalysis. There were two problems central to the study of the mechanism of the catalytic formation of ammonia from the elements: The structure of the surface of the catalyst, and the formation of intermediate compounds during the course of the ammonia reaction. Following work done by Mittasch and Hochheim, Frankenburger first chose lithium nitride as a model for the catalytic formation of ammonia.[72] In continuation of the studies of the intermediate products, Frankenburger and his collaborators investigated the decomposition of ammonia at low temperatures. They were able to attain certain intermediates, such as imides and nitrides of the catalyst tungsten, and studied the kinetics of this reaction.[73]

In studies done during 1927 and 1928, Frankenburger tried to investigate the active centres of the iron catalyst. He surmised that iron atoms or small aggregates of iron atoms were the active parts of the catalyst and therefore studied the formation of iron-nitrogen and iron-hydrogen compounds in the gaseous state.[74] Though not completely successful, Frankenburger managed to study the aggregation of hydrogen and iron and the formation of dispersed iron layers.

Frankenburger published extensively in the field of heterogeneous and homogeneous catalysis, and photochemistry.[75] He emigrated in 1938 or 1939 to the United States, and changed his name to Frankenburg. From 1942 he was director of research at the

---

[71] A. Mittasch and W. Frankenburger, "Zur geschichtlichen Entwicklung und zur Theorie der Ammoniakkatalyse," *Zeitschrift für Elektrochemie*, **35** (1929), 920-927, p. 923.

[72] Mittasch, op. cit. (60), pp. 82-83. W. Frankenburger, "Chemische Bindung von Stickstoff an Lithium," *Zeitschrift für Elektrochemie*, **32** (1926), 481-491.

[73] A. Mittasch and W. Frankenburger, "Zur geschichtlichen Entwicklung und zur Theorie der Ammoniakkatalyse," *Zeitschrift für Elektrochemie*, **35** (1929), 920-927, p. 927.

[74] W. Frankenburger, "Studien über atomar verteiltes Eisen," *Zeitschrift für Elektrochemie*, **35** (1929), 590-600.

[75] For example: W. Frankenburger, "Neuere Forschungen auf dem Gebiet der heterogenen Katalyse," *Zeitschrift für angewandte Chemie*, **41** (1928), 523-530; W. Frankenburger, "Zur Reaktionskinetik der heterogenen Katalyse," *Zeitschrift für Elektrochemie*, **35** (1929), 278-283; W. Frankenburger, *Katalytische Umsetzungen in homogenen und enzymatischen Systemen* (Leipzig: Akademische Verlagsgesellschaft, 1937); W. Frankenburger, "Zur Kenntnis der angewandten Photochemie," *Zeitschrift für angewandte Chemie*, **43** (1930), 797-800.

General Cigar Company, Lancaster, Pennsylvania.[76]

**Phase rules**

In heterogeneous systems where several materials are present the existence and variability of these substances can be determined by the application of the phase rules. For the Ammonia Laboratory the existence of phases was especially important to the chemistry of fertilisers, which were often mixtures of various salts. In September 1920, Ernst Jänecke, then professor at the Hannover Technische Hochschule, joined the staff of the Ammonia Laboratory to study the phenomenon from its theoretical side. He began with investigations on ammonium sulphate and nitrate, and the sodium chloride/ammonium nitrate system. His group members (in 1922 four scientists) did more technical work in cement and salt chemistry.[77] In Germany, theoretical work on the existence of phases did not play a major role at the universities. With the work of Jänecke, I.G. Farben had access to the theoretical underpinning of the important fertiliser business. Jänecke retired in 1935.[78]

**The end of expansion in the ammonia laboratory**

The economic crisis of the 1930s also affected the Ammonia Laboratory. In contrast to the Central Research Laboratory, many scientists had to leave the Ammonia Laboratory due to cutbacks (see table 3). Mittasch, originally a defender of 'free' research, in 1930 installed a control group which evaluated the various research projects.[79] In consequence, Mittasch held the opinion that the first and principal task of a research laboratory in the chemical industry must be the support of existing production processes, and the foundation of new manufacturing departments and production methods from the chemical side. The investigation had to be finished when the compound or process was sufficiently understood from the technical point of view. Scientific questions were not important in this respect. If research was done to accumulate pure scientific knowledge it was restricted to specific topics such as catalysis, phase rules, photochemistry, and fibre research. These topics were of interest for industry and of technical value. According to Mittasch, basic research could be done to the extent of 5-10 per cent of the total.[80]

---

[76] Poggendorff, op. cit. (70), vol. VII a, part 2, 98-99. See Manfred Gill and Peter Löhnert, *Jüdishe Chemiker aus Dessau in der Filmfabrik Wolfen* (Dessau: Moses Mendelssohn, 1997), pp. 25-26.

[77] Mittasch, op. cit. (60), pp. 29, 71-74.

[78] See E. Jänecke and W. Eissner, R. Brill, "Über das System $NH_4NO_3$-$(NH_4)_2SO_4$-$H_2O$," *Zeitschrift für anorganische und allgemeine Chemie*, **160** (1927), 171-184; E. Jänecke, "Etwas über Phasengleichgewichte bei Sulfaten," *Zeitschrift für allgemeine und anorganische Chemie*, **151** (1926), 289-308. Poggendorff, op. cit. (70), vol. VII a, part 2, 603-604.

[79] Mittasch, op. cit. (60), pp. 486, 491.

[80] Mittasch, op. cit. (60), pp. 498-500.

Most of the laboratory scientists who left the Ammonia Laboratory in 1931 were transferred to other I.G. Farben departments. To assist the control group and to move forward in the direction of planned research, Mittasch and his deputy director Grimm inaugurated seven commissions for the evaluation of the research projects of Sparte (division) I at I.G. Farben, which included the high pressure work.[81]

In 1934, Hans Georg Grimm succeeded Alwin Mittasch as manager of the Ammonia Laboratory. Grimm, who had joined the Ammonia Laboratory in 1929, was formerly a professor in Würzburg. He was an expert in the field of the chemical bond and in 1934 proposed a periodic system of simple compounds. Due to illness he retired at the end of 1938.[82] He considered basic research as one way of introducing the scientific spirit to the laboratory and of solving specific problems in those cases where empirical knowledge was ahead of science.[83]

**Conclusion**

The basic research programmes of the Central Research Laboratory and the Ammonia Laboratory both started in the mid twenties, when the economic situation of the company allowed extensions of research, including in the field of traditional applied research. They were restricted to topics of relevance for the actual and future technical development of I.G. Farben, especially to catalysis and polymer research. Some leaders, such as K.H. Meyer and Herman Mark, thought that basic research was a necessary complement of applied research. Others, such as Alwin Mittasch and H.G. Grimm, saw basic research as a possibility for introducing the scientific spirit into their laboratory and as a sort of scientific problem-solving. Consequently the Central Research Laboratory and the Ammonia Laboratory undertook projects that resembled similar attempts at Du Pont. But, probably because of the decentralized structure of I.G. Farben, its basic research projects were on a much smaller scale than those at Du Pont.

The reasons for the commencement of basic research projects at the two I.G. Farben laboratories were economic and scientific-technological. In the 1920s, the traditional markets of I.G. Farben, dyestuffs, pharmaceuticals, and intermediates, were in decline; many of the organic chemists were superfluous. To explore the new sectors of fertilisers, fibres, and related products, experience in inorganic and physical chemistry was necessary. Fundamental processes, such as catalysis, and important product groups, such as fibres, needed a scientific foundation not available at the universities and Kaiser-Wilhelm institutes. Only in 1925-26, after the formation of I.G. Farben, and a certain economic recovery, was the management willing to invest in basic research projects, mainly restricted to the above areas.

---

81 Grimm, op. cit. (60), pp. 513-516.

82 Ibid.

83 Ibid., pp. 528 ff.

In the discussion sessions of the ESF workshop a historiographic problem emerged: Did the protagonists of the basic research programmes, above all Herman Mark, overestimate the role of basic research, and their own contribution to it, in their reminiscences? Are these sources relevant for historical research? These questions are closely related to the intellectual distinction between science and technology, or basic and applied research. I hope that I have shown that there are differences between these two endeavours, and that it is not enough to seek these differences in the motivation and goals of science and technology, respectively. Certainly the scale of basic research at I.G. Farben was very limited; even Mark contributed extensively to applied chemistry. In contrast to the basic research programme at Du Pont, the similar projects at I.G. Farben were never more than an afterthought to the overwhelming amount of applied research done in the industrial research departments. But this emphasizes only the mainstream direction of industrial research in this period, that is, applied research. Nevertheless, the work of Meyer and Mark on polymers was as outstanding as the investigations of Wallace Hume Carothers at Du Pont. Both Meyer and Mark managed to continue their scientific careers after their departures from I.G. Farben. Also, their collaborators and the researchers in the Ammonia Laboratory contributed to basic science in their respective fields.

It is, however, pertinent to seek the reasons for the end of this unique period. It was not lack of scientific or technological success; the period of basic research came to an end with the economic crisis of the 1930s. The management of I.G. Farben treated the research activities in the same manner as normal investments and, when necessary, cut down on the expenditure. With the departure of Meyer and Mark the pursuit of classical, technological-oriented chemistry returned to the Central Research Laboratory; that the period I have described was no more than an interruption is clear when we compare the projects of their collaborators before and after 1930.

One example of the role of basic research in the 1930s is the work of Bernd Eistert, one of the earliest proponents of mesomerism in Germany, for which purpose he collaborated with his former teacher Fritz Arndt. Eistert joined BASF in 1929, and he took appointments in Oppau, at the Central Research Laboratory, at the triphenylmethane dyes department, and the patent department. It seems, however, as if his theoretical studies were done in a rather private way, and were not related to his tasks in the industrial departments.[84] In the years after 1932, and especially with the appointment of Walter Reppe in 1938, the investigations at the Central Laboratory were directed towards pioneering technological research, not towards a deeper understanding of corporate-related science and technology.

**Acknowledgements**
I wish to thank the staff members of the BASF Company Archives, Ludwigshafen,

[84] M. Regitz et al. "Bernd Eistert (1902-1978)," *Chemische Berichte*, **113**, 2 (1980), xxix-lviii.

especially Mrs. Ruth Fromm and Dr. Lothar Meinzer for free access to the sources and for their hospitality. Discussion with the participants of the workshop has improved my article. Finally it is a pleasure to acknowledge the linguistic contributions of Anthony S. Travis, who improved in a decisive way on the readability of this contribution.

# AMBROS, REPPE, AND THE EMERGENCE OF HEAVY ORGANIC CHEMICALS IN GERMANY, 1925-1945[1]

PETER J. T. MORRIS
*Science Museum*
*Exhibition Road*
*South Kensington*
*London SW7 2DD, UK*

We saw in acetylene a new chemical basis useful in many types of chemical synthesis, and in a number of our laboratories we specifically directed research into the fields of acetylene.
*Fritz ter Meer, testimony at Nürnberg, 1948.*[2]

The mass production of synthetic rubber must also be organized and achieved with the same urgency. From now on there must be no talk of processes not being fully determined and other such excuses. It is not a matter of discussing whether we are to wait any longer; otherwise time will be lost and the hour of peril will take us all by surprise. Above all, it is not the job of the institutions of Government to rack their brains over methods of production. This has nothing whatever to do with the Ministry of Economics. Either we possess today a private industry, in which case its job is to rack its brains about methods of production; or we believe it is the Government's job to determine methods of production, and in that case we have no further need of private industry.
*Adolf Hitler, memorandum on the Four Year Plan, 1936.*[3]

---

[1] This contribution is derived from "Strategy and System: Reppe and the Development of Organic Chemicals in I.G. Farben," a paper presented at a meeting of the Society for the History of Technology, Madison, Wisconsin, 3 November 1991. Subsequently it has been reworked and presented at various seminars. It is based largely on my doctoral thesis, "The Development of Acetylene Chemistry and Synthetic Rubber by I.G. Farbenindustrie Aktiengesellschaft, 1926-1945," Oxford University D.Phil., 1982.

[2] *Trials of War Criminals before the Nuremberg Military Tribunals under Control Law No. 10* ("Green Books") Case Six ( Washington DC: US Government Printing Office, 1953), volume VII, p. 860.

[3] Adolf Hitler's secret memorandum on the Four Year Plan, August 1936, translated in Jeremy Noakes and Geoffrey Pridham, eds, *Documents on Nazism, 1919-1945* (London: J. Cape, 1974), p. 406.



*A.S. Travis et al. (eds.),*
***Determinants in the Evolution of the European Chemical Industry, 1900–1939*, 89–122.**

**Introduction**

This paper delineates the careers of Otto Ambros and Walter Reppe, two very different chemists who worked at the Ludwigshafen works of the German behemoth I.G. Farben. Otto Ambros (1901-1990) became the chairman of I.G. Farben's Zetko (intermediates commission) in 1936 and four years later he became the youngest member of the main board of directors (Vorstand).[4] The son of a professor of agriculture, he was a cheerful and engaging Bavarian. His rise through the corporate hierarchy was assisted by his marriage to the daughter of BASF's Social Director. Both amiable and persuasive, Ambros was friendly with nearly all who met him, whether they were Jewish chemists (such as his former supervisor Richard Willstätter) or members of the Nazi hierarchy. He was adept at piloting the organic chemicals sector of I.G. Farben through the dangerous reefs of the Third Reich, and on one famous occasion even dissuaded Hitler from using nerve gas.[5]

In 1945, Ambros was arrested and eventually charged with various war crimes alongside other members of the I.G. Farben Vorstand. He was found guilty and sentenced to imprisonment in 1948, but released from American custody in 1952. Ostracized (at least in public) by the successor companies, Ambros became a consultant, most notably to the chemical engineering firm W. R. Grace & Co., and assisted the conversion of the synthetic rubber works at Hüls to petrochemicals in the mid-1950s.[6] He was also the Chairman of the Supervisory Board of Knoll AG, and a member of the Supervisory Board of several other companies including Hibernia and Telefunken.

Walter Reppe (1892-1969), according to a 1946 British intelligence report, "contributed more to the advancement of chemical science than probably any other I.G. Farben employee."[7] The son of a Thuringian schoolteacher, he was a very dedicated chemist,

4 There is very little on the life of Otto Ambros. See Jens Ulrich Heine, *Verstand & Schicksal* (Weinheim: VCH, 1990), pp. 172-174 and *Brown Book. War and Nazi Criminals in West Germany* (Dresden: Verlag Zeit im Bild, 1965), p. 36.

5 *Trials of War Criminals*, volume VII, pp. 1044-1045.

6 Paul Kränzlein, *Chemie im Revier – Hüls* (Düsseldorf: Econ Verlag, 1980), pp. 107-108. Raymond G. Stokes, *Opting for Oil: The Political Economy of Technological Change in the West German Chemical Industry, 1945-1961* (Cambridge: Cambridge University Press, 1994), pp. 201-202.

7 Reppe's curriculum vitæ can be found in Edmund Tilley, FIAT (Field Information Agency Technical) interrogation report on Dr. Walther [sic] Reppe, 23 August 1946, B/TP 200/888, Imperial War Museum, pp. 9-12; and in *J. C. Poggendorf Biographisch-Literarisches Handwörterbuch der exakten Naturwissenschaften*, Band VIIa, Teil 3: L-R (East Berlin: Akademie-Verlag, 1959), pp. 736-737. For a list of his publications and patents, see H. G. Hummel, "Walter Reppe, 1892-1969," *Chemische Berichte*, **117** (1984), i-xxi. Also see John W. Copenhaver and Maurice H. Bigelow, *Acetylene and Carbon Monoxide Chemistry* (New York: Reinhold, 1949), pp. xi-xii. For the most complete treatment of Reppe's career, see Nikolaus von Kupetow and H. Stark, "W. Reppe und die Reppe-Chemie," unpublished typescript, 73 pp., 1977. This is held by the Unternehmensarchiv, BASF AG and I am indebted to Dr. Lothar Meinzer for a copy. Quotation taken from Tilley, FIAT interrogation report, 21. For Reppe's research, see Morris, "Acetylene Chemistry and Synthetic

who spent long hours in the laboratory. He was very ambitious and clearly hoped to rise to the top of I.G. Farben on the strength of his scientific research. Two US Chemical Corps officers, John Copenhaver and Maurice Bigelow, commented in their postwar book that Reppe "had one ambition, namely to build a new chemical industry, and in so doing, to become the chemical leader of Germany."[8] He was also rather irascible and lacked social skills, which hindered his progress up the management hierarchy. He may have been envious of the rapid rise of the younger Otto Ambros, who was also educated at Munich University. In postwar interrogations, Reppe bitterly expressed annoyance at his lack of prestige. During interrogation by a British officer, Major Edmund Tilley, Reppe declared that he:

> had not been the recipient of honours for what he considered his important contributions in the field of chemistry, that the honorary title of 'Professor' had not been bestowed upon himself as upon many of his colleagues, in fact the only distinction he had received was the worthless 'Kriegsverdienstkreuz' [Distinguished War Service Cross].[9]

When BASF was revived after the war, Reppe became the company's research director until his retirement in 1957.

---

Rubber," Von Kupetow and Stark, "W. Reppe und die Reppe-Chemie;" Alfred von Nagel, *Äthylen, Acetylen* (Ludwigshafen: Firmenarchiv der BASF, 1971); and Copenhaver and Bigelow, *Acetylene and Carbon Monoxide Chemistry*, op. cit. Copenhaver and Bigelow's superb monograph is largely based on Walter Reppe's postwar typescript report on his acetylene and carbon monoxide chemistry available at Werksarchiv Hüls and Unternehmensarchiv, BASF. Reppe's own monographs are less informative, but see Walter Reppe, *Neue Entwicklungen auf dem Gebiete der Chemie des Acetylen und Kohlenoxyds* (Berlin: Springer, 1949) and Walter Reppe, *Chemie und Technik der Acetylen-Druck-Reaktionen* (Weinheim: Verlag Chemie, 1951; expanded edition, 1952); also see Otto Hecht and Hugo Kröper, "Neuere Entwicklungen auf dem Gebiete der Chemie des Acetylen und Kohlenoxyds" in Karl Ziegler, ed., *FIAT Review of German Science*, **38**, *Preparative Organic Chemistry*, part I, pp. 1-209. For other analyses of Reppe's work in the context of German heavy organic chemicals, see Stokes, *Opting for Oil*, pp. 35-39, and Basil G. Achilladelis, "Process Innovation in the Chemical Industry," University of Sussex D. Phil., 1973, 75-90. FIAT, CIOS (Combined Intelligence Objectives Subcommittee) and BIOS (British Intelligence Objectives Subcommittee) were the intelligence teams that scoured Germany for technical information, scientists, and documents in 1945 and 1946. Their published and unpublished reports are held at the Imperial War Museum's archives at Duxford, alongside the seized FD (Foreign Documents) papers. See John Gimbel, *Science, Technology, and Reparations: Exploitation and Plunder in Postwar Germany* (Stanford: Stanford University Press, 1990), and Peter Spitz, *Petrochemicals: The Rise of an Industry* (New York: John Wiley, 1988), pp. 1-17.

8 Copenhaver and Bigelow, *Acetylene and Carbon Monoxide Chemistry*, p. vi.

9 Tilley, FIAT interrogation report, p. 4.

**The formation of I.G. Farben**[10]

I.G. Farben's prehistory can be traced back to the German dyestuffs firms founded in the 1860s.[11] An enduring factor in the history of the German dyestuffs industry was the need to broaden its profit basis. Initially, this had been achieved within dyestuffs, but the rate of discovery of novel classes of dyes fell off after the mid-1880s. The development of the vat dyes, including synthetic indigo and indanthrenes, at the turn of

---

10 I.G. Farben's full name was I.G. Farbenindustrie Aktiengesellschaft. This name was derived from the title of the earlier dyestuffs cartel, the Interessengemeinschaft der deutschen Teerfarbenfabriken (community of interests of the German coal tar dye industry). Interessengemeinschaft was a common term for the more formal cartels in Wilhelmine Germany, with the same "public interest" connotations as the American 'trust.' Curt Duisberg (son of Carl Duisberg) insisted to me that the I.G. in I.G. Farben was purely a historical relic and should never be spelt out in full. Within the company, the full name was rarely used, I.G. Farben even more rarely, and 'Farben' – favoured by American writers in the period following 1945 – never, as it simply means colours or tints. It was nearly always called I.G. (pronounced ee-gay).

11 For general overviews of the history of the European chemical industry in the late 19th and early 20th centuries, see L. F. Haber, *The Chemical Industry during the Nineteenth Century. A Study of the Economic Aspect of Applied Chemistry in Europe and North America*, (Oxford: Clarendon Press, second edition, 1969); Haber, *The Chemical Industry 1900-1930. International Growth and Technological Change* (Oxford: Clarendon Press, 1971); L. F. Haber, "The chemical industry, A general survey" in Trevor I. Williams, ed., *A History of Technology*, volume VI, *The Twentieth Century, c.1900 to 1950*, Part I (Oxford: Clarendon Press, 1978), pp. 499-513; Paul M. Hohenberg, *Chemicals in Western Europe, 1850-1914: An Economic Study of Technical Change* (Chicago: Rand McNally, 1967). For the development of the German dye industry, see J. J. Beer, *The Emergence of the German Dye Industry* (Urbana: University of Illinois Press, 1959), and A. S. Travis, *The Rainbow Makers: The Origins of the Synthetic Dyestuffs Industry in Western Europe* (Bethlehem, Penn.: Lehigh University Press, 1993). For the evolution of modern industrial research in the German dye industry, see Ernst Homburg, "The emergence of the research laboratory in the dyestuffs industry, 1870-1900," *British Journal for the History of Science*, **25** (1992), 91-111, and Georg Meyer-Thurow, "The industrialization of invention: A case study from the German chemical industry," *Isis*, **73** (1982), 362-381. Peter Hayes, *Industry and Ideology: I.G. Farben in the Nazi Era* (Cambridge: Cambridge University Press, 1987) is the standard work on I.G. Farben during the Third Reich. Also see Gottfried Plumpe, *Die I.G. Farbenindustrie AG: Wirtschaft, Technik, Politik, 1904-1945* (Berlin: Duncker and Humblot, 1990), and Helmuth Tammen, *Die I.G Farbenindustrie AG (1925-1933). Ein Chemiekonzern in der Weimarer Republik* (Berlin: Helmuth Tammen, 1978). For an insider's view, see Fritz ter Meer, *Die I.G. Farbenindustrie AG. Ihre Entstehung, Entwicklung und Bedeutung* (Düsseldorf: Econ Verlag, 1953). I.G. Farben's research and development is discussed in Morris, "Acetylene Chemistry and Synthetic Rubber," pp. 38-55; Ulrich Marsch, "Strategies for success: research organisation in German chemical companies and I.G. Farben until 1936," *History and Technology*, **12** (1995), 23-77, and John E. Lesch, "Chemistry and biomedicine in an industrial setting: The invention of the sulfa drugs," in Seymour H. Mauskopf, ed., *Chemical Sciences in the Modern World* (Philadelphia: University of Pennsylvania Press, 1993), pp. 158-215. For Bayer, see F. Jacobi, ed., *Beiträge zur hundertjahrigen Firmengeschichte 1863-1963* (Leverkusen: Bayer, 1964), and Erik Verg (with Gottfried Plumpe and Heinz Schultheis), *Meilensteine. 125 Jahre Bayer 1863-1988* (Leverkusen: Bayer, 1988), translated by John Wicks: *Milestones: The Bayer Story, 1863-1988* (Leverkusen: Bayer, 1988). For Hoechst, see Ernst Bäumler, *Ein Jahrhundert Chemie* (Düsseldorf: Econ Verlag, 1963), translated by David Goodman: *A Century of Chemistry* (Düsseldorf: Econ Verlag, 1968); Ernst Bäumler, *Farben, Formeln, Forscher: Hoechst und die Geschichte der industriellen Chemie in Deutschland* (München: Piper, 1989). BASF has been less well served, but see the excellent series of short histories published in-house, especially Gerhart Wolf, *Die BASF: Vom Werden eines Weltunternehmens* (Ludwigshafen: Firmenarchiv der BASF, 1970). Also see [Wilhelm Roggersdorf], *Im Reiche der Chemie* (Düsseldorf: Econ Verlag, 1965), and [Lothar Meinzer, *et al.*], *125 Jahre BASF: Stationen ihrer Geschichte* (Ludwigshafen: BASF, 1990).

the century marked the peak of the traditional synthetic dyestuffs industry. The dye companies initially diversified into pharmaceuticals, a sector close to dyestuffs in terms of raw materials, technology and scale of production. By the early 1890s, Hoechst, Bayer, and Kalle had moved into pharmaceuticals, BASF into inorganic chemicals (vertical integration), and Griesheim into electrochemicals. Agfa moved into photoproducts at a somewhat later date. Nevertheless, dyes, and particularly the export of dyes, still provided the bulk of the major German firms' large profits on the eve of World War I.

In 1913, the German dye industry accounted for 88 per cent of the world's synthetic dyestuffs (by value).[12] This monopoly was shattered in 1914. The Allied powers were forced to develop, or at least expand, their own organic intermediates and dye industries. After 1918, these strategic industries were sheltered from German competition by a number of protectionist measures. This led naturally to a fall in German dye exports, and the German share of world synthetic dyestuffs production declined to a mere 39 per cent in 1921.[13] At the same time, the war had accelerated the movement of the German dye companies into new sectors, particularly synthetic ammonia, synthetic petrol and synthetic rubber.

Largely as result of these changes, the German dye firms, notably Badische Anilin und Soda Fabrik (BASF), Farbenfabriken Bayer (Bayer), and Meister, Lucius und Brüning (Hoechst), merged in December 1925 to form the world's largest chemical company, I.G. Farbenindustrie AG. Carl Duisberg of Bayer had insisted on a confederal structure for the new company, with the grand name of "regional decentralized centralization". The I.G. Farben factories were therefore organized into "works communities" (Betriebsgemeinschaften). Though intended to be regional, and given regional titles, these were effectively the old companies under new names. Most routine research and production decisions were initially taken by these communities. Major issues, particularly economic and financial questions, were the province of the central Technical Committee, always called TEA, which was originally made up of the technical directors of the predecessor companies; yet another example of how I.G. Farben tried to dodge the problems raised by the merger. This committee was serviced by the TEA-Büro, which also administered several specialized committees that came under TEA, including the dyes committee (Coloristic Kommission), and the important intermediates committee (Zetko).

Dyestuffs largely retained their importance in absolute terms, but their standing relative to other organic chemicals declined. The turnover of the traditional sectors of the German dye industry (dyes, pharmaceuticals and agrochemicals) in I.G. Farben

12 Plumpe, *Die I.G. Farbenindustrie AG*, p. 52.

13 Ibid., p. 129.

declined from 39 per cent in 1926 to 24 per cent in 1939.[14] By contrast, the turnover of the new areas (polymers, rayon, metals, and synthetic fuels) rose from practically nothing in 1926 to 24 per cent in 1939, and continued to expand enormously during World War II.

These changes placed I.G. Farben at a potential disadvantage. Its executives and chemists were more familiar with the 'classical' chemistry of the universities, especially aromatic chemistry. The scale factor was also significantly different. Experience of large-scale production came from the manufacture of inorganic chemicals, notably sulphuric acid, which was used in dye manufacture. Most dyes and pharmaceuticals were produced on the scale of a few tons in batch processes. Furthermore, little was known about the behaviour of large volumes of these organic chemicals. To be efficient, the new plants had to be large, but the markets for the new chemicals lagged way behind the technology and took several years to establish. The new products, solvents, polymers and various specialities, were quite different from dyes and pharmaceuticals and this naturally generated problems on the marketing side. Moreover, disposing of surpluses of organic chemicals which were by-products or co-products of the manufacture of other chemicals could sometimes be difficult.

Needless to say, the Germans were not allowed to tackle these technical difficulties in an otherwise tranquil world. Apart from the political and economic problems that faced Germany after the debacle of 1918, including the Great Inflation and the occupation of the Rhineland, the German dye industry faced fierce competition in its traditional markets from the foreign newcomers. I.G. Farben attempted to outflank this competition by reaching agreements with its German and foreign rivals, and by moving into new sectors, including novel organic chemicals. Even here, however, the German combine faced a formidable challenge from American firms.

In contrast to the United States, the interwar German organic chemicals industry was almost entirely based on coal, the only fossil fuel found in significant quantities within Germany's borders. Fermentation-based routes never found much favour in Germany, with its temperate climate and large population. It is noteworthy that wood-distillation had previously been a substantial industry. The Holzverkohlungs-Industrie AG (HIAG) was the 77th largest German company in 1913, but it was weakened by the industrial synthesis of its major product, methanol, by Matthias Pier of BASF in 1923, and was taken over by Degussa in 1931.[15] I.G. Farben's chemists distrusted empirical processes that could not be defined through standard chemical equations, a feature of many early

14 Ter Meer, *Die I.G. Farbenindustrie AG*, end table.

15 Alfred D. Chandler, Jr., *Scale and Scope: The Dynamics of Industrial Capitalism* (Cambridge, Mass.: Belknap Press, 1990), pp. 278-279, 483, 583, 698 (appendix C.1); [Mechthild Wolf], *Im Zeichen von Sonne und Mond: Vom der Frankfurter Münzscheiderei zum Weltunternehmen Degussa AG* (Frankfurt: Degussa, 1993), pp. 159-163.

petrochemical routes. Significantly, successive German governments would not support the use of imported petroleum in the chemical industry by lowering or eliminating duty.[16] Germany's own petroleum deposits were comparatively scarce and were in great demand for motor fuel. Consequently, I.G. Farben did not participate in the important shift to petrochemicals (organic chemicals made from petroleum and natural gas) that occurred in a number of American firms. Nevertheless, and quite remarkably, its interwar research and development paid off handsomely in a very different world of international trade and petrochemicals created by the Nazi defeat in 1945.[17]

**Diversification into organic chemicals**

The technological front-line for the new combine was the high-pressure hydrogenation of coal and heavy oils to gasoline, which was based at Leuna near Merseburg in eastern Germany. The story of how this new technology faltered, and nearly sank I.G. Farben, is well known and will not be repeated here.[18] The older dye factories of Ludwigshafen, Hoechst and Leverkusen endeavoured to create a counter-weight to this high-pressure chemistry, by developing new organic chemicals, particularly solvents and polymers. These included large scale production of nitrocellulose and investigations into synthetic rubber.

The early 1920s witnessed the rise of the motor-car, which was transformed from being the toy of the rich into the everyday vehicle of the middle classes. The number of cars built in North America increased almost sixty-fold from 63,000 in 1908 to 3,735,000 in

---

16 Raymond G. Stokes, *Opting for Oil*, pp. 99-106; Alfred von Nagel, *Äthylen, Acetylen*, pp. 12-13.

17 Stokes, *Opting for Oil*. Also see Detlef Mittmann, *Die chemische industrie im nordwestlichen Mitteleuropa in ihrem Strukturwandel* (Wiesbaden: Franz Steiner, 1974) for a geographical study and Raymond G. Stokes, *Divide and Prosper: The Heirs of I.G. Farben under Allied Authority, 1945-1951* (Berkeley: University of California Press, 1988) for the more general political and corporate background.

18 In addition to Hayes, *Industry and Ideology, passim*, see Plumpe, *Die I.G. Farbenindustrie AG*, pp. 69-82, 203-296, and Tammen, *Die I.G Farbenindustrie AG*, pp. 29-59, 108-112. Also see Wolfgang Birkenfeld, *Der Synthetische Treibstoff: 1933-1945. Ein Beitrag zur nationalsozialistischen Wirtschafts- und Rüstungspolitik* (Göttingen: Musterschmidt, 1964), and Wolfgang Birkenfeld, "Leuna 1933," *Tradition*, **3** (1963), 97-111; W. Greiling, "Leuna: 25 Jahre im Kampf um Deutschlands Freiheit," 27 April 1942, Werkarchiv, Leuna.; Karl-Heinz Streller and Erika Massalsky, *Geschichte des VEB Leuna-Werke "Walter Ulbricht," 1916-1945* (Leipzig: Deutscher Verlag für Grundstoffindustrie, 1989) (This book, produced only months before the downfall of the communists, is mainly about communist activity in Leuna, but contains very clear technical diagrams and a wealth of data); Thomas P. Hughes, "Technological momentum in history: Hydrogenation in Germany, 1898-1933" *Past and Present*, **44** (1969), 106-132; Arnold Krammer, "Fuelling the Third Reich," *Technology and Culture*, **19** (1978), 394-422; Alfred von Nagel, *Stickstoff* (Ludwigshafen: Firmenarchiv der BASF, 1970); von Nagel, *Methanol, Treibstoffe* (Ludwigshafen: Firmenarchiv der BASF, 1970); Raymond G. Stokes, "The oil industry in Nazi Germany, 1936-1945," *Business History Review*, **59** (1985), 254-277; Anthony N. Stranges, "Friedrich Bergius and the rise of the German synthetic fuel industry," *Isis*, **75** (1984), 643-667; Stranges, "The conversion of coal to petroleum: Its German roots," *Journal of Processing Technology*, **16** (1987), 205-225; Anthony S. Travis, *The High Pressure Chemists* (Wembley: Brent Schools & Industry Project, 1984).

1925; even in Germany it rose from 4,600 to 39,000.[19] This surge in production created a massive demand for gasoline (hence the oil-from-coal process) and various organic chemicals. The metal bodies of the cars had to be painted with stoving lacquers, made from nitrocellulose and organic solvents. The upholstery used synthetic leather cloth, another outlet for nitrocellulose and organic chemicals. The huge quantity of rubber consumed by car tyres, and the growing need for a petrol-resistant rubber, increased the market for rubber chemicals and also spurred research into synthetic rubber.[20]

The new organic solvents required for automobile lacquers created a growing demand worldwide for butanol in the early 1920s. The Commercial Solvents Corporation at Terre Haute, Indiana, which used the Weizmann fermentation process, saw its butanol production treble from 2,000t in 1923 to 6,500t a year later; by 1929, the United States produced nearly 23,000t of butanol.[21] Heavily populated Germany did not share the surplus food production of the United States, and in any event the German dye companies were uncomfortable with non-chemical processes. Nor was petroleum an attractive starting point, given the small size of the German refining industry and the paucity of native deposits. The German response was based on coal-derived acetylene and ethylene.

The well known dimerisation of acetylene-derived acetaldehyde to aldol, which could be readily converted into butanol, offered an attractive alternative to petrochemical and fermentation routes. The challenge of developing an industrial process was taken up by BASF's new solvents laboratory, led by Gerhard Steimmig and Curt Schumann.[22] When they succeeded in 1925, the future of acetylene was secured. The production of butanol was not very great, only 1,693t in 1929, but, within a few years, their process had become the starting point for the synthesis of butadiene, the building block of synthetic rubber, which was subsequently produced on a vast scale. Furthermore, one junior chemist in the butanol project made the key breakthrough in this butadiene synthesis; he was, of course, Walter Reppe.

---

19 United Nations Statistical Office, *Statistical Yearbook. Annuaire statistique* (New York: United Nations, 1948); *Statistical Abstract of the United States, 1938* (Washington DC: US Government Printing Office, 1939).

20 For the link between the development of the motor-car and organic chemicals see W. J. Reader, *Imperial Chemical Industries: A History*, volume 2, *The First Quarter-Century, 1926-1952* (Oxford University Press, London: 1975), pp. 13-14, and L. F. Haber, "The chemical industry, A general survey," p. 501.

21 Frank Greenaway, *et al.*, "The chemical industry: Part II. Raw materials for organic chemicals" in Trevor I. Williams, ed., *A History of Technology*, Volume VI, *The Twentieth Century, c.1900 to c.1950* (Oxford: Oxford University Press, 1978), p. 543; and Franz Karl Steinberger, *Die Acetylen-Chemie in Höchst* (Frankfurt-Höchst: Hoechst AG, c.1948), p. 206 (figure 27).

22 Alfred von Nagel, *Äthylen, Acetylen*, pp. 7-16. Alfred von Nagel, *Methanol, Treibstoffe*, pp. 30-32.

After the butanol research was successfully completed, Reppe worked on the manufacture of ethylene oxide. This was also connected with the rise of the motor car. The earliest anti-freezes were natural glycerol, and methanol (still obtained in America from wood distillation in this period). However, glycerol was replaced by the superior ethylene glycol, one of the first major petrochemical products, launched by Union Carbide as Prestone in 1927. Ethylene glycol presented problems for I.G. Farben, which lacked an abundant source of ethylene, apart from the expensive fermentation-based ethanol. Even so, the growing demand for anti-freeze, and the fear of competition from Union Carbide, compelled I.G. Farben to develop its own route to ethylene glycol between 1926 and 1928. Glysantin, I.G. Farben's response to Prestone, was launched in the cold winter of 1928-29.

BASF had prepared ethylene for one version of the indigo synthesis by the dehydration of fermentation ethanol even before World War I. Reppe began his work on ethylene, under the direction of Gerhard Steimmig, by looking for a better catalyst than German bauxite. He introduced a catalyst made from synthetic aluminium oxide, but the dehydration of ethanol was clearly too expensive for the mass production of ethylene glycol. The solvents laboratory then studied the cracking of petroleum middle oils. It soon became clear, however, that it was necessary to separate the ethylene from the mixture of olefins formed during the cracking process before it could be converted into ethylene oxide. The help of the liquefaction specialist, Gesellschaft für Lindes Eismachinen, was sought, and a low temperature separation process was developed in 1928. The cracking of middle oils was abandoned when the government refused to remove the petroleum duty on the feedstock, which made cracking economically uncompetitive with ethanol-based ethylene. Fortunately, the Linde process could also be used to extract ethylene from coke oven gas. Consequently, in June 1930, I.G. founded Chemische Fabrik Holten (in the Ruhr) in collaboration with two local firms, Ruhrchemie and Theo. Goldschmidt, to make ethylene oxide from coke oven ethylene.

Reppe was also working on the preparation of ethylene oxide. The first step was the reaction of ethylene with chlorine and water to form ethylene chlorohydrin (2-chloroethanol). He was able to convert the unsatisfactory batch process, which had been used for indigo manufacture, into a continuous process. This was called the "tower process" (Turm-Verfahren), because a mixture of ethylene and a carefully controlled amount of chlorine was driven up a tower filled with water. The dilute solution of ethylene chlorohydrin was drawn off at the top and then heated, without being isolated, with lime water to form ethylene oxide. Ethylene oxide was chiefly converted into the glycol by adding water, but it could also be used to make diethylene glycol[23] (diglycol),

[23] Diethylene glycol is widely used in the textile and printing inks industries, as a humectant, and can be used to make polyurethanes and polyesters. In World War II it was used in the manufacture of explosives. Infamously it was used in 1986 as an additive in Austrian white wine, as it has a sweet taste. It is used as an anti-freeze in, for instance, sprinkler systems, but not to be confused with the more poisonous ethylene glycol used as an anti-freeze in motor-cars. For the tragic case of the use of diethylene glycol in "Elixir Sulfanilamide" by the S. E. Massengill Company in 1937, see James Harvey Young, "Sulfanilamide and diethylene glycol" in John Parascandola and James C. Wharton, eds, *Chemistry and Modern Society*, ACS

and esters and ethers of ethylene glycol.

Even while he was busy with the manufacture of ethylene oxide, Walter Reppe was also involved with the development of Buna synthetic rubber. In 1926, the newly formed I.G. Farben decided to embark on the industrial synthesis of rubber, despite the poor quality of the 'methyl' rubber made during World War I.[24] This time, however, it was agreed that butadiene would be used. Several routes to butadiene were investigated, including decyclization of cyclohexene (a retro-Diels-Alder reaction), but the so-called four-step process (Vierstufen Verfahren) soon won out. This was partly because it used acetylene, and hence surplus carbide from cyanamide manufacture, but also because it drew on the steps – and hence the momentum – of the BASF butanol synthesis. As a member of the former butanol group, Reppe was a natural candidate for the four-step process project.

The first step, hydration of acetylene to acetaldehyde, was a well-established industrial process by 1926.[25] The dimerization of acetaldehyde to aldol (3-hydroxybutanal), catalyzed by dilute alkali, had been developed at Hoechst and it was converted into a continuous process (using a tubular reactor) patented by Martin Müller-Cunradi and Kurt Pieroh of the Oppau ammonia laboratory in June 1927. The third step was the reduction of aldol to 1,3 butylene glycol (butane-1,3-diol). Kurt Meisenburg of Bayer had suggested anodic reduction when he first postulated the four-step process in 1916, but this was displaced by catalytic hydrogenation. However, this hit an unexpected snag when it was found that the nickel-catalyzed reaction patented by Ulrich Hoffmann and Johann Müller in July 1927 infringed a patent of Otto Dalmer. Subsequently the nickel catalyst was replaced by a copper catalyst patented by Müller-Cunradi and Hans Sönksen in October 1927. A senior director of I.G. Farben, Arthur von Weinberg,

---

Symposium Series 228 (Washington DC: American Chemical Society, 1983), pp. 105-125.

24 For the early history of synthetic rubber, see Bettina Löser, "Der Einfluss der Arbeiten zur Strukuraufklärung und synthese der Kautschuks auf die Herausbildung der makromolekularen Chemie," Karl-Marx-Universität, Leipzig, Ph.D. thesis, 1983; Morris, "Acetylene Chemistry and Synthetic Rubber," pp. 143-15; G. S. Whitby and M. Katz, "Synthetic rubber," *Industrial and Engineering Chemistry*, **25** (1933), 1204-1218; and "The early days of synthetic rubber at Farbenfabriken Bayer. Some historical recollections," *Technical Notes for the Rubber Industry*, **36** (1966), 18-27 (published by Bayer). For the synthetic rubber research in I.G. Farben between 1926 and 1945, see: Morris, "Acetylene Chemistry and Synthetic Rubber," pp. 156-170, 178-192, 198-225; Claus Heuck, "Ein Beitrag zur Geschichte der Kautschuk-Synthese: Buna-Kautschuk I.G. (1926-1945)," *Chemiker-Zeitung*, **94** (1970), 147-157; Erich Konrad, "Uber die Entwicklung des synthetischen Kautschuks in Deutschland," *Angewandte Chemie*, **62** (1950), 423-426; Heino Logemann and Gottfried Pampus, "Buna S – Seine grosstechnische Herstellung und seine Weiterentwicklung – ein geschichtlicher Úberblick," *Kautschuk und Gummi, Kunststoffe*, **23** (1970), 479-486.

25 Morris, "Acetylene Chemistry and Synthetic Rubber," pp. 156-160, 198-203; Alfred von Nagel, *Äthylen, Acetylen*, pp. 44-55.

trumpeted a month later "that the synthesis of rubber was now an achieved fact."[26]

This boast was somewhat premature, because the most difficult step of the four-step process was still unsolved. The dehydration of butylene glycol to the desired butadiene, as Meisenburg had remarked eleven years earlier, "presented great difficulties, because apart from butadiene, other products of the dehydration were also formed, which gave rise to heavy contamination."[27] I.G. Höchst had developed an alum catalyst, based on earlier work at Bayer, but Richard Leopold of Höchst ruefully reported in April 1927 that one of the hydroxyl groups resisted dehydration, giving rise to allyl carbinol and crotyl alcohol as by-products.[28]

Reppe was entrusted with the challenge of finding a highly specific dehydration catalyst for the production of butadiene without excessive contamination, a task so important that it was nicknamed the "key to Buna" (Schlüssel zum Buna). This was obviously connected with his contemporary work on the dehydration of ethanol to ethylene. Reppe and Ulrich Hoffmann discovered that phosphoric acid was a good catalyst for the reaction, an observation made by Meisenburg some fifteen years earlier. They found that the best results were produced by using sodium hydrogen phosphate, which was patented in December 1927. This breakthrough placed the four-step process on a firm footing and it was used to make butadiene at Hüls in West Germany until the mid-1950s, and Schkopau in East Germany up to 1990.

As a result of the concurrent progress on the polymerization side, Ludwigshafen and Leverkusen agreed in July 1929 to build a semi-technical works plant for Buna at Knapsack, alongside the carbide works.[29] This plan was blocked by Carl Krauch of Oppau, largely because he wanted to wait until Oppau's methane-to-acetylene electric arc process was ready. A few months later, the Buna program was effectively halted by the onset of the Depression, which soon reduced natural rubber prices to minimal levels. When the production of synthetic rubber was revived in Hitler's Third Reich, the weak Buna, which was a sodium-polymerized polybutadiene, had been displaced by the superior copolymers of butadiene with styrene (Buna S) and acrylonitrile (Buna N or Perbunan).

Walter Reppe's breakthrough in the butadiene synthesis was central to the new

26 "Synthetic Rubber," *New York Times*, 20 November 1927, press cutting held at Bayerarchiv 151/16.2.

27 K. Meisenburg, "Gewinnung von Erythren aus Acetaldehyd" (21 March 1916), Bayerarchiv 103/7.1.

28 Minutes of the second "rubber conference," April 1927, Bayerarchiv 153/3.1.

29 Ernst Struss, "Zur Gründung der Buna-Werke der I.G." NI-7241, 1-2. This is one of the "Nuremberg Industrialists" documents, available at the National Archives, Washington DC, Imperial War Museum, London, and the Bayerarchiv (151/7).

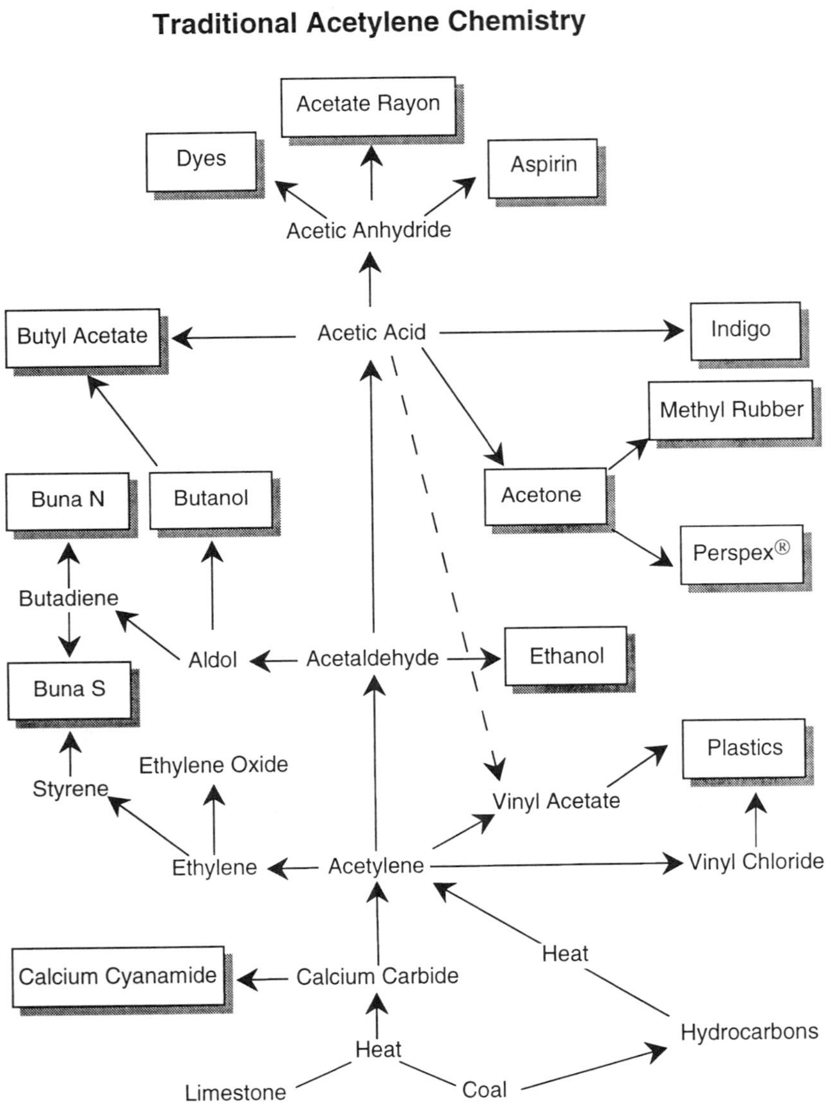
Traditional Acetylene Chemistry
Acetate Rayon
Dyes
Aspirin
Acetic Anhydride
Butyl Acetate
Acetic Acid
Indigo
Methyl Rubber
Buna N
Butanol
Acetone
Perspex®
Butadiene
Aldol
Acetaldehyde
Ethanol
Buna S
Plastics
Ethylene Oxide
Styrene
Vinyl Acetate
Ethylene
Acetylene
Vinyl Chloride
Heat
Calcium Cyanamide
Calcium Carbide
Hydrocarbons
Heat
Limestone
Coal

## Advanced Acetylene Chemistry

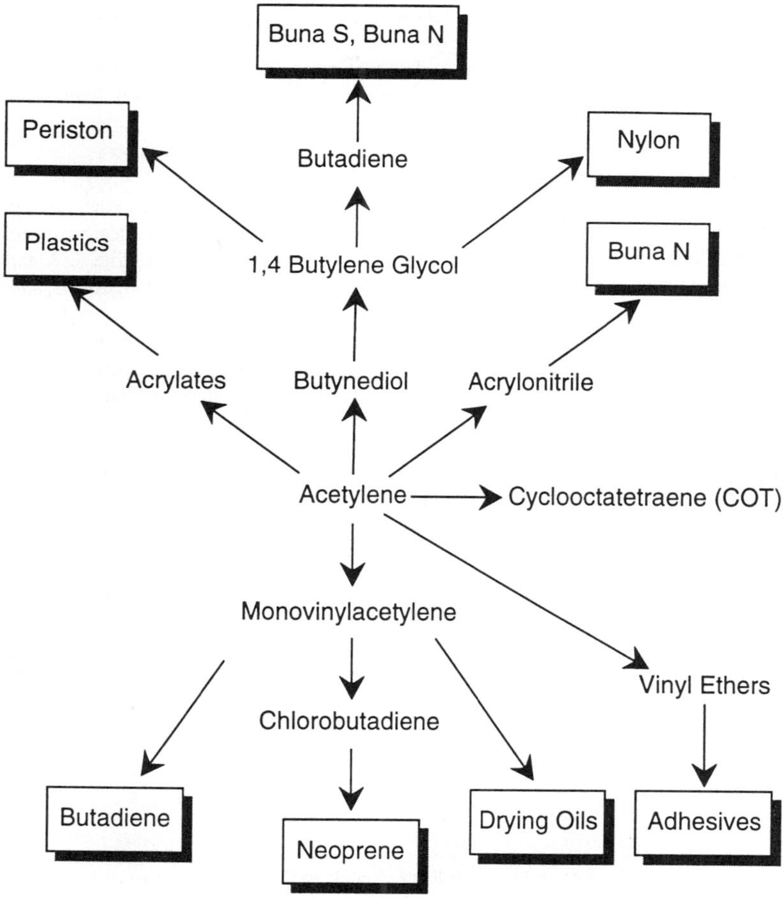

programme. When he finished this work, Reppe turned to the synthesis of monomers for a new class of promising polymers, the vinyl ethers.[30] It was thought that the polymers of these ethers would be resistant to hydrolysis and be less prone to side-reactions than the hitherto troublesome polyvinyl chloride (PVC). Reppe disproved the accepted wisdom that vinyl chloride would not react with sodium salts of alcohols, and went on to discover that acetylene would react directly with alcohols to form the corresponding vinyl ethers. This was a remarkable achievement, not least because it involved the handling of acetylene under pressure, an operation that was banned in welding as inherently dangerous.

**New structure, new problems**

In the summer of 1929, as the new I.G. Farben headquarters in Frankfurt was nearing completion, Carl Bosch decided to weaken Duisberg's concept of decentralized centralized regionalism by creating three production divisions or Sparten. Fritz ter Meer was made the leader of Sparte II, which encompassed the traditional dyestuffs and pharmaceuticals sectors as well as the new organic chemicals and polymers.[31] Whereas Sparte I (high-pressure chemistry) and Sparte III (fibres and photographic) had relatively simple product-lines and were confined to a few plants, Sparte II was intricate. It covered numerous factories across Germany and inherited no less than twenty-two committees.

In the early 1930s, ter Meer was in a strong position to increase the importance of organic chemicals within I.G. Farben, particularly after he became chairman of TEA at the beginning of 1933. The strenuous efforts of BASF, spearheaded by Carl Bosch and his deputy Carl Krauch, to move away from dyes and seize the opportunities offered by the new high-pressure chemistry had made the combine equally dependent on the success of synthetic fertilisers and fuels. It is not a little ironic that the continuing profitability of dyes and pharmaceuticals enabled I.G. to survive the almost total commercial failure of high-pressure chemistry during the Depression. The rapid

---

[30] Morris, "Acetylene Chemistry and Synthetic Rubber," pp. 113-117; John W. Copenhaver and Maurice H. Bigelow, *Acetylene and Carbon Monoxide Chemistry*, pp. 32-90; Friedrich Hölscher, *Kautschuke, Kunststoffe, Fasern* (Ludwigshafen: Firmenarchiv der BASF, 1972), pp. 79-84; Alfred von Nagel, *Äthylen, Acetylen*, pp. 40-41.

[31] Fritz ter Meer was the son of Edmund ter Meer who had founded a small firm (Weiler-ter Meer of Uerdingen) that became part of I.G. Farben. For ter Meer see Jens Ulrich Heine, *Verstand und Schicksal*, pp. 108-110. For the history of Weiler-ter Meer, see Wolfgang Scheinert, "Joseph Wilhelm Weiler, Julius Weiler und das Anilin: zur Entwicklungsgeschichte der deutschen Teerfarbenindustrie und der chemischen Technik vor dem ersten Weltkrieg," *Zeitschrift für Unternehmensgeschichte*, **33** (1988), 217-231.

## The Structure of I.G. Farben in 1931

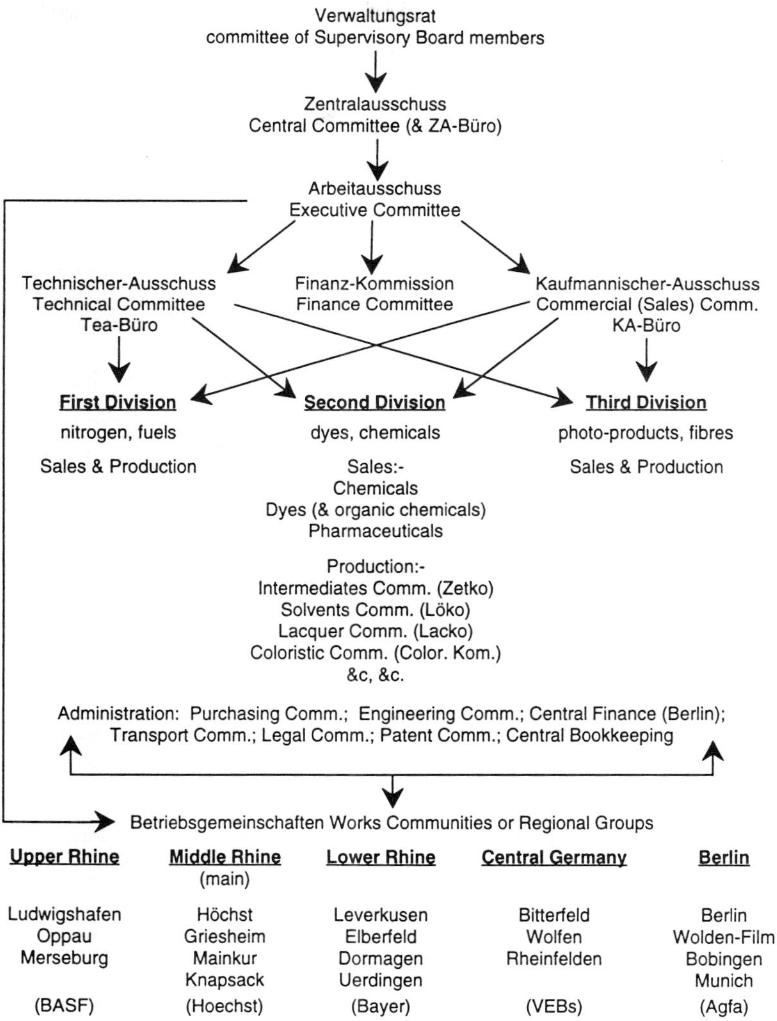


Source:
I.G. Kommissionen-Handbuch (1931)
Bayerarchiv-Leverkusen

development of the embryonic heavy organic chemicals[32] and polymers sector became all the more urgent. Ter Meer maintained the existing strategy of diversifying away from dyes by sponsoring a number of new areas, namely synthetic rubber, other polymers, acetylene chemistry, and pharmaceuticals. In a very real sense, ter Meer and his colleagues saw Buna rubber as standing in the same tradition of industrial synthesis as alizarin and indigo. It might have its teething problems, it might be struggling to match natural rubber in price, but eventually, even if it took another generation or so, a superior form of Buna rubber would displace the rubber tree as inexorably as synthetic alizarin had overthrown the madder plant.[33]

When Adolf Hitler came to power in 1933, he was determined to expand German self-sufficiency and create a 'defence' economy *(Wehrwirtschaft)*. Hitler's plans for German autarky dovetailed with, and yet significantly altered, ter Meer's original strategy.[34] I.G. Farben was forced to concentrate its research and development on the creation of strategic materials at the expense of other potentially more profitable lines of research. For instance, in the field of synthetic rubber, the profitable specialty Buna N was displaced by mass produced Buna S, which was more suitable for tyres. Given the inevitable technological lag between inception and production, Hitler's ideas had relatively little impact on Sparte II's production in the early years of the Third Reich. The more immediate result was a quickening of the pace of research; the amount spent on research in Sparte II more than doubled between 1934 and 1937.[35]

As the new fields of synthetic rubber and acetylene chemistry burgeoned, a separate laboratory became necessary for the development of the chemical intermediates. Reppe became the director of the new Intermediates and Plastics Laboratory (*Zwischenprodukte-Kunststoff Laboratorium* or ZK-Labor) on 1 January 1934. The first task of the new laboratory was the development of a continuous process for the production of methyl vinyl ether. A pilot plant was erected in 1935, and two full-scale plants were operating in Ludwigshafen during the war. A discontinuous polymerization process was developed by Otto Schlichting of the ZK-Labor in 1936 and this was first

[32] The term "heavy organic chemicals" may sound strange to non-chemists and non-native English speakers, hinting at the production of very dense liquids. It derives from "heavy chemicals," originally applied to inorganic chemicals (sulphuric acid, soda) made in vast quantities and extended to organic chemicals made on the scale of thousands of tonnes per plant in contrast to organic chemicals produced by the kilogram.

[33] Carl Krauch of I.G. Farben used the parallel with indigo to reassure Hermann Göring in 1936. Testimony of Carl Krauch, *Trials of War Criminals*, volume VII, p. 1003.

[34] Morris, "Acetylene Chemistry and Synthetic Rubber," Chapter Six, summarized in Peter J. T. Morris, "Synthetic rubber: Autarky and war" in Susan T. I. Mossman and Peter J. T. Morris, eds, *The Development of Plastics* (Cambridge: Royal Society of Chemistry, 1994), pp. 57-59.

[35] *United States of America v. Carl Krauch, et al.*, ter Meer Defense Documents, exhibit no. 37. National Archives Microfilm M892, roll no. 74.

brought to the industrial scale two years later. The polymers were not as widely applicable as first hoped, but they were used for a variety of purposes including adhesives, lacquers and waxes. One remarkable application was the use of poly(isobutyl vinyl) ether as a chewing gum substitute.[36]

Nonetheless, to Hitler's mounting irritation, ter Meer was slow to scale up the production of synthetic rubber. I.G. Farben had intended to use Buna N to meet the state's demand for synthetic rubber and sell any excess production as a profitable oil-resistant rubber in competition with Du Pont's neoprene. Unfortunately for I.G. Farben, in May 1934, General Tire of Akron reported that its tests demonstrated that Buna N was very hard, destroyed its rubber-processing machinery and was "definitely inferior" to natural rubber. Furthermore, in July 1934, Continental Gummiwerke discovered that Buna N could not be blended satisfactorily with natural rubber, which ruled it out as a tyre rubber. I.G. Farben was very impressed with neoprene, and ter Meer travelled to Wilmington, Delaware, in October 1934 to negotiate a neoprene licence with Du Pont. However, the talks ended in stalemate. While Buna S was suitable for tyres, it lacked the oil-resistance that made neoprene and Buna N profitable. At this stage Ambros and ter Meer were content to delay the scaling up of synthetic rubber because they were also keen to replace the cumbersome four-step process with a two-step process based on the dimerization of acetylene to monovinylacetylene, used by Du Pont to make neoprene. This process was possibly more economic than the four-step process, which would make it a useful carrot for the deadlocked negotiations with Du Pont. Unfortunately, it also turned out to be explosive, and two chemists were killed in an explosion at Höchst on 27 August 1935.[37]

**Ambros and the Buna system**

The situation changed radically when Adolf Hitler forced I.G. Farben to press ahead with erection of large Buna factories by his stunning – and wholly unexpected – announcement at the opening of the Nürnberg rally on 11 September 1935 that the construction of Germany's first synthetic rubber factory was underway. To make up for lost time, I.G. was forced to abandon the old methods of lengthy deliberation and compromise between the different regional communities, and create a centralized

---

36 R. Owens and A. W. Johnson, *The Acetylene Industry and Acetylene Chemistry in Germany during the period 1939-1945* (London: HMSO, 1951) [BIOS Surveys Report no. 30], p. 35.

37 Morris, "Acetylene Chemistry and Synthetic Rubber," summarized in Morris, "Transatlantic transfer of Buna S synthetic rubber, 1932-1945," in David J. Jeremy, ed., *Transfer of International Technology: Europe, Japan and the USA in the Twentieth Century* (Worcester: Edward Elgar, 1992), pp. 63-65. Also see David A. Hounshell and John K. Smith, Jr., *Science and Corporate Strategy: Du Pont R&D, 1902-1980* (Cambridge: Cambridge University Press, 1988), pp. 205-209. The late Franz Broich, who worked for I.G. Farben at Schkopau, gave me much personal information about the development of the two-step process, for which see Morris, "Acetylene Chemistry and Synthetic Rubber," pp. 209-214.

## Chart of the Committee Structure for Plastics and Synthetic Rubber

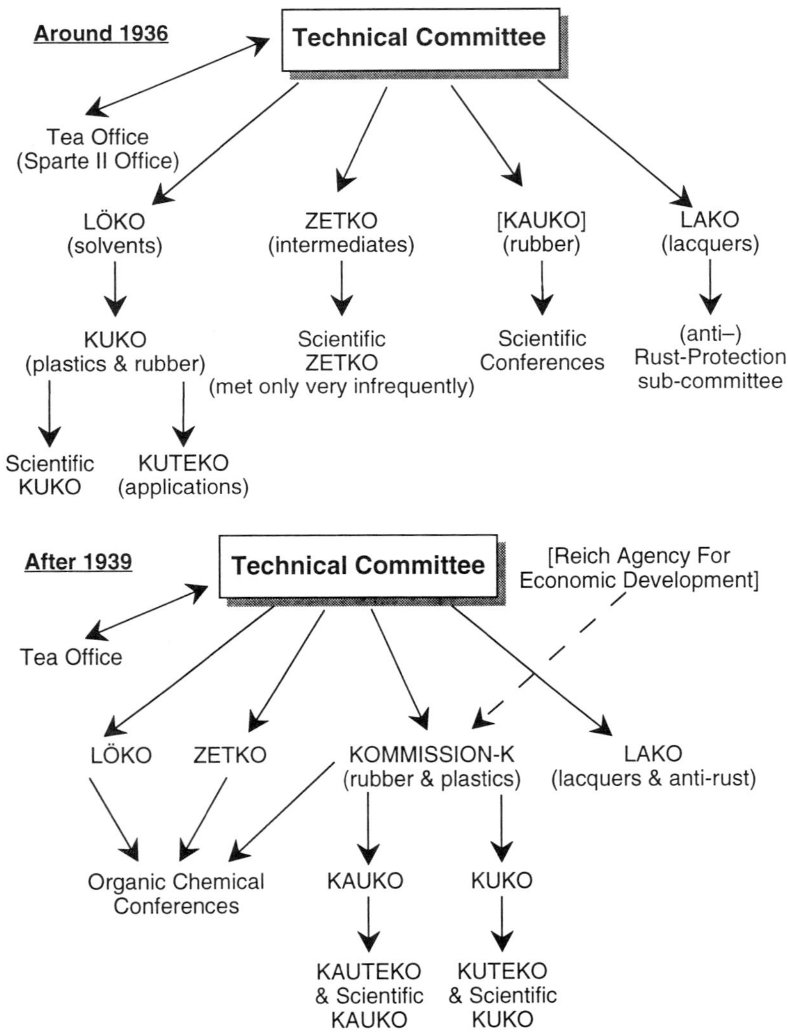

command structure.

As the chief co-ordinator of the synthetic rubber programme, Otto Ambros threw himself whole-heartedly into the creation of the "Buna system".[38] The desired end-product of the new system was clear: the production of Buna S on a large scale, and its conversion into tyres and other rubber goods. To do this, Ambros had to bring four distinct components together. The first was the manufacture of acetylene, which had hitherto been mostly associated with welding technology. The production of acetylene from coal and lime, via calcium carbide, was greatly improved. Furthermore, the synthetic rubber project stimulated research on the manufacture of acetylene from hydrocarbon gases. During 1937, Paul Baumann brought this new arc process for acetylene to the industrial scale at Leuna using the experience he had gained working with Jersey Standard at Baton Rouge in the United States. The second element of the system was the development of the intermediates that linked acetylene with butadiene, and which could form the basis of alternative peacetime outlets for acetylene. This organic chemical research was coordinated by Zetko, and linked Reppe with chemists in other factories, especially Otto Nicodemus of Höchst. This facet of the programme was I.G.'s strong point, stemming from its traditional field of dyestuffs chemistry. However, the firm was never at ease with the new petrochemistry that was becoming increasingly important in the United States. Ambros saw this field as the major challenge for I.G. Farben if Germany won the war.

Thirdly, organic synthesis had to be combined with the new field of polymer chemistry to produce the best possible Buna S rubber, an emulsion-polymerized co-polymer of butadiene and styrene. The earlier synthetic rubber research at Bayer and BASF had been largely empirical, but the foundations of a more scientific approach had been laid at Ludwigshafen in the late 1920s by Reppe's former thesis supervisor, Kurt Hans Otto Meyer, and Hermann Mark. Their pioneering work was largely based on the study of cellulose (I.G. was very interested in rayon during this period), but applied to synthetic plastics by Hans Fikentscher at Ludwigshafen and to synthetic rubber by Richard Ludwig at Leverkusen. Ludwig improved the emulsion polymerization process, in 1936, by switching from soap to detergent and by adding a 'modifier' (linoleic acid) which produced a polymer that was easier to process. A year later, the veteran rubber chemist Kurt Meisenburg of Leverkusen discovered better modifiers (mercaptans), but they were not adopted by I.G. until 1943. Ironically, dodecyl mercaptan – fondly known as OEI (one essential ingredient) – had already been adopted by the American industry in March 1942.[39]

---

[38] This section is based on Morris, "Acetylene Chemistry and Synthetic Rubber," pp. 281-370. For an independent, but similar account, see Peter Hayes, *Industry and Ideology*, pp. 148-151, 186-193, 339-340, and 347-358.

[39] For the American synthetic rubber programme, see Peter J. T. Morris, *The American Synthetic Rubber Research Program* (Philadelphia: University of Pennsylvania Press, 1989), and Vernon Herbert and Attilio

On its own, however, a good polymer was not enough. It had to be processed into a stock that could be made into tyres. This final element was perhaps the single greatest bottleneck in the development of a viable synthetic rubber at the outset of the Four Year Plan in 1936. The successful change-over from natural rubber tyres to a natural-synthetic mixture was a complex task which required diplomacy to handle the differences between I.G. Farben and the rubber companies, who did not welcome the new ersatz rubber. The task was bedevilled by rivalries within the rubber industry. In 1937, I.G. and the large rubber concern Continental Gummiwerke of Hannover introduced a crude process called thermal degradation (thermisches Abbau) that softened the rubber by heating it in air. A major research and testing laboratory was built in the late 1930s for Erich Konrad, the leader of rubber research at Leverkusen, and Paul Stöcklin, the chief rubber technologist.

The different aspects of the development of synthetic rubber have been reviewed in some detail to show the disparate and often immature components of the programme that Ambros was forced to weld into a single system in a remarkably short time-span. The actual construction of the factories was another important element of Ambros's planning. The first plant, essentially a large pilot plant, was built in 1937 at Schkopau, located between Halle and Leuna at the centre of I.G.'s new network of energy and raw materials, and the main factory was inaugurated on 20 April 1939. This was followed by a second factory at Hüls on the northern edge of the Ruhr in August 1940. The third factory was initially planned for Rattwitz (near Breslau), but was eventually erected between Ludwigshafen and Oppau in 1943. The infamous fourth factory at Auschwitz was never completed.

As a senior manager at Ludwigshafen, Ambros was familiar with Walter Reppe's research, which had already held out great promise for his strategy. His trust in Reppe was rewarded in September 1937, with the discovery of an entirely new three-step reaction, one which could be used to make butadiene. Reppe found that two molecules of formaldehyde (derived from methanol) could be united with acetylene, without loss of the triple bond, to form 2-butyne-1,4-diol. The most striking feature of the new process was its employment of the detonating compound copper acetylide as the catalyst for a reaction which also used acetylene under pressure. By retaining the triple bond, in contrast to most acetylene reactions, the new reaction conserved the considerable quantity of electricity needed to manufacture acetylene. Furthermore, it used only one molecule of acetylene to make a molecule of butadiene, and calculations suggested that the new Reppe process would only consume a third of the coal required by the four-stage process.

This breakthrough appeared to vindicate Ambros's aversion towards the four-stage

---

Bisio, *Synthetic Rubber: A Project That Had to Succeed* (Westport, Conn.: Greenwood Press, 1985).

Otto Ambros laying the foundation stone of Schkopau on 25 April 1936. Two days earlier Ambros had become the Chairman of the Intermediates Committee (Zetko). (Courtesy Dow BSL-Olefinverbund.)

Ambros (on the left) and Reppe (second from right) observing the construction of the new butynediol building at Ludwigshafen, c.1942. (Author's collection.)

The Reppe process (butynediol) building at Ludwigshafen. The open construction was adopted to minimize the damage from flying debris in the (likely) event of an explosion. (Author's collection.)

process. He decided as early as June 1938 to use the three-stage Reppe process in the first two Buna factories, but, as major technical problems soon arose, he withdrew this proposal in the spring of 1939. Solutions for these problems were found by Georg Niemann of Ludwigshafen by the end of 1939, and Ambros decided to use the Reppe process in the next synthetic rubber factory (Buna III). This was an expensive project, with research and development costs of 2.6 million Reichmarks (roughly US $1 million) in 1941-43 alone. Franz Reicheneder successfully operated a 200t/m plant in Schkopau at the end of 1941. The full-scale 2,500t/m plant, between Ludwigshafen and Oppau, started up in the spring of 1943, but never operated at full capacity. It only produced 19,000t of Buna S, before it was closed down by Allied air-raids in December 1944.[40]

The operation of a high-pressure acetylene plant that employed a percussive compound as a catalyst in the midst of aerial bombardment was a feat of technical skill, not to mention sang-froid. A leading British acetylene chemist, Alan W. Johnson, recalled the amazement of Allied chemists when they first heard about this plant:

> A small experimental plant was erected in the early war years at the Dyestuffs Division of I.C.I. Ltd. at Blackley, Manchester ... and after much toil some 3-4 lbs. of [butynediol] was prepared, the workers (and I was one of them) satisfied under the illusion that this was by far the largest quantity of [butynediol] that had ever been prepared. How completely this illusion was shattered may be judged when, in 1945, they learnt that the Ludwigshafen [butynediol] plant had a capacity of [30,000] tons a year and had been producing well over 60 per cent of this amount in 1944. ... Approximately 45 tons of the copper acetylide-silica catalyst were in use at once and the whole project represented an outstanding feat of chemical engineering.[41]

**The impact of World War II**

If the outbreak of war in 1939 wrecked most of I.G. Farben's overseas markets, it also promoted the explosive growth of heavy organic chemicals, which had begun in 1937. The most obvious example was Buna S, the production of which rose over fifty-fold between 1937 and 1943, and thereby spurred the production of equally vast quantities of butadiene and styrene. The large-scale production of synthetic rubber gave an

---

40 Morris, "Acetylene Chemistry and Synthetic Rubber," pp. 123-126, 185-186, 215-218; John W. Copenhaver and Maurice H. Bigelow, *Acetylene and Carbon Monoxide Chemistry*, pp. 91-104; Alfred von Nagel, *Äthylen, Acetylen*, pp. 48-51.

41 A. W. Johnson, *Some Applications of Acetylenic Compounds in Organic Synthesis,* Meldola Medal Lecture (London: Royal Institute of Chemistry, 1948), p. 4. The original quotation gives a figure of 50,000 tons, but this is almost certainly a misprint for 30,000; see A. W. Johnson, "Acetylene chemistry," in Owens and Johnson, *The Acetylene Industry and Acetylene Chemistry in Germany* (written in 1948) for the latter figure, which agrees with Copenhaver and Bigelow, *Acetylene and Carbon Monoxide Chemistry*, p. 98.

## Routes to Butadiene

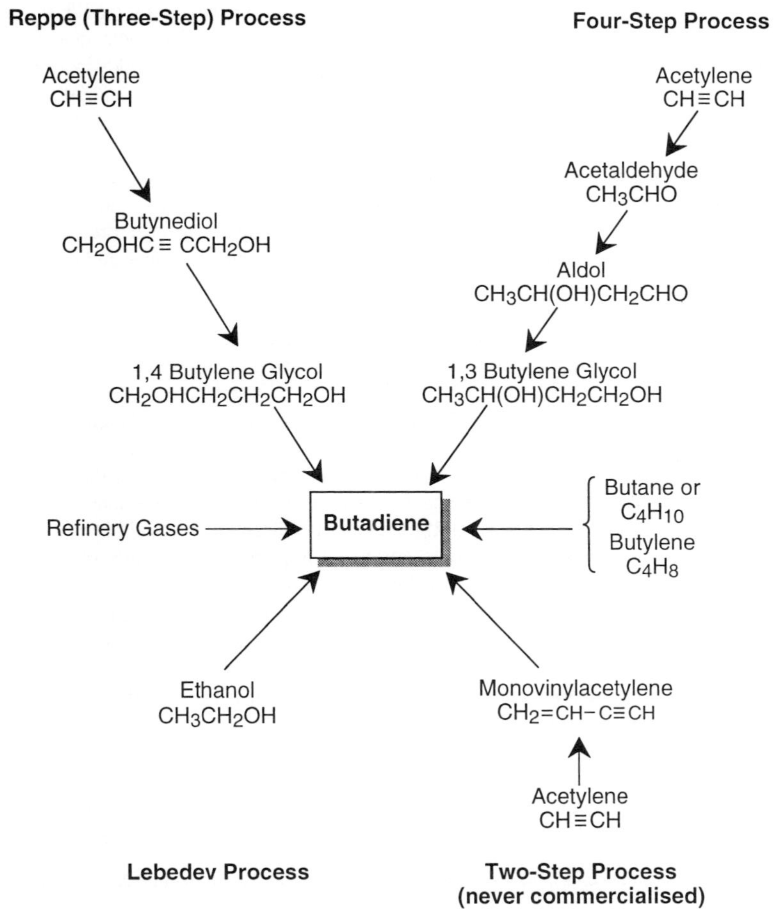

immense impetus to the production of calcium carbide and acetylene: thus Buna accounted for only a thousand tonnes of carbide in 1935, but 330,000t in 1943. The Buna factory at Schkopau was the largest and most modern carbide factory in Germany. German carbide production nearly doubled between 1936 and 1943 to 1.3 million tonnes or about half of total world production. In the same period, the quantity of carbide used in chemical manufacture increased sixfold to about 850,000t in 1943. In addition to the 280,000t of acetylene generated from this carbide, 50,000t were produced in 1943 by the hydrocarbon-cracking arc process at Hüls.

If vast quantities of acetylene had to be generated for synthetic rubber manufacture, it made sense to find other end-uses for it, if only to cover the anticipated postwar dip in demand for Buna. Otto Ambros always regarded the synthetic rubber project as the vanguard of a larger programme of diversification into organic chemicals and polymers. The production of these non-strategic materials would dilute the state-directed character of these factories, hopefully increase profits, and make the factories more like the rest of I.G. Farben. For Ambros the Buna factories were not an end in themselves, but were the 'seeds' for future organic chemical complexes scattered across Europe.

The first of these complexes was obviously Schkopau. As early as November 1935, Ambros reported that "Schkopau is, therefore, a good site for a new factory, that can become a seed for powerful growth."[42] From the outset, he was anxious that the new factory should not be limited to Buna production and rejected sites that would be restricted in this way (such as Piesteritz). In a letter to Bürgin of Bitterfeld in September 1936, he said "It now lies within the purpose of this purely private enterprise to also place a few key inorganic industries in Schkopau next to the acetylene chemistry."[43] During the war Schkopau produced vinyl chloride (and some polyvinyl chloride), acetic acid, phthalic anhydride, diglycol, polyethylene (as an oil), and other organic chemicals such as tetrahydrofuran (THF). Since 1945, Schkopau has developed into a large 'combinat' of five factories and numerous plants producing a range of polymers and chemicals.[44]

---

[42] Otto Ambros, "Die Projekte der Bunafabrik und des Lösungsmittelbetriebes in Mitteldeutschland," 16 November 1935, 6, I.G. Zentralarchiv file 370 (section 2), Firmenarchiv Hoechst AG.

[43] Letter from Ambros to Bürgin, 22 September 1936, I.G. Zentralarchiv file 370, Firmenarchiv Hoechst AG.

[44] For the history of Schkopau see Morris, "Acetylene Chemistry and Synthetic Rubber," pp. 281-297, 313-320; Peter Hayes, *Industry and Ideology*, pp. 148-151, 186-193, 339-340; *VEB Chemische Werke Buna, 1945-1965* (Schkopau: VEB Chemische Werke Buna, 1965); *Zahlen, Fakten, Ereignisse: Chronik zur Geschichte der Arbeiterwegung des Kombinates VEB Chemische Werke Buna*, vol. 1 (1936-1974), (Schkopau: Kreisleitung der SED des Kombinates VEB Chemische Werke Buna, 1975), vol. 2 (1975-1981), (Schkopau: Kreisleitung der SED des Kombinates VEB Chemische Werke Buna, c.1981). For the production range at Schkopau, see R. Owens and A. W. Johnson, *The Acetylene Industry and Acetylene Chemistry in Germany*, pp. 20, 22-23; CIOS Reports XXII-20, XXII-22 and XXVIII-29; and *Buna: Traditionsreich, Leistungsstark und Zukunftsorientiert* (VEB Chemische Werke Buna, 1984).

Carbide furnace at Schkopau. In 1943, Schkopau produced nearly 300,000 tonnes of calcium carbide and had become the largest carbide factory in Germany. (Courtesy Dow BSL-Olefinverbund.)

In the case of the second chemical factory, Hüls, Ambros even suggested that it should be started as a factory for organic chemicals and only converted to synthetic rubber at a later date. This was unacceptable to the state authorities. However, Hüls also produced ethylene oxide which was converted into ethylene glycol (in part for anti-freeze), diethylene glycol, and the co-products chlorine and sodium hydroxide.[45] Ambros

[45] For the history of Hüls, see Morris, "Acetylene Chemistry and Synthetic Rubber," pp. 297-304, 313-320; Paul Kränzlein, *Chemie im Revier – Hüls* (Düsseldorf, 1980); Raymond Stokes, *Opting for Oil*, pp. 84-89, 197-216; Franz J. Wünsch, "Das Werk Hüls. Geschichte der Chemischen Werke Hüls AG in Marl von 1938 bis 1949," *Tradition*, **9** (1964), 70-79; Wünsch, "Vom Bunawerk zum Werk der Grosschemie" *Der Lichtbogen* (the Hüls house magazine) (2/1963), pp. 41-72; and Wünsch's excellent unpublished history, "Chronik der Chemische Werke Hüls" (in three volumes), Werkarchiv, Hüls AG. For the wartime

intended that Hüls would be an intermediates producer for the rest of I.G., but this was prevented by the war and the heavy air-raids of 1943.[46] When the Allies banned Buna production for a period after 1945, the manager Paul Baumann diversified into a wide range of organic chemicals and plastics. With the help of the holding company, VEBA, Hüls has since grown from being a relatively small synthetic rubber plant into a major chemical corporation, with a growing presence in the United States.

The synthetic rubber factory at Rattwitz, near Breslau, was designed by Ambros to be a centre of Reppe chemistry, producing vinyl ethers and acrylates in addition to Buna S. Construction was halted in July 1940 and the synthetic rubber plant was eventually erected at Ludwigshafen in 1943 with the co-production of butynediol.[47] Ambros saw Auschwitz as a good site for combining the production of methanol and iso-octane (hitherto Sparte I products) with carbide-based acetylene and synthetic rubber, thereby creating a suitable site for "a nucleus for modern chemistry".[48] In common with its sister factories, the present-day Oswiecim Chemical Factory (Zaklady Chemiczne Oswiecim) produces chemicals and plastics in addition to synthetic rubber.[49]

There is no specific evidence for the production of organic chemicals at the foreign Buna factories at St. Fons, close to Lyon (Rhône-Poulenc-Ugine-I.G.), and Ferrara (SAIGS with I.G. assistance), partly because both plants were using (or in the case of St. Fons planning to use) alcohol-based butadiene. However, it seems probable that St. Fons would have produced polystyrene and possibly ethylene compounds; the Ugine carbide works at Anglefort in the Alps had an area set aside for the working up of by-products, presumably acetic acid and acetone.

---

production range at Hüls, see R. Owens and A. W. Johnson, *The Acetylene Industry and Acetylene Chemistry in Germany*, pp. 20, 22; CIOS Reports XXII-21, XXVI-51 and XXXI-75;

46 Letter from Otto Ambros to author, 27 October 1980.

47 For the history of Rattwitz and Ludwigshafen synthetic rubber plants, see Morris, "Acetylene Chemistry and Synthetic Rubber," pp. 322-329; Peter Hayes, *Industry and Ideology*, pp. 347-348; Alfred von Nagel, *Äthylen, Acetylen*, pp. 48-50.

48 Letter from Otto Ambros to author, 27 October 1980. For the history of the Auschwitz Buna factory, see Morris, "Acetylene Chemistry and Synthetic Rubber," pp. 330-345; Peter Hayes, *Industry and Ideology*, pp. 347-368; the excellent thesis by Joseph Robert White, "The Politics of Labor Utilization: I.G. Farben, the SS and Auschwitz," Georgia State University MA thesis, 1989, and the less original work of Robert Simon Yavner, "I.G. Farben's Petro-Chemical Plant and Concentration Camp at Auschwitz," Old Dominion University MA thesis, 1984. Unfortunately, Debórah Dwork and Robert Jan van Pelt did not consult this literature (except for Hayes) for the chapter on I.G. Farben in their *Auschwitz, 1270 to the Present* (New York: W. W. Norton, 1996). For a personal account, see Primo Levi, *If this Man is a Man?*, a translation of *Se questo è un uomo* by Stuart Woolf (New York: Orion, 1959), also published as *Survival in Auschwitz*.

49 Frank X. Didik, *Eastern European Business Directory*, 1st edition (Detroit: Gale Research, 1992), p. 847.

Reppe was rewarded for the three step process with the directorship of the Main Laboratory (Hauptlaboratorium) at Ludwigshafen on 1 January 1938. Copenhaver and Bigelow, in their book on Reppe chemistry, commented that "This was the largest research institute of the I.G., and its research activities extended into all branches of chemical science and technique."[50] The director of the Main Laboratory was effectively head of research at Ludwigshafen, and his new position gave Reppe greater freedom to pursue his ideas and to investigate the potential of the ethynylation reaction for compounds other than butadiene. It is noteworthy that nearly all Reppe's proposed syntheses used only the original two compounds, although the ethynylation reaction could be used to yield many different products. This was partly a reflection of the economies of scale anticipated in the use of butynediol in the Reppe process and also fears of post-war overcapacity following the likely cut-backs in Buna S production.

The most important of these compounds were the monomers of nylon 66, hexamethylene diamine (HMDA) and adipic acid. During World War II, I.G. Farben (and Du Pont) prepared both compounds from coal tar products, but Reppe and his co-workers devised several new routes from propargyl alcohol and butynediol. Similarly, caprolactam, the monomer of the rival nylon 6, was also synthesized from butynediol.[51]

Butanediol was also used as an intermediate in the production of polyurethanes. These novel plastics had been discovered by Otto Bayer, Reppe's counterpart at Leverkusen, in 1937.[52] They were formed by the condensation of di-isocyanates (derived from diamines) with alcohols containing two hydroxyl groups. One early polyurethane,

---

50 Copenhaver and Bigelow, *Acetylene and Carbon Monoxide Chemistry*, p. xii.

51 Nylon 6,6 was developed by Wallace Hume Carothers at Du Pont. Hounshell and Smith, *Science and Corporate Strategy*, pp. 207, 257-273. When I.G. Farben became aware of nylon 6,6 in 1937, the company trawled the technical literature and its own research, and decided to develop caprolactam-based nylon 6. The key research was carried out at the I.G. factory in Berlin-Lichtenberg by Paul Schlack, who had previously championed research into polyamides without success. See Basil G. Achilladelis, "A study in technological history: Part I. The manufacture of 'Perlon' (nylon 6) and caprolactam by I.G. Farbenindustrie," *Chemistry and Industry*, 5 December 1970, pp. 1549-1554. For the production of nylon monomers by the Reppe process, see BIOS 350, translation of a report written by Reppe in 1941. Also see CIOS XXXIII-50, Part II, pp. 534-547 and Walter Reppe, *Neue Entwicklungen auf dem Gebiet der Chemie des Acetylens und Kohlenoxyds* (Berlin: Springer, 1949), pp. 46-48, 110-112, 140. Dehydrogenation of the saturated butanediol yielded butyro-lactone, an intermediate and solvent, which was used for the manufacture of Periston, an effective blood-plasma substitute developed by Reppe and introduced in 1942. Copenhaver and Bigelow, *Acetylene and Carbon Monoxide Chemistry*, pp. 152-166, 81-82; FIAT 945; Walter Reppe, *Polyvinylpyrrolidon* (Weinheim: Verlag Chemie, 1954).

52 Owens and Johnson, *The Acetylene Industry and Acetylene Chemistry in Germany*, pp. 84-85; John M. Debell, William C. Goggin, and Walter E. Gloor, *German Plastics Practice* (Springfield, Mass: DeBell and Richardson, 1946), pp. 289-290, 300-316. Verg, *Milestones*, pp. 284-289; V. M. Walsh, J. F. Townsend, B. G. Achilladelis and C. Freeman, "Trends in Invention and Innovation in the Chemical Industry," report to the Social Science Research Council, June 1979 (obtainable from the Science Policy Research Unit, University of Sussex), pp. 3.49-3.52.

called Perlon-U, was made from butanediol and the di-isocyanate of HMDA. The Leverkusen plant was on the point of coming into full-scale operation when captured in 1945. Tetrahydrofuran (THF) was a by-product of the Reppe process, and was used by I.G. as a solvent and, in a polymerized form, as a lubricant.[53] The trihydric alcohol 1,2,4 butanediol could be used as a substitute for glycerol, and its trinitrate was employed as an explosive. Reppe patented its preparation by a multi-stage process from butynediol in September 1940, and a pilot plant was later erected.[54] A full-scale 6,000t/y plant was planned, but never built. It is unlikely that butanetriol was ever competitive with the simpler glycerol substitute, diethylene glycol.

Walter Reppe also used his new base to expand the chemistry of acetylene. His first major breakthrough, in the summer of 1939, was the addition of carbon monoxide to acetylene in the presence of alcohols (or water) and a nickel catalyst to form acrylates.[55] Carbon monoxide had attracted attention for many years as a readily available, cheap and reactive carbon compound. I.G. Farben employed it in the Pier methanol synthesis, Ruhrchemie used it in the Fischer-Tropsch synthetic petrol process, and Du Pont had carried out research on the addition of carbon monoxide to olefins at very high pressure and temperatures.[56] Additional impetus for the use of carbon monoxide in acetylene chemistry was provided by the introduction of covered carbide furnaces at I.G. Farben's Knapsack plant in 1938, which permitted the collection of by-product carbon monoxide. The polymers of acrylic esters were already used for treating leather and for paint, but acrylic acid was made from ethylene oxide, and consequently was rather expensive. Reppe's process reached the pilot plant stage by 1945, and was subsequently used on a large scale by BASF and its American partners.

Reppe was told about Otto Roelen's OXO process by Ruhrchemie in the spring of 1940, which gave him the idea that the carbonylation reaction could be extended to ethylene.[57]

---

53 Modern polyurethanes fibres (such as Lycra®) use tetrahydrofuran as a starting material rather than butanediol.

54 Copenhaver and Bigelow, *Acetylene and Carbon Monoxide Chemistry*, pp. 138-139; BIOS 1708.

55 Morris, "Acetylene Chemistry and Synthetic Rubber," pp. 128-130; Copenhaver and Bigelow, *Acetylene and Carbon Monoxide Chemistry*, pp. 246-294; Alfred von Nagel, *Äthylen, Acetylen*, pp. 58-63.

56 For the Pier methanol synthesis, see von Nagel, *Methanol, Treibstoffe*, pp. 7-11. For the Fischer-Tropsch process, see the references to oil-from-coal chemistry above (note 18) and Stokes, *Opting for Oil*, pp. 217-230. For the use of carbon monoxide at Du Pont, see Reader, *Imperial Chemical Industries: A History*, volume 2, p. 321.

57 The OXO process, patented by Roelen in 1938, involved the addition of carbon monoxide and hydrogen to olefins using a cobalt catalyst at 150°C and 200 atmospheres. See BIOS 355, 7 for the Reppe-Ruhrchemie link. For the OXO process, see Milton Orchin, "The OXO Reaction," in Benjamin T. Brooks, Cecil E. Boord, Stewart S. Kurtz, Jr., and Louis Schmerling, eds, *The Chemistry of Petroleum Hydrocarbons*, volume 3 (New York: Reinhold, 1955), pp. 341-361; FIAT 1000; Klaus Weissermel and Hans-Jurgen Arpe, *Industrielle organische chemie* (Weinheim, 1976), translated by Alexander Mullen, *Industrial Organic*

With Hugo Kröper, he demonstrated that carbon monoxide and ethylene formed propionic acid at 300°C and 200 atmospheres under the influence of nickel carbonyl. A silver-lined pilot plant for a continuous process was erected in 1943, but, owing to the war and the French occupation, a semi-works plant was not ready until 1951.[58] Even before the outbreak of war in 1939, Reppe had used nickel carbonyl to carbonylate methanol to acetic acid, a reaction previously only possible under extreme conditions of pressure and temperature. The war (and corrosion problems) delayed its development for over a decade and a full-scale plant was not built until 1957. It is still used at Ludwigshafen and by Borden at Geismar, Louisiana.[59]

A wholly new reaction was discovered by chance by Reppe and Tim Toepel in the winter of 1940-41. Attempting to prepare 3-hexyne-1,6-diol by adding ethylene oxide to acetylene (presumably with the aim of preparing the monomers of nylon 66) they found that acetylene could undergo cyclization in the presence of nickel cyanide to form the elusive cyclooctatetraene (COT). This compound had been prepared just before World War I by Otto Ambros's former professor Richard Willstätter using a lengthy synthetic sequence, but his claim had come under attack by American chemists in 1939. Despite the difficulties created by wartime conditions, particularly air-raids and shortage of trained staff, Reppe carried out considerable fundamental research on COT, and confirmed that Willstätter had indeed managed to synthesize it. When Reppe's work was published in 1948, it became the basis for later academic research, but the early hopes for industrial uses for COT turned out to be ill-founded.[60] Wilhelm Keim, a leading organometallic chemist, has remarked "although this reaction has not, as yet, found commercial application, it has nevertheless played a decisive role in the development of theoretical chemistry."[61]

Contrary to the impression given by some histories of the chemical industry, Reppe chemistry was still largely embryonic in 1943, with the partial exception of the vinyl ethers. Most of the new Reppe reactions were still at the laboratory, or at best at the pilot plant, stage. In fact, Reppe chemistry was responsible for less than 2½ per cent of

---

*Chemistry* (Weinheim, 1978), pp. 111-123. Also see Manfred Rasch, "Ruhrchemie AG, 1945-1951: Wiederaufbau, Entnazifizierung und Demontage," *Technikgeschichte*, **54** (1987), 101-120.

58 Dieter Franz, "Propionic Acid" *Die BASF* (English version) **18** (October 1968), 51-58; Weissermel and Arpe, *Industrial Organic Chemistry*, pp. 123-4.

59 Weissermel and Arpe, *Industrial Organic Chemistry*, pp. 156-157.

60 Peter J. T. Morris, "The technology-science interaction: Walter Reppe and cyclooctatetraene chemistry," *British Journal for the History of Science*, **25** (1992), 145-167.

61 Wilhelm Keim, "Nickel: An element with wide application in industrial homogeneous catalysis," *Angewandte Chemie International Edition in English*, **29** (1990), p. 236.

all the acetylene consumed by the German chemical industry in 1943. Somewhat surprisingly, plastics as a whole only accounted for about 60,000 tonnes per year of acetylene (18 per cent) in 1943, despite a vast increase in production since 1936. The production of Igelit PCU (PVC), for instance, rose over a hundredfold in this period to 18,383 tonnes in 1943, but it still consumed less than 7,000 tonnes per year of acetylene.

After a final desperate attempt by a task-force led by Georg Niemann to get production started at I.G. Auschwitz in the autumn of 1944, Ambros concentrated on Project 'Cornerstone' (Eckstein). This was the conversion of an abandoned underground aircraft plant in Schwandorf in the Upper Palatinate into a chemical factory based on Reppe chemistry using equipment salvaged from Ludwigshafen and perhaps diverted from Auschwitz. Presumably this was intended – if Ambros ever regarded it as other than a fantasy to keep the Nazis happy – to supply Hitler's last stand in his Bavarian redoubt with Buna and other chemicals. Interestingly, Schwandorf is not far from Ambros's birthplace in Weiden. At this time, Reppe was still working on the elaboration of his new chemistry, especially the manufacture of acrylic acid, and the carbonylation of olefins and methanol. On 21 September 1944, Ambros and Reppe made their way to the Anorgana factory at Gendorf[62] in Upper Bavaria, partly to escape from the bombing and advancing Americans, and partly to prepare for the starting up of Schwandorf in the autumn of 1945. In the modest laboratory at Gendorf, Reppe concentrated on the formation of ethanol and propanol from the corresponding olefins using water under high pressure. The US Army occupied Gendorf on 6 May and detained Ambros and Reppe on 10 June. Reppe was released but taken in July 1945 to 'Dustbin,' the Allied interrogation centre at Kranzberg Castle, near Frankfurt, for interrogation.

**Conclusion**

What does the relationship of Ambros and Reppe tell us about innovation in the chemical industry? They were linked by a strategy of diversification and modernization. I.G. Farben's leaders (and the leaders of Sparte II in particular) wanted to get away from dyes, on the one hand, and the ruinously expensive high-pressure chemistry, on the other. During the Third Reich, ter Meer and Ambros also sought to diversify away from 'political' products such as Buna S towards potentially more profitable consumer goods, such as oil-resistant Buna N, nylon and polyurethanes.

I.G. Farben (and Ambros in particular) sought to modernize production methods, with a

62 The Gendorf factory had started up in May 1941, producing ethylene glycol, diethylene glycol and mustard gas; ethylene dichloride was produced as a by-product. Anorgana (presumably from the German for 'inorganic') was set up as a 'shadow' company to produce war gases in January 1940. Ernst Bäumler, *Die Fabrik im Grünen* (Gendorf: Hoechst AG, c.1990); Stokes, *Opting for Oil*, 122; Debell, Goggin, and Gloor, *German Plastics Practice*, pp. 131-139. Testimony of Otto Ambros, *Trials of War Criminals*, volume VII, pp. 1260-1263.

strong movement towards continuous automatically controlled processes. Thus, during his visit to I.G.'s works in June 1936, Du Pont's Elmer K. Bolton noted research on automation which Ambros assured him was used to improve product quality rather than save labour. There are numerous examples of the conversion of batch reactions into continuous processes by I.G., including Reppe's tower process for ethylene oxide, the nickel chloride-THF process for COT, and the employment of continuous processes throughout the manufacture of Buna S, including the polymerization stage. This development was assisted by the involvement of research scientists (including Reppe) in the scaling-up stage. Bolton commented that the semi-works in the neighbouring Oppau works was "very well-equipped and has a very large amount of duplicate equipment, thereby permitting a mass attack on any problem."[63]

Modernization and diversification also implied the introduction of new products. Many of the new products associated with Reppe were innovated as a result of an existing demand. The expansion of motor-car ownership was one important driving force. Butanol and butyl acetate were required as solvents for stoving lacquers used on car bodies. Ethylene glycol was first used as an anti-freeze, and oil-resistant rubbers, such as Buna N, were needed for petrol hoses and gaskets. There was also a strong consumer demand for plastics, which were highly fashionable in this period, in contrast to their tawdry image in the 1950s and 1960s. The Third Reich demanded a wide range of strategic materials and substitutes.

However, I would argue, the demand factor provides us with an incomplete picture of technological innovation in the organic chemical industry. The chemical industry plays a crucial role in 'mediating' between an existing (or a potential) demand, scientific knowledge and technical know-how. It perceives technical advances or new products and uses them to meet external demands. The development and production of the new products requires a system which promotes innovation and links it to the development of the industrial plant, thereby creating a production system with its own momentum. The result of this system can be seen in the diversified organic chemical complexes erected during and after World War II. In this sense, Schkopau and Hüls can be regarded as the forerunners of the petrochemical complexes built in Germany in the late 1950s and early 1960s.

The chemical industry also mediates in another important way, namely between abundant (and hence cheap) raw materials, perhaps even the by-product of another process, and the desired products. Hence, modernization can also involve a change of feedstock, or at least its source. Carl Heinrich Krauch of Hüls AG (and the son of I.G.'s Carl Krauch), has argued that the organic chemical industry has been driven by

---

[63] Report from Bolton to Crane, 11 June 1936, E. I. du Pont & Co. Inc. records, series II, part 2, box 1038, Hagley Museum and Library. I thank David Hounshell for drawing my attention to this document.

feedstocks, rather than by demand or technology.[64]

As in any form of mediation, the result of this frequently complex process is often a compromise. The raw material may be scarcer or dearer than is desirable or, more often, the final product has to struggle to become profitable, as was the case with Buna S. In such cases, the industrial chemist may look for a more suitable raw material, as when Reppe strove to replace fermentation-based ethanol with middle oil as the feedstock for ethylene. Alternatively, he may retain the original starting material, but search for a reaction with a higher yield (the "key to Buna"), or one with a more desirable outcome, such as the Reppe process, which conserved the dearly won triple bond. The Reppe process also featured the partial replacement of an expensive feedstock, since one of the costly acetylene molecules was replaced by two cheap formaldehyde molecules. It is thus clear that Reppe's work was strongly influenced by feedstock considerations.

Although a feedstock, or a particular feedstock source, is initially taken up because it is produced in excess or is even a waste product, its success can result in the need to manufacture it specially for that process, thereby losing the rationale for its original adoption. One can see this progression at work with coal-tar (in particular phenol[65]) and synthetic dyes, with coke-based ammonia and the Solvay process, with calcium carbide and acetylene-based chemicals, and, more recently, with refinery gases and petro-chemicals. Conversely, butanol became such a major by-product of butadiene production that I.G. could only sell a quarter of it by 1943.[66]

Furthermore, Reppe's research is a good illustration of the "founder effect" (or "bandwagon effect"), whereby positive feedback tends to segregate competing technologies.[67] Acetylene was adopted as the feedstock for BASF's butanol synthesis,

---

[64] Peter Spitz, *Petrochemicals: The Rise of an Industry* (New York: John Wiley & Sons, 1988), p. 514. V. M. Walsh, J. F. Townsend, B. G. Achilladelis and C. Freeman, "Trends in invention and innovation in the chemical industry," reached similar conclusions about the importance of demand and feedstocks (see page 5.20). By contrast, Achilladelis played down the importance of feedstocks in his thesis; see "Process Innovation in the Chemical Industry," p. 245. The Krauch family form an impressive dynasty. Carl Heinrich's grandfather, also called Carl, was a pharmacist and production manager at Merck (Darmstadt). See Heine *Verstand & Shicksal*, p. 98, and H. Benniga, *A History of Lactic Acid Manufacture* (Dordrecht: Kluwer, 1990), pp. 129-134, 153.

[65] Phenol is used in the manufacture of explosives and plastics, as well as dyes. See Travis, "Modernizing industrial organic chemistry," this volume.

[66] Folder entitled "Lösungsmittel," section 7, chart of butanol production and turnover, FD 3295/45 (corrected from FD 3296/45), Imperial War Museum.

[67] The 'founder' effect was introduced by Brian Arthur. See W. Brian Arthur, "Competing technologies, increasing returns and lock-in by historical events," *Economic Journal*, **99** (1989), 116-131. The concept was first applied to heavy organic chemicals in Germany by Ray Stokes. See Stokes, *Opting for Oil*, p. 245. This is probably a better model for heavy organic chemicals than Thomas Hughes's concept of technological momentum, originally applied by Hughes to the oil-from-coal process in his classic paper "Technological momentum in history: Hydrogenation in Germany, 1898-1933," *Past and Present*, **44** (1969), 106-132.

partly because of the well-known aldol condensation but also because calcium carbide production was in considerable excess of demand in the early 1920s owing to the collapse in the market for calcium cyanamide. Because it was in place for the butanol synthesis, aldol (and hence acetylene) enjoyed a head start in the competitive struggle for the route to butadiene. Butanol acted as a founder-effect for butadiene and, owing to the sheer scale of butadiene production (and hence acetylene production) in the early 1940s, it influenced the production of nearly all other organic chemicals, even ethylene oxide and the monomers of nylon. Had acetylene not been adopted for butadiene production, some acetylene-based processes would have still flourished (acetic acid and PVC spring to mind), but they would have been more isolated and vulnerable to erosion by the processes adopted for butadiene. The fate of acetylene after 1945 shows that this erosion would have been assisted by the high cost of carbide-based acetylene.

Even in the 1920s, ethylene was regarded as an attractive alternative for many important products. It can be argued that the post-1945 switch in Germany and Britain to petrochemicals is an illustration of the founder-effect of polyethylene. Because ethylene was needed in such huge amounts for polyethylene, far beyond the capacity of fermentation alcohol, coke-ovens or even acetylene, it influenced the production of other organic chemicals in a manner similar to butadiene a decade or so earlier.

From today's perspective, Ambros and Reppe may appear to be technological dinosaurs, prisoners of the Nazis' autarky policies left stranded by the postwar petrochemical industry. Nonetheless, their joint efforts laid the foundations for the switchover of the West German chemical industry to petroleum in the 1950s and early 1960s. Both men participated in this process, as a consultant and as a research leader, respectively. The uneconomic nature of coal-based synthetic rubber in a world market dominated by the American petroleum-based product led to the abandonment of the four-stage process in West Germany in 1956.[68] Nevertheless, Hüls AG and BASF have remained key players in industrial acetylene chemistry, despite a strong trend away from acetylene (even petroleum-based acetylene) in West Germany and the United States in the 1970s and 1980s. Schkopau retained a heavy commitment to acetylene chemistry until the Communist regime fell in 1989. Thus in 1984, about 55 per cent of its products were derived from the use of calcium carbide as the feedstock, which made Schkopau one of the largest carbide works in the world.[69] This is in keeping with the findings of Basil Achilladelis that technological traditions – such as acetylene chemistry, high pressure chemistry and their combination in Reppe chemistry – played an important role in the development of the post-1945 chemical industry. He concluded that the large number of innovations and patents produced in the 1950s and 1960s partly resulted from "the establishment of technological traditions within companies which were responsible for early market successful radical innovations." He also

---

68 Stokes, *Opting for Oil*, pp. 197-216.

69 *Buna: Traditionsreich, Leistungsstark und Zukunftsorientiert*, p. 5.

remarked that these corporate traditions "acquire their own momentum, so that over periods ranging from 15 to ... 40 years a disproportionate number of innovations results from the same technology... Such traditions characterize strongly innovating companies and provided many of their most successful products and processes."[70] Many chemical firms have recently abandoned, or at least severely curtailed, corporate traditions dating from the period covered by this paper. The long-term effect of this break with the past is still uncertain, but will doubtlessly provide material for a sequel to this paper to mark the centenary of Reppe chemistry in 2032.

**Acknowledgements**
I would like to thank Ernst Homburg, Raymond G. Stokes, Anthony S. Travis, Basil G. Achilladelis, Ulrich Marsch and J. Eric Elliott for their criticisms and suggestions; any mistakes that remain are, of course, wholly my own.

[70] Basil Achilladelis, Albert Schwarzkopf and Martin Cines, "The dynamics of technological innovation: The case of the chemical industry," *Research Policy*, **19** (1990), 1-34; quotations on pages 27 and 28. It should be noted that the 'bandwagon' effect mentioned on page 16 is different from the 'founder' effect introduced by Arthur and Stokes (see note 67). Also see Achilladelis, Schwarzkopf and Cines, "A study of innovation in the pesticides industry: Analysis of the innovation record of an industrial sector," *Research Policy*, **16** (1987), 175-212, and Achilladelis, "Process Innovation in the Chemical Industry," pp. 244-248.

# THE DEVELOPMENT OF CHEMICAL INDUSTRY IN SWEDEN AND THE CONTRIBUTION OF ACADEMIC CHEMISTRY AFTER 1900

ANDERS LUNDGREN
*Department for History of Science and Ideas*
*Uppsala University*
*Slottet, Ingång A0*
*S- 752 37 Uppsala*
*Sweden*

## 1. Introduction

Until recently historians, chemists and industrialists invariably viewed technology as applied science, with chemical technology as the supreme example of this. Reflecting on the Swedish chemical industry in 1916, the chemical engineer Alfred Larson subscribed to this viewpoint. He used the classical example of Liebig and agricultural chemistry, but he never suggested how science was actually transferred into industrial practice.[1] The step just seemed so smooth and self evident. The science of chemistry was the indisputable driving force behind the growth of chemical industry.

That chemistry was essential for production of goods had been emphasized for well over a century.[2] And doubtless, chemistry was, and remained, part of many production processes even though what went on in those processes was not always understood to the extent that science-based control, and even improvement, was possible. The prime example of how chemistry created a new industry is of course the German dye industry during 1869-1900. During the first three decades of the 20th century, academic chemistry contributed towards a number of developments, including fertiliser production (Haber's work on the ammonia synthesis), plastics, and pharmaceuticals. In all cases the successful use of science for production purposes had been taken for granted. It naturally followed that the development of academic chemistry played a

---

1 Alfred Larson, *Sveriges kemiska industri* (Stockholm: Sveriges teknologförening, 1916).

2 See for example T. Thomson, *A System of Chemistry*, 5 Vols. (Edinburgh, 1810), Vol. 1, p. 3, where chemistry is "intimately connected with all our manufactures."



*A.S. Travis et al. (eds.),*
***Determinants in the Evolution of the European Chemical Industry, 1900–1939,*** 123–141.

prominent role in histories of chemical industry.[3] This was not always the case, however, as Robert Bud and Gerrylynn Roberts have pointed out. Thus in England the dependence on academic chemistry in the dye industry was "unusual, not typical" and rarely is there found a simple causal chain from science to technology.[4]

It seems plausible to assume that if chemical studies were of use for industry then the relevant work would have been undertaken in research and development laboratories. However, except in the case of Germany not much has been said on laboratories in industry.[5] A preliminary glance at Swedish material from around 1900 reveals that if a laboratory did exist at an industrial plant, it was usually badly equipped, and certainly not well suited to scientific research. Production processes rarely seem to have been directly influenced by science.[6] In production, the laboratory, as well as the science, often disappears from the story, but remained part of its rhetoric. This paper is an attempt, using Swedish examples, to look at scientific activities in industry. In particular, it will enter inside the industrial laboratory in order to observe what kind of work was done there. Hopefully this will lead to an increased awareness of the problematic relations between science, technology and industry, and on the role of chemistry in the development of chemical industry.

Examples will be taken from those chemical industries that according to historiographic and/or hagiographic traditions had been strongly supported by science during crucial phases in their development. For sake of completeness, the prior history of three relevant developments will be described. Firstly, in phosphate production through the most important producer, Stockholms superfosfat fabriks aktiebolag (hereafter Stockholms superfosfat); secondly, in pulp and paper, perhaps the most important chemical industry in the Swedish economy; and, finally, in the electrochemical industry which is considered a pioneer in the use of science for production purposes.

---

3 For a contemporary viewpoint see Ernst von Meyer, *A History of Chemistry from Earliest Times to the Present*, 2nd English edition (London: MacMillan, 1898), pp. 554-580; for a recent version see Fred Aftalion, *A History of the International Chemical Industry* (Philadelphia: University of Pennsylvania Press, 1991), p. xxiii, "chemical manufacturers cannot do without the scientific knowhow provided by research chemists."

4 Robert Bud and Gerrylynn Roberts, *Science versus Practice: Chemistry in Victorian Britain* (Manchester: Manchester University Press, 1984), p. 165.

5 Cf. Seymour M. Mauskopf, "Introduction," in Mauskopf, ed., *Chemical Sciences in the Modern World* (Philadelphia: University of Pennsylvania Press, 1993); Ernst Homburg, "The emergence of research laboratories in the dyestuff industry, 1879-1900," *British Journal for the History of Science*, **25** (1992), 91-111. The many often excellent studies on laboratory life we have today rarely relate academic work to industrialization.

6 Cf. Bud and Roberts, op. cit. (4), pp. 33, 95.

## 2. The rise of chemical industry in Sweden

### *a. Phosphates*

The need for phosphate and other fertilisers in Sweden was originally satisfied by imports, but from 1850 also by domestic production. Many small short-lived production units were established, most of which closed after a few years in operation. The outcome was that in 1892 imports were ten times greater than exports.[7]

An example of domestic production is Klippans superfosfatfabrik, which operated from 1857 to 1874. The factory was primitive, built "with very simple arrangements for mixing and storing" ["med mycket enkla anordningar för blandning och magasinering"] and "after a description in a German brochure." ["efter en beskrivning i en tysk broschyr."][8] Its founder and owner C.F. Waern originally imported fertilisers from England but from 1859 marketed a phosphate of his own manufacture, made from bone charcoal and sulphuric acid. The company employed C.J. Tranberg, a teacher of chemistry at Chalmers Technological Institute (Chalmers teknologiska institut) in Gothenburg, to analyse both its own products as well as those from competing companies. Analyses were done at Chalmers but Tranberg's successor, the 18 years old A.T. Wedelin, organized a new laboratory, so that they could be done "without interference or insights from unknown chemists." ["utan inblandning eller insyn från främmande kemister."][9]

The most important phosphate producer was Stockholms superfosfat, founded in 1871 by industrialist Oscar Carlson (1844-1916), who had qualified in the field of mining, an activity which bore little relationship to the industry he was to run.[10] No laboratory seems to have existed at his factory, at least at the start. There was no need for laboratory activities, routine or otherwise, since well-tried production methods were imported from England.[11] After a fire in 1889 the factory was rebuilt to include an "office building, also containing a larger laboratory ... in which research was done, which was unusual during the time." ["kontorsbyggnad, även inrymmande ett större

---

[7] For an overview see *Stockholms superfosfat fabriks aktiebolag 1871-1946. Minnesskrift* (Stockholm: Stockholms superfosfat fabriks aktiebolag, 1946); Erik Karlström & Agne Wenzel, "Superfosfat - 100årigt gödselmedel från moderna fabriker," *Svensk kemisk tidskrift*, **69** (1957), 547-561. Statistics from *Svensk kemisk tidskrift*, **5** (1893), 183.

[8] *Stockholms superfosfat fabriks aktiebolag*, op. cit. (7), p. 40; Gösta Bodman, "Klippans superfosfatfabrik 1857-1875. C. Fr. Waern & Co gödselmedelfabrik vid Göteborg," *Daedalus*, **1947**, 41-68.

[9] Bodman, op. cit. (8), 61f.

[10] Biographical information from Sven Odén, "Oscar Fredrik Carlson," *Svenskt biografiskt lexikon*, **7** (Stockholm, 1927), 500-511.

[11] *Stockholms superfosfat fabriks aktiebolag*, op. cit. (7), p. 69.

laboratorium ... här började nu forskningsverksamhet, som var ovanligt på den tiden."][12] Because the results were classified business secrets there is little information on the activities in the laboratory.

Production was a dirty, foul-smelling, delicate handicraft, far from scientific laboratory work. For production of bone phosphate, in principal only two items, a tank, and a big bowl, were needed.[13] There were no possibilities for undertaking exact measurements, and the role of the laboratory was as a works laboratory where only analysis was undertaken.

Reference to scientific input nevertheless endowed considerable status and was therefore useful as a marketing ploy. When advertising a phosphate, Waern made known that one chemist from Chalmers and another from the Academy of Agriculture (Lantbruksakademien) had verified its quality. The more established the scientist who verified the quality, the more likely that sales would benefit. Stockholm superfosfat hired C.E. Bergstrand, a leading agricultural chemist, whose analytical testimony was of a particular value.

***b. Pulp and paper***

The pulp and paper industry used both the sulphite and the sulphate processes, and in the few historical sketches of its development that exist there are found some references to laboratories.[14]

Sulphate factories operated from the 1870s, often with help from English engineers. Swedish developments are connected with Alvar Müntzing (1848-1917) at Munksjö bruk.[15] In order to avoid destruction of the cellulose during the boiling of the wood Müntzing added sodium sulphate instead of caustic soda. Together with a longer boiling time, lower pressure and less reactive chemicals than usual, the result was a strong "kraft paper." Müntzing described his work as consisting of "endless experiments but also efforts on a greater scale" ["otaliga experiment och även försök i större skala"] and

12 *Stockholms superfosfat fabriks aktiebolag*, op. cit. (7), p. 74.

13 W.A. Campbell, *The Chemical Industry* (London: Longman, 1971), p. 76, gives a vivid description.

14 *Den svenska cellulosaindustriens utveckling och ekonomiska betydelse* (Stockholm: Svenska cellulosaföreningen, 1918), pp. 7-55; Alfred Larson, *Den svenska kemiska industrien*, 2 Vols (Stockholm: AB Hasse Tullberg, 1922), Vol **1**, pp. 203-282.

15 Karl Nyström, "Den norrländska trämasse- och cellulosa industrien" in *Norrländskt handbibliotek*, **10** (Uppsala, 1924), pp. 388-390; *Den svenska cellulosaindustriens*, op. cit. (14), pp. 23f.; Torsten Althin, *Papyrus 1895-1945* (Mölndal: AB Papyrus, 1945), pp. 96-98. According to Nyström (p. 404) there existed in 1889 twelve small factories, many of which went bankrupt; between 1900-1910 eight new ones were founded, and in 1923, after rationalization, there were still twenty-six sulphate factories. On Münzing see Bosse Sundin, "Alvar Müntzing," *Svenskt biografiskt lexikon*, **26** (Stockholm, 1987-89), 131-133.

not until many years after could necessary experiments "on a scale big enough" ["i tillräckligt stor skala"] be performed. It took until 1906 before he could build a boiler of $7m^3$ capacity "in all arranged as if for continuous operation." ["i allt monterad så som man tänkt sig för maskindrift."][16] If there was a laboratory at Munksjö, then it was the whole factory that was the laboratory. A specific research laboratory did not exist, and would not have solved the problems of scaling up any better than methods based on daily acquaintance and long practical experience with the actual processes.[17]

Sulphite pulp was from 1873 produced by the mill at Bergvik, and from 1881 at Korndal and Billerud.[18] At Bergvik, Carl David Ekman (1854-1904) developed the process.[19] Originally mechanical wood pulp was produced according to George Fry's process (the owner of Bergvik was the English firm I. Thomson & T. Bonard & Co.) that involved boiling wood under high pressure. The pulp was often discoloured, and Ekman, in order to avoid the oxidation which was believed to cause the colour, added sulphurous acid and magnesite.[20] This resulted in a new highly useable pulp, and his ideas were immediately and successfully applied to industrial production.

All sources, even if few and not always reliable, indicate that Ekman had at his disposal a laboratory "in which he spent all his time" ["i vilket han alltid vistades"], and in which he carried out "laboratory-like work."[21] Preserved plans of the plant show a comparatively large building named the laboratory. But the word 'laboratory' is often put within quotation marks and referred to a small production plant with the main purpose to produce sulphuric acid.[22] Only in one small part of the 'laboratory' did Ekman start "a series of laboratory experiments" ["en serie laboratorieförsök"].[23] There he used devices from other factories, for example a test boiler from Fry's, later replaced

---

16 Quoted from Elis Bosaeus, *Munksjö bruks minnen* (Uppsala: Munksjö pappersbruk, 1953), p. 68.

17 Cf. John Kenley Smith, "The evolution of chemical industry. A technological perspective," in Mauskopf, ed., op. cit. (5), pp. 137-157.

18 Also, this industry grew quickly. Fifteen factories were started between 1881-1890, twenty-four between 1891-1900, thirty between 1901-1910, and six between 1911-1920. Nyström, op. cit. (15), pp. 429f.

19 On Ekman see Torsten Althin and Karin Forsberg, "Carl David Ekman," *Svenskt biografiskt lexikon*, **13** (Stockholm, 1950), 153-157; Althin "Carl David Ekmans liv och person," *Daedalus*, **1935**, 47-66.

20 *Den svenska cellulosaindustriens utveckling*, op. cit. (14), p. 13; Althin, op. cit. (19), 52, 57.

21 Althin, op. cit. (19), 52.

22 Ewald Granström, "Beskrivning över Bergviks gamla trämassefabrik," *Svensk papperstidning*, **26**:8, (154-158); cf. Gunnar Eriksson, *Kartläggarna* (Umeå: Umeå universitetsbibliotek, 1978), p. 80.

23 George Spaak, *Männen kring Carl David Ekman och tillkomsten av världens första sulfitcellulosafabrik* (Stockholm: Svenska cellulosa- och trämasseföreningarna, 1957), pp. 44-57.

by Swedish test-boilers capable of performing experiments at half the original scale. Ekman's laboratory work aimed at "scaling up" and was based on a thorough knowledge of actual processes, with the purpose of finding technical improvements. The industrial problem of "scaling up," is one that academic scientists do not usually meet in their daily work. Problems of "scaling up" could hardly have been solved in a normal laboratory, especially the primitive laboratories of Ekman and others. Ekman has been described in obituaries and similar accounts as a chemist only because he worked in a laboratory, not because of what he did there.

Carl Waldemar Flodquist (1847-1894), at Korndal, was a civil engineer, who had earlier worked in the transport business and mechanics, and who tried to introduce Alexander Mitscherlich's (son of Eilhard) process for cellulose production "although he had no education in chemistry." ["trots att han saknade kemistutbildning."][24] He was helped by a young man, H.A. Gustafsson, "who wasn't a chemist but who through books knew about the contemporary history of production of sulphite cellulose." ["som ej var kemist, men genom bokliga studier bekant med den dåtida sulfitcellulosametodens historia."][25] At Billerud, Victor Folin (1850-1915) also tried Mitscherlich's method. Folin was educated at a lower technological school, and his knowledge of paper production came directly from factory work.[26] There seems to have been no laboratory facilities, neither at Billerud nor at Korndal, but Flodquist's and Folin's activities, based on long experience, are often described with words such as 'experimental,' at the same time as their 'primitive' character is stressed. The experiments often failed, and Folin brought the project close to economic ruin.[27]

In summary, in the pulp and paper industry there were places called laboratories in the sense that there was a room or a reserved area where some kind of 'experiments' were going on. But these experiments seem to have depended little on contemporary science, and were more based on daily acquaintance with industrial routines.

***c. The electrochemical industry***

The Månsbo fabric, founded in 1893 and owned by Stockholms superfosfat, has been called the first Swedish science-based electrochemical industry.[28] The amount of electricity needed made it necessary to move from the Stockholm area to Månsbo, about

---

24 George Spaak, "Carl Wilhelm Flodquist," *Svenskt biografiskt lexikon*, **16** (Stockholm, 1964-66), 213.

25 Spaak, op. cit. (24), 213. See also Myllyntaus, this volume, pp. 348-350.

26 Spaak, "Victor Folin," *Svenskt biografiskt lexikon*, **16** (Stockholm, 1964-66), 255-257.

27 *Billerud 1883-1983* (Säffle, 1993), p. 21. This is an extremely hagiographic work.

28 The historian will of course find it meaningless to look for the 'first,' which is only a question of definition. It is enough here that Månsbo was both early and comparatively big.

200km northwest of Stockholm, where sufficient water power was available.[29] Laboratory experiments had been performed in Stockholm, and "as soon as there were results from laboratory experiments" ["så snart laboratorieförsöken ... gett resultat"], the Månsbo plant started to operate.[30] At Månsbo there was a laboratory, not surprisingly called the first industrial laboratory in Sweden.

There is little information on the experiments done in Stockholm, but there are some details of what went on in the Månsbo laboratory. Laboratory worker Julius Gelhaar, employed from 1898, described the early laboratory as small and primitive (water was carried in buckets from the river, etc.). The work load was not too heavy, and there was always time to do other things. Gelhaar and his colleagues, on their own initiatives, experimented with production of new chlorates.[31] In 1901 a new laboratory specifically dedicated to research was built, and successively enlarged, but its appointed director, Birger Carlson (son of Oscar Carlson) complained that resources did not meet scientific demands.[32] Gelhaar in 1910 also worked in an ammunition laboratory established "under exceedingly primitive circumstances." ["under synnerligen primitiva former."][33]

**3. Science in early industry**

There were thus laboratories, for various purposes, at some Swedish chemical factories. They were often run by individuals with little training in chemistry, and the role of science seems to have been exclusively restricted to analysis.

To the outsider, however, chemistry and its application seemed like a magical enterprise. The image of alchemy hovered over much of chemical industry, and everything the chemist touched "seemed to turn into gold." ["tycktes ... förvandlas till guld."][34] Descriptions of the laboratory at Bergvik foster a picture of the chemist as magician: "Everything related to production of acid is veiled in a severe isolation, from the beginning and many years onwards, hidden under the secret name of Laboratory."

---

29 On the use of water power, see Staffan Hansson, *Porjus: En vision för industriell utveckling i övre Norrland* (Luleå: Tekniska högskolan i Luleå, 1994), p. 331. One of the first buyers of power from northern Sweden was Stockholms superfosfat, but the attempts to create an electrochemical industry in northern Sweden, however, failed. See Hansson, pp. 226f., 245, 252-260.

30 *Stockholms superfosfat fabriks aktiebolag*, op. cit. (7), pp. 44-46, 101. See also Bosse Sundin, *Ingenjörsvetenskapens tidevarv* (Umeå: Umeå universitetsbibliotek, 1981), pp. 31ff.

31 Julius Gelhaar, "45 år på ett industrilaboratorium," *Harald Nordenson 60 år* (Stockholm, 1946), pp. 105-116; *Stockholms superfosfat fabriks aktiebolag*, op. cit. (7), p. 46.

32 *Stockholms superfosfat fabriks aktiebolag*, op. cit. (7), p. 103; see also Sundin, op. cit. (30), p. 32.

33 Gelhaar, op. cit. (31), p. 111.

34 Eriksson, op. cit. (22), p. 165.

["allt som hade med syraberedning att göra, är insvept i en sträng avskildhet från början och under många år framåt och dolt under det hemlighetsfulla namnet Laboratoriet."][35] Such descriptions are not unique, and mask the fact that science in fact contributed very little, thereby indirectly preserving the picture that industry was developed by the help of science. The works laboratory was a place where activities were described in jargon belonging to science, as 'experiment,' "developed in the laboratory," and thereby evoking the status of science.

When the real activities in the laboratory are established another, contradictory, myth is exposed: the bad equipment and primitive state of the laboratory. To the examples already given can be added Alfred Nobel, whose first laboratory was his kitchen; conditions in his first factory were "fairly primitive." ["tämligen primitiv/a/."][36] Intriguingly the myth of the primitive laboratory does not exclude financial success. Of course it is part of the pioneer myth, stressing hard work, little support, reduced circumstances, and eventual success. The contribution of scientific research was minimal. True, similar myths turn up when academic chemists describe their careers.[37] But if the images of the university laboratory and the industrial laboratory at certain levels were similar, there were important differences. Theory was not part of the industrial myth, which underscores the fact that modern science had few practical applications in early chemical industry.[38]

One example outside the branches treated here was agriculture. The chemist L.F. Nilson, at the Agricultural Academy, in 1890 noticed that practical men did not like Liebig's ideas since they "lead to consequences that do not conform with ... experience." ["leder till konsequenser som icke öfverenstämma med ... erfarenhet."][39] Chemistry in agriculture and in the food industry evolved by doing analysis, and not through the study of chemical processes.[40] In early phosphate production, science was

---

35 Spaak, op. cit. (23), p. 53.

36 *Nitroglycerin aktiebolaget* (Gyttorp, 1964), pp. 9, 24. The metaphors "the laboratory is a kitchen," and "chemistry is cooking" are still commonly in use, especially among organic chemists. The significance of this is itself a worthy field for study.

37 Cf. Peter Klason, "Några drag ur en kemikers lefnad," in *Festskrift till Peter Klason* (Stockholm: Norstedts, 1910), pp. xliiiff, on the laboratory in Lund during the 1860s.

38 A. Lundgren "Between science and industry. The formation of the Swedish Chemical Society" (in press).

39 "Om förloppet af näringsämnenas upptagande och dess betydelse för kulturväxterna," *Svensk kemisk tidskrift*, **2** (1890), 53-67.

40 Cf. the situation in Denmark, Helge Kragh and Hans Jörgen Styhr Pedersen, *En nyttig videnskab. Episoder fra den tekniske kemis historie i Danmark* (Koebenhavn: Gyldendal, 1995), p. 193; see also Bud and Roberts, op. cit. (4), pp. 59, 109.

far away, especially when the farmer was advised to produce phosphate himself "on stamped clay or earth on the floor of his own barn ... and mixing can be done with a shovel." ["i stampad lera eller jord på sitt eget loggolv ... och att omröringen kan ske med skovel."][41] Commenting on the production of sodium cellulose, engineer and teacher of chemistry K.P. Peterson, noticed that one should not "strictly keep to the rules" ["ängsligt fasthålla vid föreskrifterna"] but could "almost at pleasure vary the factors /strength of lye, boiling time etc./ and still obtain a good result" ["nästan efter behag variera de ovannämnda faktorerna /lutens styrka, koktid m.m./ och ändock erhålla ett godt resultat."][42] Alvar Müntzing did so when he obtained "kraft paper." As late as 1910 Peter Klason, professor of chemical technology at the Royal Institute of Technology (Kungliga tekniska högskolan), somewhat drastically concluded "indeed any theory on the boiling of wood according to the sulphite process seems fairly superfluous, because during decades cellulose has been produced according to this method, without any theory, and sulphite cellulose has nevertheless become an even bigger article with a world-wide market." ["visserligen synes en teori för vedens kokning enligt sulfitmetoden vara tämligen överflödig, ty man har nu i decennier framställt cellulosa enligt denna metod nästan utan hvarje teori, och sulfitcellulosa har dock för varje år blifvit en allt större världsartikel."][43] Experiments of interest to industry were of little interest for an exact science.

Even at Månsbo, science was clearly in the background. According to Gelhaar the work was "very much empirical ... there was no theory for chlorate electrolysis" ["synnerligen empiriskt ... någon teori för kloratelektrolysen fanns icke"], there were no attempts to "to place the question in a broader context" ["att sätta in frågan i ett större sammanhang"],[44] and the work to find new chlorates was done outside normal factory working hours. Production was one thing, scientific understanding another, as formulated by Gelhaar in 1915, "even if there are no difficulties in producing ammonium chlorate it seems as if ... the salt is very little studied." ["ehuru det ej erbjuder några större svårigheter att framställa ammoniumklorat synes det ... att saltet är mycket litet studerat."][45] If a production process worked there was no commercial need to study it scientifically. Birger Carlson noticed that during one early step in chlorate production, "more chloride was formed than required by theory" ["mer klorid bildades än teorin fodrade"] but the attempts to avoid this, by using magnesia instead of caustic

---

41 Elisabeth Waern-Bugge, *En gammal herrgård. Familjepapper ur Baldersnäs arkiv* (Stockholm, 1920), p. 177.

42 K.P. P[eterson], "Natroncellulosa," *Svensk Kemisk Tidskrift*, **2** (1890), 9, 70-75.

43 Klason, "Oregelmässigheter i kokprocessen vid framställning av sulfitcellulosa och dess orsaker," *Arkiv för kemi, mineralogi och geologi*, **4**:1 (1910), 1.

44 Gelhaar, op. cit. (31), pp. 105-116; cf *Stockholms superfosfat fabriks aktiebolag*, op. cit. (7), p. 103.

45 "Några iakttagelser öfver ammoniumklorat," *Teknisk tidskrift*, **1915**:2, 1-12.

soda, could not be economically motivated and "therefore the theoretically elegant magnesia process was abandoned." ["därför uppgafs ... den teoretiskt eleganta magnesiumprocessen."][46] The scientifically best method was not necessarily practically the best, nor the most profitable. On another occasion Gelhaar, to his great surprise, did an experiment that produced a high yield of fine barium chlorate, but he failed to repeat it: "many chemists have experienced that an experiment had given a very surprising result that never has occurred once again." ["mången kemist /har/ gjort denna erfarenhet vid sina försök att resultatet blivit något högst oväntat, som aldrig inträffat senare."][47] Unrepeatable experiments are of no interest to industry, however promising they may be.

At the same time Swedish industrial products were of little interest to scientists. Ekman's boiler, with a capacity of 120 litres, could hardly be used for delicate scientific experiments. To the academic chemist, industrial processes were either theoretically not interesting, as in the case of phosphate, or too complicated to be understood, as in the case of pulp and paper. In both cases chemistry could not significantly change production processes. The knowledge embedded in a thorough experience of actual production was still much better than science for development of more economical processes.

**4. Process improvements**

Industrial laboratory work focused on technological improvement of existing processes, for which a chemical understanding was not necessary.[48] Of course there are no sharp lines between chemical and technical improvements, but developments in Swedish chemical industry were predominantly directed towards the latter. One reason was that process details for the phosphate, paper and pulp, and electrochemical industries had been largely imported, often from England, so there was no need to develop new ones. All that was needed was the ability to accommodate already operational technologies to existing local circumstances.

To make changes to a technical process required use of sophisticated trial-and-error methods, in the form of parameter variation, based on many years of direct experience, rather than on scientific reasoning based on theory. One example is Ekman's innovation of covering the inside of his boiler with lead to prevent corrosion.[49] That science and

---

46 Carlson, "Den elektrokemiska kloratindustriens uppkomst och utveckling med särskild hänsyn till Månsbo-fabriken," *Festskrift till Peter Klason*, op. cit. (37), pp. 170-171. According to Carlson, scientifically 'wrong' patents had been accepted, which implies that they actually did work (p. 184).

47 Gelhaar, op. cit. (31), p. 107.

48 For an earlier example, see the relation between Gay-Lussac and John Glover concerning production of sulphuric acid. Cf. W.A. Campbell, "Industrial chemistry," in C.A. Russell, ed., *Recent Developments in the History of Chemistry* (London: Royal Society of Chemistry, 1985), p. 243.

49 *Den svenska cellulosaindustriens*, op. cit. (14), p. 13.

technical improvements were conceived as two different things was clear to the many chemical-technical bureaus established towards the end of the century (see below), and which in advertisements emphasized their capacity to do "real *technical* experiments." ["verkligt *tekniska* försök."][50]

Problems in pulp and paper were mostly technical.[51] One of Müntzing's main difficulties was of how to mechanically separate the cellulose from the lye, once the chemicals had done their job. To do this he used a technology he learned when working in a sugar refinery.[52] These improvements were based on accumulated knowledge gathered over long periods of time, and consisting of small adjustments and changes until the optimum outcome was reached: "It seems as if day after day, year after year, the persistent collected observations and incremental improvements by [J.E.] Lundström, Ljungquist and Müntzing, towards the end of the 90s had brought the specialized production of asphalted cardboard at Munksjö to its optimum state." ["Det förefaller som en dag för dag, år efter år, av Lundström, Ljungquist och Müntzing ihärdigt samlade rön och genomförda detaljförbättringar framemot 90-talet bragt Munksjö tillverkning av asfaltpapp fram till ett för denna specialitet optimalt stadium."][53] Production was "much of an art of handicraft" ["mycket av konsthantverk"] based on constant observation in order to determine the exact time to take the next step.[54] Small technological changes could be done without the help of academic chemistry.[55]

The role of chemistry in Swedish industry before 1900 was mainly analytical, regardless of all the rhetoric stating the opposite. The knowledge then needed for industrial development in chemical industry has to be distinguished from the knowledge needed by the chemist doing science in a laboratory, not because industrialists did not see the possibilities in chemistry, but because chemistry did not in practice offer any possibilities.

But from Gelhaar it can be seen that chemistry slowly was becoming more important, that scientists *became* interested. Once the chlorate electrolysis had been initiated at Månsbo "a large number of researchers threw themselves into this subject." ["ett stort

---

50 Advertisement for *Kem. tekn. experimental-laboratorium* (J.E. Hällstrand & J. Landin), in *Svensk kemisk tidskrift*, **1** (1889), emphasis in original.

51 *Den svenska cellulosaindustrien*, op. cit. (14), p. 13.

52 Bosaeus, op. cit. (16), pp. 475f.

53 Bosaeus, op. cit. (16), p. 516.

54 Bosaeus, op. cit. (16), pp. 491, 531.

55 A branch which might constitute an exception to this general pattern could be mining and metallurgy, which, however, is not treated here.

antal forskare kastade sig in på detta ämne."][56] Here industrial development stimulated scientists to undertake research, and, as a consequence, the results of the scientists' work in the long run was bound to have repercussions on the industrial process itself. Eventually, however, chemistry did have a growing influence on chemical industry.[57]

**5. Towards a changing role for chemistry**

Here I tentatively suggest some perspectives from which it might be fruitful to study the process whereby chemical science infiltrated into chemical industry. Hopefully this can contribute to discussions on the relations between science and industry with reference to the history of chemical industry.

***a. The consultants***

While chemical analysis in works laboratories was mostly done by individuals employed by the industry, circumstances around 1900 led to the appearance of a new type of intermediary between science and industry, the consultant. He usually worked in independent technical-chemical bureaus, was temporarily hired by industry, and often educated as a chemical engineer, sometimes in Germany, more often at the Royal Institute of Technology.[58] Stockholm superfosfat engaged, but did not employ on a regular basis, academic chemists as consultants. Among these were such well-known scientists as Klason, Nilson and Svante Arrhenius.

That academic chemists acted as industrial consultants mirrors the increasing role of science. Klason and Ludwig Ramberg (professor of chemistry at Uppsala university) had many such assignments, and also did much work on governmental committees as experts.[59] But the expert was still mostly required for analytical purposes. Existing committees often had the practical purpose of controlling goods on the market (food, arsenic, etc.), not of engaging in any form of scientific research.

Consultants functioned as a bridge, over which science spread into practical fields and into technological developments, such as when Ramberg developed a new microanalytical method for determination of arsenic. Certainly analysis took up most of the consultant's work, but from 1900 he could also advise on developments of

---

[56] Gelhaar, op. cit. (31), p. 105.

[57] Without drawing too far-reaching conclusions, there was a similar situation in medicine. Here the increased scientific understanding of body mechanisms did not lead to any therapeutic break through until the end of the century. Cf. W.F. Bynum, *Science and the Practice of Medicine in the Nineteenth Century* (Cambridge: Cambridge University Press, 1994).

[58] For the Institute and the relationship between science and industry see Lundgren, op. cit. (38).

[59] Ramberg thus took part in the important Arsenic Committee, Thomas Hillmo, *Arsenikprocessen. Debatt och problemperspektiv kring ett hälso- och miljöfarligt ämne i Sverige 1850-1919* (Linköping: Linköpings universitet, Linköping Studies in Arts and Science, vol. 102, 1994).

processes (this was an important part of Arrhenius's work). A popular consultant was A.W. Cronquist. He was a member of many committees, and had his own bureau, which carried out more than 43,000 analyses. At the same time he published more than 100 articles, many of them in scientific journals.[60]

The consultant brought science from academic laboratories and institutions into contact with industrial practice, not by taking science into industry, but by bringing pieces of industry into science. He brought industry into the laboratory, not the other way around.

***b. Technical and scientific institutions***

Scientific work related to the branches of chemical industry studied here was done in Sweden, but it was neither initiated nor undertaken by industry. Industry relied on the universities or the Royal Technological Institute, Stockholm. This was part of the ordinary job of professional chairholders and others, and was undertaken with the resources that went with the position. Nevertheless these resources were usually too small to initiate any major project of interest to industry. One notable exception, however, was research into wood and paper, headed by Klason at the Royal Technological Institute and financed by a rare governmental allocation from the Swedish Diet.[61]

Klason began his career as a promising academic chemist with a degree from the university of Lund,[62] but in 1890 he accepted the post of professor of technical chemistry at the Royal Technological Institute. Even so, he once denied the value of theory in pulp production. Nevertheless he saw that even if theory could not be applied directly, an increased knowledge of processes in actual use could help to explain why something went wrong.[63] Science thereby facilitated improvements of existing processes, but did not find new ones. It was an 'error-finder' and 'corrector' of processes. This work, done with the intention of assisting industry, did not receive any financial help from industry.

Other industries, operating smoothly and profitably thanks to reliance on long experience, slowly opened to science, but not on their own initiatives. It was a grant from the Royal Academy of Science (Kungliga vetenskapskademien), not money from the soda industry, which gave the possibility for professor Fredrik Ekman at the Royal

60 Eriksson, op. cit. (22), pp. 89f.

61 Sundin, op. cit. (30), p. 131. The question of as to what extent governmental money gave more space to science than industrial money needs to be studied in greater depth.

62 In fact his teacher in Lund, Wilhelm Blomstrand, wanted to keep him at the university, cf. Blomstrand to L.F. Nilson, April 11 1889, *The Archives of the Swedish Royal Academy of Science*, Nilson papers.

63 Klason, "Bidrag till kännedom om de kemiska processerna vid sulfitcellulosa-tillverkning," *Teknisk tidskrift*, **1893**, 49.

Institute of Technology and Alfred Larson, to travel abroad to study the manufacture of soda.[64]

So if science slowly did prove its industrial usefulness in fields apart from analysis, it was not carried out in industry itself, but at research institutions.[65] The work was therefore dependent on individual initiatives from persons such as Klason, and on money existing either within the framework of these organisations, or in rare cases from the government. Science slowly proved its potential, and in 1918 the first industrial research institute was created by the paper industry.[66]

### *c. Education of industrial chemists*

A great part of the knowledge essential for chemical industry could not be learned at schools, but only in daily practice. The impression that chemistry played a minor role in the development of early chemical industry is strengthened by reviewing the education of individuals involved in industry. In the laboratories of the phosphate industry it is difficult to identify any employee with a chemical education.[67]

The majority of those working in chemical industry had not been educated in, had no theoretical training in, nor begun their careers in, the field in which they achieved most. Müntzing had an elementary technical education, and some experiences from a sugar refinery prior to entering the pulp industry. Ekman was a chemical engineer, but his first job had been with a distillery and he had no experience with pulp and paper before Bergvik. Müntzing, Ekman, Oscar Carlson, Gelhaar and others all possessed a direct knowledge of industrial practice, which was superior to scientific knowledge for the efficient running of an industry.

Consultants and academics interested in industrial problems received more scientific training. Klason has already been mentioned, but after 1900 academically trained researchers more often directly found their way into industry. Scientific education often became a prerequisite for employment. E.L. Rinman received a Ph.D. in 1902 from Uppsala, though not in the field in which he later worked industrially, and a docentship in 1904.[68] He was employed by the sulphate and sulphite mill at Skutskär where he developed processes to recover and reuse the chemicals wasted in the lye. He was

64 Larson, op. cit. (1), pp. 131ff.

65 Nyström, op. cit. (15), p. 420.

66 Sundin, op. cit. (30), pp. 154-162.

67 As for A.W. Friesteds tekniska kemiska industri, see *Sveriges industriella etablissementer*, 2 Vols (Stockholm, 1872-79), vol 1, p. 16

68 E. Rinman, *Om triazol, bitriazol och tritriazol-föreningar* (Uppsala, 1902); on Rinman see A. Lundgren, "Erik Ludvig Rinman," *Svenskt biografiskt lexikon* (in press).

assisted by engineer Gösta Ekström, who had graduated as Diplom-ingenieur from the Darmstadt Technische Hochschule. Ekström also used his knowledge to develop new methods "for production of ethyl alcohol." ["för framställning av sulfitsprit."][69] Analysis was no longer the only task, although it still dominated laboratory work. Part of the scientific effort was directed towards development of industrial processes by studying the process itself. Ekström's work was soon commercialized, and in 1909 the firm AB Ethyl emerged.

Birger Carlson, educated at the Charlottenburg Technische Hochschule, received a scientific training that made him a prominent co-worker to his father, whom he helped "in the scientific development of his [the father's] ideas and suggestions in the electrochemical field." ["vid det vetenskapliga utarbetandet av dennes idéer och uppslag på elektrokemins område."][70] As far as the father was concerned, it was practical work, not science, which gave rise to new ideas. The elaboration of these ideas, however, more and more needed science, and was a task for the younger generation.

Scientific education slowly became more important in chemical industry, but scientific education was not enough to establish an industry. Chemical industry still needed practical knowledge – and adequate sources of finance.

***d. Economy and chemical industry***

Economics have been implicit in much of what has been said above. And chemists were aware that technical development was related to economics, "all technical questions are questions about the state of the market." ["alla tekniska spörsmål är konjunkturfrågor."][71]

The cost of works laboratories where routine analysis was conducted was marginal. The laboratory in Bergvik cost just 0.25 per cent of the total budget.[72] By contrast, investments into full scale production with new processes were so high that it was deemed financially more satisfactory to continue with already successful processes, or to make just minor changes in them. Swedish chemical industry in general saw science as too expensive and unprofitable. It was one of the principle goals of the Swedish Chemical Association (Svenska Kemistsamfundet), founded in 1883 by chemical engineers, to convince industry that the opposite was true.[73] Alfred Larson in 1916

69 Nyström, op. cit. (14), p. 446; on Ekström see "Gustaf Ekström," *Svenska män och kvinnor*, **2** (Stockholm, 1944), 377.

70 "Birger Carlson," *Svenska män och kvinnor*, **2** (Stockholm, 1944), 33.

71 *Svensk kemisk tidskrift*, **1** (1899), 123.

72 Spaak, op. cit. (25), p. 98.

73 See Lundgren, op. cit. (38).

described the history and present state of chemical industry with the quite deliberate intention of increasing support for it. Compared to the strong and highly profitable German chemical industry, the Swedish "did not cooperate." ["sakna/de/ sammanhållning"] and had to give up "pettiness and jealousy" ["småsintheten och afundsjukan."][74] Larson saw economic competition and business secrets as obstacles for a more rational use of science and thereby for the development of chemical industry.

A few large firms, such as Stockholms superfosfat, did financially support scientific research, even if it did not automatically result in increased production. Sometimes scientific results actually could not be used at all. Full chlorate production capacity at Månsbo was never achieved, partly because of international regulations, and partly because the price of chlorates would fall if production was increased.[75] And why put more science into production if the results did not increase profits? Inevitably, as elsewhere, after 1900 cooperation between chemistry and industry resulted in profitable endeavours, as in the case of the reused lye at AB Ethyl. But the prime driving force was economic, not scientific.

From the scientist's point of view economics tended to isolate the chemist. Basically all information on production processes were business secrets, at least until a patent arrangement had been reached. Before that, research results had to be kept within the firm. Ekman kept his journals secret and warned of the dangers of industrial espionage.[76] The factory in Munksjö was described as an isolated place, due to "jealous vigilance of that time" ["den dåtida svartsjuka vaksamheten"] behind which there was economic competition.[77] This secrecy prevented scientific interchange but also, by isolating the chemist, helped sustain the image of him as a magician.

It seems safe to conclude that the economic influence on science prevented an exchange of scientific results and did not encourage new ideas; at the most it contributed to the development of existing ones. By contrast there followed a concentration of research in certain areas where a high standard was reached. It is not by chance that Swedish chemists (Klason, Erik Hägglund, Holger Erdtman, etc.) during the 20th century became international authorities on wood chemistry.

It should be added also that the Swedish government at the end of the 19th century was

---

74 Larson, op. cit. (1), pp. 53f.

75 Carlson, op. cit. (46), p. 246.

76 Eriksson, op. cit. (22), p. 60; Spaak, op. cit. (25), p. 89. This causes evident difficulties for historians seeking sources. The problem of patents in relation to the spread of scientific results is critical in this context, but is not treated here.

77 Bosaeus, op. cit. (16), pp. 471ff.

reluctant to support science for applied purposes.[78] It was deemed necessary in Sweden only for industries critical to the economy, such as, and notably, Klason's project.

***e. The emergence of laboratories***

Over the years, buildings incorporating laboratories, or even special laboratory buildings, became more common at industrial sites. In general, however, the laboratory as a distinct feature of the industrial landscape was a late arrival in Sweden. The space needed for a works laboratory was small, and since consultants had their own laboratories outside of industry there was no need to build one at the plant.

The Rosendahl paper mill had no laboratory in 1895. Half a century later there was a small structure close to the office building called the laboratory.[79] A photo taken around 1950 from Munksjö shows laboratory director Sven Grahn in his laboratory.[80] But in the photo he is speaking on the phone, and he is not wearing a white laboratory coat; behind him there only are a few instruments for analysis, test-tubes, one microscope and one balance. Nowhere does the text tell us what he is doing, or even when the laboratory was built. The existence of a laboratory at Billerud is not mentioned until the 1920s, and then as a works laboratory; in 1946 it was described as an "extensive laboratory." ["omfattande laboratorium."][81]

More diversified and bigger laboratories can be found in industries where development of new processes were essential. For AB Ethyl, analysis was part of routine. The goal to use waste lye to produce ethyl alcohol demanded the study of processes and therefore of different laboratory equipment. Photos of this laboratory taken around 1920 show a larger laboratory, with more advanced instruments and distillation apparatus as a central feature.[82]

## 6. Final words

From this brief survey it follows that the idea of technology as applied science in the case of the prime example, chemistry, has to be substantially modified. Chemistry as a science could not, at least in Sweden, do much for industry apart from analyse its products. Even if analysis was a difficult task itself, which needed knowledge, skill and experience, it did not aspire towards the more traditional task associated with science,

---

78 Eriksson, op. cit. (22), *passim*; Sundin, op. cit. (30), p. 31.

79 Althin, op. cit. (15), pp. 200f.

80 Bosaeus, op. cit. (16), p. 531.

81 *Billerud 1883-1983*, op. cit. (27); I. Schyman, *Christian Storjohann: Mannen som byggde Billerud* (Stockholm: Bok och bild, 1968), pp. 60f. Both sources are hagiographic and should be checked before use.

82 *Minnesskrift utgiven med anledning av Aktiebolagets Ethyls 10-åriga tillvaro* (Stockholm: Aktiebolaget Ethyl, 1920).

that of understanding nature. Analysis just controlled and guaranteed the quality of the final product. The knowledge required in industry was that related to economic profit, for which the chemical understanding of nature at the end of the 19th century could help little. When the early works laboratory became involved in anything beside analysis it was to ameliorate existing processes mechanically, or to scale up, and not to verify or discuss theories. Process improvements were done by parameter variations, which represented a level of knowledge many times better for the purpose in hand than scientific knowledge. It was invariably a trial-and-error method, though a systematic one, requiring experience and knowledge which could not be formulated scientifically. When scientific chemistry entered industry at the beginning of the 20th century it brought with it an increased chemical knowledge of the industrial processes.

It is possible to describe these changes with the help of Melvin Kranzberg's more general thesis that research carried out for a certain mission results in small technological changes, whereas technological breakthroughs have their foundations in scientific non-mission-orientated research outside industry, and purchased by industry when developed enough.[83] The thesis seems confirmed by developments in Sweden, where the role of a scientific laboratory in industry was small; it did not begin to play a more prominent role until around 1900. However, science done with the purpose of being useful was carried out at other institutions, even if the initiative did not come from industry. The fact that chemists employed by industry increasingly had academic educations, as in the case of Rinman, shows an enhanced awareness among industrialists that chemistry could be useful.

During the period examined here there was no drastic increase in chemical industry production due to the application of science, apart from a few exceptions, such as AB Ethyl. According to Kranzberg's thesis this is because the chemical industry used scientific research in a mission-oriented way, not for the production of pure scientific knowledge. The development of the first significant international chemical industry in Sweden, that of pharmaceuticals during the 1940s, points towards the validity and fruitfulness of Kranzberg's thesis. Here the industry was built on research conducted outside industry, but, once the research moved into industry, patent regulations and economic interests narrowed its scope.[84] I see no reason to suppose that the situation was very much different forty years earlier.

Early chemical industry used science either to find out why something went wrong, or to make small improvements in existing processes, rather than to introduce new ones.

[83] Kranzberg's thesis had been applied to the synthetic rubber industry. Peter J.T. Morris, *The American Synthetic Rubber Research Program* (Philadelphia: University of Pennsylvania Press, 1989), p. 89, from which Kranzberg is quoted.

[84] A. Lundgren, *Vetenskap till vardags. En historia kring Xylocain* (Stockholm: Stiftelsen Bengt Lundquists minne, 1995). The expression "patent research" was actually used by the scientists involved.

Even if the line between small improvements and science-based breakthroughs is bound to be diffuse, and more so the further we go into the 20th century, it is still fruitful to separate these two kinds of changes when discussing science-industry relations. This is supported by Aaron Ihde's classic *The Development of Modern Chemistry* (1964). When treating chemical industry, Ihde often says, for example, that discoveries at the universities were commercialized, made "ready for commercial production," or "quickly brought into commercial production."[85] A full analysis along this line must probably make a distinction between different kinds of chemical industries.

Although science grew in importance from 1900, it never supplanted technical innovation, and even if the gap between science and industry was continually closing there was, and remains, a gap. Ramberg, whose work on the development of microanalytical methods for the arsenic committee was, in obituaries, described as scientifically successful, in 1918 privately yearned for an end to the "damned arsenic reports ... so that finally there can be some science done." ["förbannade arsenikbilagorna ... så att man äntligen åter får syssla med vetenskap."][86] The knowledge needed to use laboratory chemistry for practical purposes is not by necessity the same as the knowledge needed to do laboratory chemistry as science. Any history of chemical industry needs to take this into account.

---

[85] A. Ihde, *The Development of Modern Chemistry* (New York: Dover, 1964, reprinted 1984), quotations from pp. 679, 683, 685, respectively.

[86] Ramberg to Bror Holmberg, 1918, *Archives of Lund University Library*, Holmberg papers.

# SELLING SCIENCE: DUTCH DEBATES ON THE INDUSTRIAL SIGNIFICANCE OF UNIVERSITY CHEMISTRY, 1903-1932*

GEERT J. SOMSEN
*Institute for the History of Science*
*Universiteit Utrecht*
*Nieuwegracht 187*
*3512 LM Utrecht*
*The Netherlands*

## 1. Introduction

The intensive contacts between academic and industrial chemistry, which grew in Germany after the mid-19th century, remained unique to that country for quite a long time. As a number of contributors to this volume suggest (Petri on Italy; MacLeod and Travis on Britain; Kragh, Lundgren and Myllyntaus on Scandinavia), the ties between academia and the corporate world were considerably less strong in many other European nations right up to World War I. At the start of the century the situation in the Netherlands was hardly different: university research and industrial production were quite unrelated activities and a Ph.D. in chemistry generally did not pave the way to employment in a private company. In 1908, industry employed less than 10 per cent of university graduates.[1] So while Dutch science was going through its "Second Golden Age" and Van't Hoff had ushered academic chemistry into prosperity too, the chemical industry virtually sat by and watched from a distance.

At times, the gap was so wide that some could not even *see* the other side, such as in 1904, when Utrecht professor Ernest Cohen and his Delft colleague Sebastiaan Hoogewerff

* I thank participants at the ESF Workshop in Strasbourg for stimulating discussions, and Ernst Homburg and the members of my own Institute for their comments on an earlier version of this paper.

[1] *De Toekomst der Academisch Gegradueerden. Rapport van de Commissie ter Bestudeering van de Toenemende Bevolking van Universiteiten en Hoogescholen en de Werkgelegenheid voor Academisch Gevormden* (Groningen, Batavia: J.B. Wolters, 1936) chapter 7, p. 173. The figures in this extensive employment report were based mainly on the yearbooks of the universities, a polytechnic, the Dutch Chemical Society, two secondary school teacher associations, and on the governmental almanac for the Dutch East Indies and the announcements in *Chemisch Weekblad*.



*A.S. Travis et al. (eds.),*
***Determinants in the Evolution of the European Chemical Industry, 1900–1939,*** 143–168.

claimed the virtual non-existence of a national chemical industry, and asked themselves why this was. Their claim was certainly peculiar, for in fact Holland possessed a wide variety of chemical manufacturers, a richness that was not ignored by contemporary foreign observers. But Cohen and Hoogewerff had basically restricted their definition of what *counted* as chemical industry such that it included only companies making intensive use of academic expertise and personnel. Their non-observation of this type of enterprise *was* quite correct, and thus rightly indicates the feeble university-industry relations at that time.[2]

Some thirty years later the situation had changed dramatically. Informal and organized contacts provided a strong bond between the chemical industry and academia. Many companies had drawn up contracts with university consultants, while some professors even joined industrial firms to take up leading positions in their research departments. Conversely, new professors were often recruited from corporate laboratories, while former students working in those laboratories kept close contact with their *alma mater*. In Utrecht, alumni pursuing all sorts of career tracks came back to the university every month to gather at a colloquium to maintain their old "family relations." All of this had been primarily made possible because of an enormous increase in the estimation of university trained employees by the chemical industry; by 1938 almost 50 per cent of university chemistry graduates worked in chemical firms, a four fold increase on the situation at the beginning of the century.

How did this change come about? The foremost factor in the formation of the link is not hard to find. It was the emergence during the intervening period of a number of large industrial research laboratories, which were staffed by university graduates. In 1914, the Philips electrical corporation created a sizeable laboratory, for electro-technical as well as chemical research, which became the largest employer of academic chemists before World War II. Second in line was the Bataafsche Petroleum Maatschappij (predecessor of Royal Dutch/Shell) which opened a laboratory in 1902 and started extensive chemical research after 1927. Enka and the Hollandsche Kunstzijde Industrie (predecessor of Akzo), fast growing producers of artificial silk, established high level research facilities during the course of the 1920s. And in the same period, the laboratory of the Jurgens margarine factory (later Unilever) in the town of Oss provided ample academic employment.[3] Beside these big companies, some smaller

---

[2] H.A.M. Snelders, *De Geschiedenis van de Scheikunde in Nederland. Van alchemie tot chemie en chemische industrie rond 1900* (Delft: Delftse Universitaire Pers, 1993) ch. 8. Ernst Homburg, *Chemische Techniek in Nederland in de Negentiende Eeuw* (Zutphen: Walburg Pers, 1994) esp. ch. 7. Homburg's volume demonstrates the rich variety of Dutch chemical companies in the 19th century, which included fertiliser producers, the white lead industry, sulphuric acid and pharmaceutical manufacturers.

[3] Ernst Homburg, "De overgang naar een moderne chemische industrie," *Chemisch Magazine*, (1989), 741-743; and Ibid., (1990), 311-314. J.J. Hutter, "Nederlandse Laboratoria 1860-1940, een kwantitatief overzicht," *Gewina*, **9** (1986), 150-174. E.S.A. Bloemen, "Bezieling en 'esprit d'équipe.' Industriële research in Nederland in het interbellum," in P. Boomgaard *et al.*, eds, *Exercities in ons Verleden. 12 opstellen over de economische en sociale geschiedenis van Nederland en koloniën 1800-1950* (Assen: Van Gorcum, 1981), pp. 153-167. Philips's chemical investigations were linked to its argon production (for light bulb fillings) and

ones opened joint laboratory facilities, which were usually organized by branches and often government supported. Together, these industrial laboratories forged the link between the corporate world and the university, simply because of their constant need of knowledge, expertise and highly educated personnel trained in doing research work.

The next question is: where did those research laboratories originate? Again, it seems easy to find an answer. The four dominant companies mentioned above were all relatively new on the Dutch scene, and, more importantly, they represented a new *type* of chemical industry, one that had not previously existed in the Netherlands.[4] Unlike their predecessors, they were export-based and internationally oriented; they manufactured on a large scale often with continuous production processes; and, most of all, they were "knowledge-intensive," requiring a substantial input of chemical and chemical-engineering expertise. In order to meet that demand they created research laboratories. Thus, it was the emergence of this new brand of chemical company that was finally responsible for the great intensification of university-industry connections.

If this was the whole story, this paper might end right here. But it is not so. True as it may be, there is also something incomplete and dissatisfying about the explanation just given. It is dissatisfying because it pictures the industrial need for scientific research and researchers as all too natural, to the point of appearing an inevitable consequence of the types and natures of the new companies. But one can very well wonder to what extent these industries were "of necessity" science-based. It is a fact that they all initially managed to operate for some time *without* reliance on research, and without the prospect of starting such a reliance. Their need of science was not self-evident – industrial leaders had to become *convinced* of this first, otherwise no laboratories would ever have been established. Accounts of the type just given ignore all the efforts it took to create such a conviction, particularly the arguments, rhetoric and persuasion employed to overcome the skepticism to science that was invariably present at the outset. Therefore, these accounts are also very incomplete, for they fail to pay attention to the advocates of industrial research, their particular reasons for making their pleas and their favoured instruments of persuasion.

In this paper I aim to make up for this deficiency. I will do so by shifting attention from the 'consumers' of research and researchers (the industries) to the producers, keeping in mind the law of economics that a supply may create its own demand. The suppliers on whom I primarily focus are the university chemists, endeavouring to bridge their gap with industry and convince industrial leaders of the value of science and the necessity

---

later the synthesis of the insulating material 'Philite' (a variety of bakelite).

4 Another possible answer would have involved the adoption of a patent law in the Netherlands, in 1910. But, as Jakob Tanner in this volume shows, a patent law was by no means a necessary condition for stimulating industrial research.

of doing research. During the period under scrutiny Dutch academics were remarkably active in this kind of self-promotional activity, employing all channels of public and private communication, and developing a strong rhetoric on the usefulness of science.[5] In the process, however, they readily encountered conflicting ideas, from the industrialists themselves and from the competing "chemical technologists" of the Delft polytechnic (I shall return later to the distinction between the Dutch university and polytechnic). These two parties often held very different views on what kinds of research and researchers were actually needed, and on who should best deliver those. Disagreements were voiced in fierce debates, and the emerging 'market' for research soon proved to be a highly contested area, claimed by university, industry and polytechnic.

In the following, I explore this contested area by analyzing three episodes in the debates on academic-industrial relations. The first episode (section 3) consists of a conflict after 1903 between university chemists and polytechnic professors concerning the goals of their educational programmes and the labour markets for their students. The second (section 4) follows a discussion after 1912 on the possibilities for starting large scale chemical industry in the Netherlands. The third episode (section 5) deals with the efforts of scientists at the Dutch Academy to establish a national research organization, modelled on institutions, such as the United States National Research Council, which were founded during World War I. In the concluding section, I review the debates and identify some of the main rhetorical strategies which were employed. But let me start by giving an outline of the professional position of academic chemistry at the beginning of the century, which is after all what the university chemists set out to change.

## 2. The state of research around 1900

The era around the turn of the century has become known as the "Second Golden Age" of science in the Netherlands (the first one being in the 17th century). It was in this period that H.A. Lorentz, J.D. van der Waals, and H. Kamerlingh Onnes did their famous work in physics; Hugo de Vries, M.W. Beijerinck, and Eugène Dubois became renowned biologists; and J.H. van't Hoff undertook his ground-breaking studies in physical- and stereo-chemistry. Historians of Dutch science have for a long time wondered what caused this sudden upheaval of individual talent and scientific achievement. A definitive answer has not yet been found (and regarding the talent may very well never be found), but it is generally agreed that the national reform of the universities, which directly preceded the flourishing period, played a major facilitating role.[6]

---

5 Chemists were certainly not the only participants in this campaign, although their contribution was perhaps the most effective. For an overview of the relative contributions from chemists, biologists and physicists, see Bert Theunissen and Frans van Lunteren, eds, *Zuivere Wetenschap en Praktisch Nut. Visies op de maatschappelijke betekenis van wetenschappelijk onderzoek rond 1900*, special issue of *Gewina*, **17** (1994), 141-205.

6 See for example K. van Berkel, *In het Voetspoor van Stevin. Geschiedenis van de Natuurwetenschap in*

In 1876, the Dutch parliament passed a bill that called for a drastic reorganization of the nation's university system. Largely following the German model, faculties were reorganized, the position of *privaatdocent* was created, and, under the banner of *Forschung und Lehre*, research was introduced as a main university task. The official aim of academic teaching was no longer merely to transfer knowledge, but primarily to educate young men (and soon women) in the autonomous practice of science. The effect of this was that a scientific training would contain a substantial amount of laboratory experience and students would be required to finish their studies by undertaking a piece of original research, reported on in a publicly defended dissertation. These novel goals required a host of extra facilities, which were largely provided during the three decades that followed the reorganization. New government funding was furnished, the academic staff was expanded (the number of science professors doubled between 1875 and 1880), and at all four universities, large and well-equipped laboratories were built, among them Van't Hoff's chemistry laboratory in Amsterdam in 1892.[7] These laboratories plus the growing number of students performing their investigations inside them, enabled scientists such as Van't Hoff to undertake research programmes of a scope and scale that had not been attainable by previous generations. Thus, the academic reform of 1876 facilitated the "Second Golden Age."

The university had become a research institution, by law, through its structure and facilities, but no less in the spirit of those who worked in it. Students and staff were well aware of the flourishing period they were going through, and they took pride in the international recognition for investigations and discoveries made in their laboratories. The new scientific leaders, such as Van der Waals, and Van't Hoff, but also his successor H.W. Bakhuis Roozeboom (founder of phase rule chemistry), were greatly admired. Collectively Dutch scientists shared an unprecedented self-assurance, which was expressed in the proud speeches made at the meetings of the new scientific association, the Nederlandsch Natuur- en Geneeskundig Congres (NNGC), a Dutch version of the BAAS, but also in a growing interest in the history of science.[8] Scientific success in modern times was believed to be only attainable through the systematic conduct of investigations, and a talent for research was regarded as the most important quality of the new university man. "The highest ideal of the chemist of the twentieth

---

*Nederland 1580-1940* (Meppel: Boom, 1985), pp. 132-143.

7 After 1876, there were four Dutch universities: Leiden, Utrecht, Groningen, and Amsterdam. In 1880, the *Vrije Universiteit* in Amsterdam was added, but this did not open a science department until 1930. There was, as described in the text, one polytechnic in the Netherlands, located in Delft.

8 The NNGC was established in 1887 and based on foreign organizations, such as the British Association for the Advancement of Science (BAAS). See R.P.W. Visser, "Het 'Nederlandsch Natuur- en Geneeskundig Congres' over de relatie natuurwetenschap en samenleving, 1887-1900," in J.J. Kloek and W.W. Mijnhardt, eds, *De productie, distributie en consumptie van cultuur. Balans en Perspectief van de Nederlandse cultuurgeschiedenis* (Amsterdam: Rodopi, 1991), pp. 37-48.

century," Bakhuis Roozeboom exclaimed in 1904, "is not to be a brilliant teacher and an excellent instructor, but a researcher, and he who does not perform in the latter capacity will seldom become significant in the former two functions."[9]

Yet despite all the pride and the undeniable achievement, the justification of large-scale scientific research remained in fact rather narrow. Research proved to be expensive, and for years after the university reform, members of parliament kept arguing that the "squandering law" of 1876 should at least be countered by the closing down of one or more universities. What, after all, were all those laboratories good for? The interested speakers of the NNGC often claimed that science possessed above all cultural value, being the epitome of civilization. Modern civilized nations, they said, were ranked according to their scientific performance, and so there should be good reason to keep that level high. But "cultural value" was of course not going to be the most solid basis in the long run, and it had certainly not been the main reason why the state agreed to sponsor research at the university in the first place. There, its function was still primarily one of education: research was undertaken because students were required to be *trained* in the practice of science. Laboratory facilities and scientific personnel all existed by the grace of the students' need for them. Many academics knew this by experience, for in requests for expansion of laboratories or staff the most effective argument was usually the inadequacy of the existing situation due to an increase in the number of students. In the eyes of its sole sponsor, the state, research existed for education's sake only. Thus the fate of the university as a research institution depended very much on its success as a centre of education.

What did an academic training in science lead to? According to the 1876 bill, it prepared students for "those professions in society for which a scientific training is required." Not a very illuminating phrase, and perhaps for good reason, for there were not too many such jobs altogether. As for the graduate chemists, industry was hardly an option (as noted above). Agricultural experimental stations were a more realistic destination, and a few Ph.D.s moved to posts in government inspection services for food, water, etc. But the main prospect for the majority of graduates (64 per cent by 1908) was to pursue a career in education. And of them by far the greater number turned to teaching at the Hogere Burger School or HBS, a type of secondary school comparable to the German Realschule, which had a large science component. Such were the expectations for the chemistry freshman around the turn of the century, and these prospects constituted the real (although perhaps not the intended) goals of academic education.

So if we were to take an extreme point of view, we could say that Dutch university

---

[9] H.W. Bakhuis Roozeboom, "In Memoriam Prof. Lobry de Bruyn. Rede, uitgesproken bij de hervatting der colleges op 26 September 1904," *Chemisch Weekblad*, **1** (1904), 957-971, esp. 966. Roozeboom's maxim would often be quoted, e.g., as late as 1941, in H.R. Kruyt, *De Weg der Wetenschap* (Amsterdam: H.J. Paris, 1941).

chemistry departments were nothing but overly luxurious teacher training institutions, maintaining research facilities and performing research only for the sake of preparing students for the teaching profession. Of course this image is somewhat distorted and unrealistic, but it does enable us to see why the university chemists, who identified so strongly as researchers, felt that the justification of their activity was frighteningly feeble. It was for this reason that they started attempts to broaden the significance of university research, by trying to increase the variety of professions that an academic training could lead to, and by endeavouring to build up closer ties with the chemical industry.

**3. University chemists vs. Delft technologists**

The endeavours of the university chemists to broaden the narrow basis for their research only really started after its main existing function was threatened with jeopardy. This threat occurred in the spring of 1903, when the government presented plans to raise the status of the Delft polytechnic to the same level as that of the universities. Traditionally, the school in Delft had been intended to educate a middle class body of engineers, who were to serve the country in technical professions, while the universities were to create the learned elite. This distinction also existed in the paths to the two sorts of schools: for Delft an HBS preparation focusing on science and modern languages sufficed, but for the universities a classical Gymnasium training was required, including the study of Latin and Greek. Now in the new plans these different admission requirements were maintained, while the degrees that the institutes offered were made equivalent. According to the newly proposed measures, the polytechnic professors would be granted the academic *jus promovendi*, and as a consequence their students would gain the opportunity to conclude their studies with a completely valid Ph.D., indistinguishable from the degree of their university counterparts.** In itself this did not trouble the university students, but they did fear new competition on the job market as a result of the intended measures. The Delft department of chemical technology embraced at least as many students as all four chemistry faculties of the country together, and at the same time more and more 'technologists' were also moving to the position of teacher at the HBS, the traditional professional field for university graduates.

On 26 June 1903, representatives of the "Natural-Philosophical Student Association" of the University of Amsterdam (the largest in chemistry) submitted a petition to the national parliament in which they raised the pressing matter. The intended measures would bring university graduates to "extraordinary unfavourable circumstances," they wrote, for they already had to battle chemical engineers and technologists on the HBS job market. Competition there would only increase, while for posts "of a more practical nature" it would become considerably more acute. There was a danger that "the future equality of title will create the idea that both categories of people also possess the same competencies, which certainly will not be the case." Most importantly, the forthcoming

** For lack of a suitable English translation, I call the Delft institute 'polytechnic,' even though it was officially renamed in 1905, from *Polytechnische School* to *Technische Hoogeschool*.

competition would be unfair, for the polytechnic students would be enabled to obtain the same degree as their university counterparts in three to four years less time: firstly, because the chemical technology study programme itself was shorter by two years; and secondly, because Delft students did not need the Gymnasium degree, which took one or two more years to obtain. For the sake of fairness, the petitioners proposed to add two more steps to the existing plans: 1. removal of the Gymnasium requirement for university chemists also (a request already made by many academics in previous years); and 2. a legal ban on HBS-teaching for polytechnic graduates.[10]

Had the discussion been closed here, the issue would probably have gained little publicity and impact. But soon the students were backed by their professors, and an unusually divisive controversy ensued in the Dutch chemical community. First, the students' request was followed by a similar petition from the assembled science professors in Amsterdam.[11] And in January 1904, a second address was sent to the Minister of the Interior (who was responsible for education) by *all* chemistry professors in the country – except the ones from Delft. The scope of the considerations was much wider this time. So far, the authors claimed, everybody had ignored the important issue of the significance of university chemistry for the chemical industry. Many academic graduates already worked "in the field of practice," while most of the current one hundred students should also seek their futures in industry. Industrial progress had always originated, as it still did, from the university, so "[w]hile the Polytechnic aims to promote industry, the Universities do exactly the same and certainly to no lesser an extent."[12] Chemical science interacted so closely with its applications, that it was unimaginable that they should exist without each other. Therefore, the professors argued, it would have been better if the chemical technology studies had been located at the universities in the first place. So in principle the Delft institute should not even have existed, but now that it was a *fait accompli*, suitable measures needed to be taken: 1. students should be able to switch between the two kinds of institutes; 2. an HBS preparatory training should suffice for university enrollment; and 3. the right to teach at the HBS should be restricted for polytechnic graduates, especially since, the petitioners emphasized, "the universities alone will already supply a sufficient number of the necessary teachers."

It was time for the Delft professors to react, and so they did, in a letter to the minister on 11 February. What had troubled them most, the polytechnic professors confessed, was

---

10 City Archives of Amsterdam. Files of the Faculty of Mathematics and Science of the University of Amsterdam, 1020 no. 196. The petition was later published as J. Olie jr. and E.J. Swaab, "Adres in zake de wet op het Hooger Onderwijs," *Chemisch Weekblad*, **1** (1903), 23-26.

11 *Chemisch Weekblad*, **1** (1904), 198-200.

12 H.W. Bakhuis Roozeboom, C.A. Lobry de Bruyn, A.F. Holleman, A.P.N. Franchimont, F.A.H. Schreinemakers, P. van Romburgh and Ernst Cohen, "Aan Zijne Excellentie den Minister van Binnenlandsche Zaken," *Chemisch Weekblad*, **1** (1904), 245-250, esp. 248.

the suggestion that their educational programme could have been taken over by the universities. Nothing was less true, for chemical technology comprised much more than purely chemical subjects, e.g. construction, knowledge of machinery, technical drawing, business, etc. The university professors had incorrectly equated chemistry with chemical technology, thus displaying a severe underestimation of the necessary autonomy of the latter field. But the chemical industry did know the distinction and preferred polytechnic graduates, and, contrary to what the previous petitioners had claimed, employed hardly any university Ph.D.s. These had predominantly become teachers at the HBS, but apparently that market was now drying up. No wonder then, that most of the current one hundred university students would "need to seek a future in industry," the Delft professors cynically remarked. But there, they would have to face normal competition with the polytechnic graduates, just as they ought to on the HBS front, where, also, no protectionist measures were justified.[13]

Once more, university men responded, and this time the stakes were set even higher than before. The core of the whole issue, Bakhuis Roozeboom and his Amsterdam colleague C.A. Lobry de Bruyn replied, was not merely jobs or petty student rivalries. It was the importance of a flourishing chemical science. That was already a matter of the highest national significance. It also constituted the true basis for a flourishing chemical industry. One had to look no further than Germany to see this. The distinction between pure science and technological science, which "the Delft gentlemen" had conjured up, was entirely fictitious: "technological problems do not distinguish themselves from scientific ones by any method followed in solving them." In the end "there is but one scientific chemistry, which is always relating, in *all* her parts, to application and industry." Therefore, Roozeboom and Lobry de Bruyn again stated the ideal of uniting the now separated chemistries, by transferring Delft's chemical technology programme to the universities. Doing so would also increase the number of university students and thus strengthen "chemical science" generally, which was, after all, a great public good. Nevertheless the two professors feared that in reality the situation was going to be exactly the opposite, for if the government plans were to be followed, *no* student would want to come to the university. Freshmen having the same prospects with respect to degree and career would naturally all choose to follow the Delft route, since that would take them three to four years less. As a consequence, the university laboratories would be left evacuated, and "chemical science" severely damaged.

The debates ended there and their effect would be quite small in the short term. In the following year the government raised the status of the polytechnic exactly as it had intended, while leaving both the HBS teaching authorizations and the university enrollment requirements untouched. Nevertheless, the controversy did have some relevant lasting consequences, by having brought out sharply the relative professional positions of the two groups of academic chemists, and their aspirations to maintain or to

[13] L. Aronstein, H. Behrens, and S. Hoogewerff, "Aan Zijne Excellentie den Minister van Binnenlandsche Zaken," *Chemisch Weekblad*, **1** (1904), 270-281.

change those positions. For us, as distant observers, it has thus become clear how in doing so each community used its own image of the science-technology relationship: chemical science as an integrated whole, according to the university representatives, against chemical technology as a separate enterprise, according to the polytechnicians. And apart from this we observe how in the debates the issues of job markets for graduates, student enrollment numbers and the fate of research, were tightly interwoven. But of course the actors made very different sorts of observations. For them, the educational *status quo* had been most ruthlessly revealed: Delft led to industry, while the university taught teachers. This had even been emphasized by the support which had come from these fields during the debates. The Association of Teachers at Secondary Schools had expressed its agreement with the university professors; whereas B. Wigersma, an influential chemical technologist, had sided with the Delft spokesmen, in a public reaction claiming the unfitness of university graduates for work in industry.[14]

The university chemists were left worried, for they saw, quite correctly, that they were drastically disadvantaged with respect to their Delft colleagues. Their study programme required a longer preparatory training, took more years itself, and led to fewer opportunities than its polytechnic counterpart. The professors resolved to undertake action on all three points. First, they continued to plead, incessantly, for a removal of the requirement of Gymnasium preparation as the only road to the university. They would succeed on this, finally, in 1917, when a new bill opened up admission for HBS students too. Second, the professors started looking for ways to reduce the length of the chemistry study programme itself. This proved to be more difficult to achieve. For example, in 1913 the Utrecht chemistry department presented ambitious plans to cut back the average time which students spent at the university from 7.75 to 5.5 years or less. Their main instrument was increased differentiation: students could choose between a theoretical and a practical bias, and, accordingly, they were allowed to skip the examinations on unrelated subjects. Nevertheless, by 1935 the average period of study for those who had entered after 1917 still turned out to be 7.6 years, which made Utrecht the worst case in the country. Polytechnicians graduated in an average of 6.4 years.[15]

---

14 N. Quint, secretary of the board of the *Vereeniging van Leeraren aan Inrichtingen van Middelbaar Onderwijs*, to Prof. Dr. R. Sissingh [president of the Amsterdam Faculty of Mathematics and Science], 12 January 1904. City Archives of Amsterdam. Files of the Faculty of Mathematics and Science of the University of Amsterdam, 1020 no. 196. B. Wigersma, "Het wetsontwerp op het Hooger Onderwijs en de opleiding der Chemici in ons land," *Chemisch Weekblad*, **1** (1904), 317-321. Wigersma was a very colourful chemist. He worked in several industries during his career, was a regular contributor to discussions on education, industry, etc., often bringing in original points of view (such as the influence of national character on industrial success). He was secretary of the Dutch Chemical Society from 1922 to 1924. In 1932 he published a book on the theory of relativity.

15 The 7.75 figure is an average taken over all chemistry graduates between 1877 and 1913. Ernst Cohen, "Non Vitae, Sed Scholae Docemus," *Chemisch Weekblad*, **10** (1913), 571-583. *De Toekomst der Academisch Gegradueerden* (note 1), ch. 7, esp. 193-195.

The university professors put most of their energy into working on the third problem: expanding their graduates' labour market by making the chemical industry a strong part of it. Already during the debates with the polytechnicians, the Amsterdam professors Bakhuis Roozeboom and Lobry de Bruyn proposed in their faculty to improve their students' preparation for a possible industrial career by appointing a lecturer in "technical chemistry." The competition with Delft lay directly behind the initiative, as indicated in a letter of explanation:

> The number of those who study chemistry has so much increased in recent years, that only a small part of them will find a future in education and the large majority will have to work in technical institutions. The conference of the professors of chemistry, which has lately been held in connection with the proposed elevation of the Polytechnic to the level of the University, has led the chemistry professors of the country to the opinion that the main principles of technical chemistry are of the greatest importance for the chemists who constitute the universities.[16]

Again, it should be pointed out that it was not only charity that moved the professors to undertake this move for their students. As Bakhuis Roozeboom added to the proposal, another main reason was fear for "depopulation of the university" and the subsequent superfluity of its chemistry departments.[17] But whatever the motives, the proposal was accepted, and in the same year Andreas Smits, chief chemist of the municipal gasworks, was appointed the new Lecturer in Technical Chemistry.

**4. Debating a new chemical industry**

While the debates of 1903/04 had initially centred on the suitability of university students as compared with polytechnic students for the HBS teaching profession, the next episode's exchanges focused immediately on their fitness for chemical industry. Although university chemists managed to increase their importance this time, they again started out by receiving a serious blow. It soon turned out that the appointment of lecturers in technical chemistry was not a sufficient measure to get more graduates to enter industrial posts and embark on a career in the corporate world. This was demonstrated sharply after 1912, when a discussion arose on the establishment of a heavy chemical industry in the Netherlands. In these debates the university chemists were still very much left to the side. This was all the more painful since the type of industry being

---

[16] R. Sissingh and P. Zeeman to the Mayor and Aldermen, May 17th, 1904. City Archives of Amsterdam. Files of the Faculty of Mathematics and Science of the University of Amsterdam, 1020 no. 197. Sissingh and Zeeman were the president and secretary, respectively, of the faculty. Since the University of Amsterdam was a municipal university, the Mayor and Aldermen were its highest executive board.

[17] Notes of the faculty meeting of May 11th, 1904. City Archives of Amsterdam. Files of the Faculty of Mathematics and Science of the University of Amsterdam, 1020 no. 4. The non-chemical colleagues immediately accepted the proposal, with only a small objection from Van der Waals, who wondered whether the university should move in the direction of Delft.

discussed was exactly the kind in which scientific chemistry should play a role, as had been claimed so often by academics: a synthetic chemical industry after the German model, based on coal-tar and, allegedly, on science. But instead of the latter, it was the presence of the former ingredient, coal, that had formed the starting point of the considerations.

Alphons Steger, a new professor at the Delft polytechnic, had pointed out in his inaugural speech that the newly discovered coalfields in Limburg provided the country with a unique opportunity to set up a large-scale chemical production system, comparable to that of the German factories. He indicated the precise possibilities of the Dutch situation and pointed at the possible obstacles on the route to such an industry. Furthermore, he declared it part of his task to apply his professorial energy to contribute to its realization.[18] The proposal was welcomed with enthusiasm, and the dream of a large-scale chemical industry turned out to be very vivid in the Netherlands.

However, some reservations with respect to the chances of success in the enterprise also appeared. In a published reaction to Steger's ideas, B. Wigersma pointed out that the determining factor would probably not be the local presence of coal (as Switzerland's synthetic chemical industries had shown), but the availability of a suitable workforce, and that was questionable in Holland. There existed no experience with this type of industry, and, moreover, the projected leaders of the factories, namely the chemical technologists from Delft, systematically lacked respect for the problems of practice and the expertise of practical workers. Wigersma himself had seen all too often how this attitude, displayed in the factory, had led to "a gigantic fiasco of the man of science." J.F.B. van Hasselt, an industrial chemist in Rotterdam, shared Wigersma's opinion, although he believed that there were reasons deeper than attitude for the incompatibility of academic knowledge with practical experience. Anybody who had ever tried to scale up a textbook chemical reaction knew that this was a very complicated matter. In fact, experience showed that "*if* one has determined a certain procedure by laboratory experiments, it can be readily assumed that that prescription will *not* be usable, as such, on a technical scale." Therefore, Van Hasselt suggested that the polytechnic education should be altered so as to include more practical training in the handling of factory scale processes.[19]

Thus, the discussion on the establishment of a large-scale industry took place primarily between industrialists and Delft professors. The polytechnical education received some

---

[18] Alph. Steger, *De Toekomst der Chemische Industrie in Nederland. Rede uitgesproken bij de aanvaarding van het Hoogleeraarsambt aan de Technische Hoogeschool te Delft den 4den Juni 1912* (Amsterdam: Holdert & Co., 1912). Steger had previously been Smits' successor as a teacher of technical chemistry in Amsterdam.

[19] B. Wigersma, "De Toekomst der Chemische Industrie in Nederland," *Chemisch Weekblad*, **9** (1912), 810-812. J.F.B. van Hasselt, "Chemische Procédé's in Laboratorium en Fabriek," Ibid., **10** (1913), 1052-1055. See also A. Vosmaer, "Iets over Techniek en Wetenschap," Ibid., **10** (1913), 142-147.

sharp criticisms, but at least its quality was regarded essential to the matter. If university training was left untouched of criticism, it was not because it was considered a better preparation for industrial work; it was because it was not considered at all. University chemistry was ruthlessly sidelined, and that position endured for as long as the heavy industry discussion lasted. After the outbreak of World War I, the issue was picked up by the government, which foresaw shortages of drugs and strategic chemicals. Its representatives wondered (as Steger had done) whether a national industry could not be established, and decided to appoint a "governmental consultant for the chemical-pharmaceutical industry" to provide advice on these matters. To fill the post, however, they, once more, did not choose a university man, nor even a polytechnic professor, but a pharmacist, and, moreover, one who shared the opinions of Wigersma on the value of academics. In a speech made shortly afterwards, P. van Romburgh, professor of organic chemistry at the University of Utrecht, publicly deplored this course of action. He claimed that the basis of autarky and also improvements in military defense, had always been provided by pure, disinterested science – especially by chemical research. Therefore, the government would better profit from the expertise of university chemists.[20]

But however bad the situation may have been, World War I would in fact also provide the conditions for a later improvement in the relations between the university and chemical industry, mainly because the special circumstances enabled an unprecedented growth of the latter. The Netherlands was a neutral country in the war, its borders were closed, and this turned out to be quite advantageous for the Dutch chemical industries which found themselves shielded from German competition for once. The hampered import of raw materials naturally gave rise to many problems, but it also gave way to some initiatives which otherwise might not have been taken. Autarky considerations led to the establishment of Hoogovens, a large-scale steel factory in IJmuiden, which would later start chemical production in collaboration with Bataafsche Petroleum Maatschappij.[21] Also during the war, the Dutch State Mines opened a chemical production plant in the coke factory known as 'Emma' (its production sites were always named after members of the Royal Family). And it was the halt of argon imports that urged Philips to start its own noble gas factory, since it needed argon to fill its light bulbs. Many of the new or expanded companies were of the novel type mentioned in the introduction of this paper, and the most important of them would establish research laboratories in the course of the 1920s.[22]

---

20 The governmental consultant was H.F. Tillema. See H.F. Tillema, "Aan Allen die Belang Stellen in den Bloei van 'Groot Nederland,'" *Chemisch Weekblad*, **11** (1914), 930-932 and 990-991. P. van Romburgh, "Chemie en Onafhankelijkheid," *Verslag van het Verhandelde in de Algemeene Vergadering van het Provinciaal Utrechtsch Genootschap van Kunsten en Wetenschappen*, (1915), 9-37.

21 J.J. Dankers and J. Verheul, *Hoogovens 1945 - 1993. Van staalbedrijf tot twee-metalenconcern. Een studie in industriële strategie* (Den Haag: SDU, 1993), pp. 19-61

22 Homburg (note 3).

But this would not happen before the university chemists had changed their courses of action too, taking advantage of the new situation and addressing themselves directly to the rising industries, emphasizing the significance of science. Most of their speeches were also made by a new type of academic chemist, one convinced of the value of research but also of the social task of being a researcher. In the case of H.R. Kruyt, a new Utrecht professor whose background usefully illustrates the situation, this social conscience had largely been implanted during his college days in the early 1900s, when a wave a socialism had spread through the Dutch universities. Students of that period not only felt compassion for the fate of the poor, they also developed strong non-elitist senses of their future functions in, and not above, society. It was this notion of a worldly professional task which stuck with someone like Kruyt, and expressed itself in his attitude to industry.[23]

When Kruyt became president of the Dutch Chemical Society in 1918, he immediately took upon himself the task of establishing "cooperation of science and industry in the Netherlands." He organized the first general meeting of the society in an industrial town, and opened the proceedings with a speech in which he directly approached the newly established or expanded firms. Kruyt acknowledged that the Dutch chemical industry had boomed during the war and was doing very well at the moment. But the flourishing period might soon come to an end when the foreign competition appeared on the stage again. In order to consolidate the present situation, Kruyt told the industrialists, it was absolutely necessary that they assured themselves of the support of science. The fate of Perkin's early factory for the aniline dye mauve showed "how even a prospering chemical industry goes down, if scientific research does not keep it up." And the contemporary German industries, as always, provided the more successful example. Cooperation was the key, and therefore Kruyt invited industrialists to contact university laboratories and ask for advice or even investigations into their particular production problems. But, more importantly, he claimed that they should employ university graduates, for they possessed exactly the qualities which industry needed at the time. Kruyt dressed his offer in a quote from Raphael Meldola's 1907 presidential address before the British Chemical Society:

> We can provide you with young men, having a general understanding of the foundations of our science, who know how to find their way ingeniously; place them in your factories and your research laboratories, give them the necessary freedom, make them familiar with the problems, which ask for solution in your company, and do not be impatient if there is no direct result; in the long run these people will always prove to be worth their salaries.[24]

---

23 Much of this has been taken from my article "'Hooge School en Maatschappij.' H.R. Kruyt en het Ideaal van Wetenschap voor de Samenleving," in Theunissen and Van Lunteren (note 5), 162-176; and my forthcoming dissertation on H.R. Kruyt.

24 H.R. Kruyt, "Samenwerking van Wetenschap en Industrie in Nederland," *Chemisch Weekblad*, **15**

Kruyt's address went down well, and in the following years he would acquire some considerable reputation for his opinions on science and industry because of his repeated speeches and publications on the subject.[25] But as he realized, these were still only words, and more solid connections remained necessary. In 1918 a perfect opportunity for this occurred when the assembled chemical companies founded a joint Association of the Dutch Chemical Industry to promote their branch of industry and further common interests. From the start of this undertaking, Kruyt was closely involved, and he managed to establish some real collaboration. In the first place, the association decided to develop ties with the Dutch Chemical Society. A policy was adopted of always appointing a representative of the chemical society on the association's board (the first being Kruyt himself), while an industrialist would normally have a place on the chemical society's executive committee. Besides, the chemical companies decided to support the society financially, and assist with reorganization and improvement of its journals. The *Recueil des Travaux Chimiques des Pays-Bas* was acquired as the society's main scientific publication. Meanwhile, the *Chemisch Weekblad* was brought under the joint auspices of the chemical society and the industrial association, while being largely expanded so as to include, among other things, special divisions for industrial news. From 1923 on, a separate section called "Chemistry and Industry" was added every week. Last but not least, the connection was officially laid down in the association's statutes, of which number 3 expressed the aim of "establishment of the bond between science and industry which is so strongly desired for the prosperity of the Dutch chemical companies."[26]

Kruyt's endeavours started to bear some real fruit, perhaps because his approach to the industrial sector had been unusually direct, especially for a university professor. Note for example how he bluntly passed over the polytechnicians and their possible contribution to the new industries, unlike his predecessors who had often tried to deal with, or defeat, the competition from Delft. Kruyt's approach was also reflected in the image of the relation between science and technology, which he chose to project to his audience. The central point in his presentation was that there existed no conflict between pure science and practical utility. As Kruyt repeatedly argued, the more fundamental that research became, the broader the significance of its outcomes would be. The same was true for education. A student thoroughly trained in the fundamentals of his field was much more employable than one who only knew of a few specific

---

(1918), 418-426. Kruyt unfolded his agenda with the Dutch Chemical Society later in "Herinneringen," *Chemisch Weekblad*, **25** (1928), 342-344.

25 For example, at a similar meeting, five years later: "Wetenschappelijk Onderzoek en Algemeen Belang," *Chemisch Weekblad*, **20** (1923), 541-544.

26 *Recueil des Travaux Chimiques des Pays-Bas*, **39** (1920), 1. "Eerste Jaarverslag van de VNCI," *Chemisch Weekblad*, **17** (1920), 2-4.

applications. In the long run, pure science would pay off best.

With this image, Kruyt sought to overcome the existing industrial skepticism toward academic research and researchers. In his function as consultant to many chemical firms, he tried to get these views realized. But also in the direction of academia, Kruyt's image of a useful pure science served to counter objections and fear that the university might sell out to the world of business. Since existing pure science was exactly what industry would ask for, there was nothing to fear. In keeping with this image, Kruyt also did not admit any straightforward industrial research in his own laboratory. He would accept a practical assignment from time to time, but then its question was to be directly translatable into a general problem of theoretical interest. Collaboration between university and industry was one good thing, but academic autonomy was another. As has been set out before, for university men like Kruyt, the bridge between science and industry served primarily to give the existing university research a broader justification.

**5. Building a national research organization**

The significance of university research was further expanded in a third episode, but again in a different way than earlier. While the value of university trained students for teaching and industry had been the issue of the first sets of debates, it was the more direct fruits of academic research that were claimed to be crucial for society now. Also, this time, the discussions were not confined to chemists alone, but all scientists (and industrialists) played a role. The impulse for these developments was once more provided by World War I.

As is well-known, in many of the belligerent countries, the war had a strong catalyzing effect on the development of scientific research. In order to cope with the problems caused by loss of imports and the new chemical warfare, the governments of both Central and Allied nations began mobilizing scientists (often at their own initiative) in newly established national research institutions. Britain inaugurated the Department of Scientific and Industrial Research, Germany largely expanded its Kaiser Wilhelm Institute, and the United States established the National Research Council. These institutes were unprecedented in their type and scale. First of all, there were enormous numbers of researchers working in them: hundreds of academically trained scientists and thousands of assistants and technicians. But also the type of work was new, not only because it was aimed at practical goals, but because it consisted exclusively of research which was not done for any educational objective. Furthermore, the available research facilities were often very good, and for many scientists their 'mobilization' in these institutes was a positive experience.[27]

---

[27] Jeffrey A. Johnson, *The Kaiser's Chemists. Science and Modernization in Imperial Germany* (Chapel Hill: University of North Carolina Press, 1990), esp. ch. 9. Roy MacLeod, "The chemists go to war: the mobilization of civilian chemists and the British war effort, 1914-1918," *Annals of Science*, **50** (1993), 455-481. Daniel J. Kevles, *The Physicists. The History of a Scientific Community in Modern America* (Cambridge, Mass.: Harvard University Press, 1995), chs 7-9. See also Roy MacLeod's contribution to this volume.

The Netherlands were not at war, but seeing the great research institutes arise in the countries around them made the Dutch academics long for something similar. Since the international circumstances had also blocked imports, and led to mobilization of the Dutch army, perhaps the time was ripe now to let university research play its larger role. In the autumn of 1917, members of the Royal Academy of Sciences discussed the question of whether researchers could not serve the country in the same ways and on the same scale as their foreign colleagues did. A new opportunity was signalled and it was decided to move to act. H.A. Lorentz, the academy's president, started an inquiry among science professors around the country, asking them to raise pressing national problems on which they thought they could give scientific advice or perform useful investigations. Having assembled several such topics (mainly on product shortages) Lorentz requested a conference with the Dutch prime minister in order to suggest what the academy could offer. A cabinet meeting was quickly arranged, and among the ministers present Lorentz found "full sympathy" for the academy's intentions and he received 100,000 guilders to start the research projects, a very substantial sum.[28]

The academy formed a committee consisting of science professors and some representatives of industry and agriculture to classify the particular practical problems into topical categories, such as foodstuffs, ammunition, chemical industry, the "N[itrogen] problem," etc. For each of these topics special subcommittees were then established, which organized the necessary research and carried it out, typically in one of the existing university or polytechnic laboratories. In itself this type of activity was nothing new: when asked, the academy had always provided advice to the government by having one of its members investigate a particular problem in his own laboratory (often on Saturdays after classes). But the sheer size of the research and its level of organization were much bigger than before, and although the name of the new body made it appear like a small gathering of academicians, the "Lorentz Committee" truly resembled its foreign sister institutions. In the opening address of the Committee on 30 March 1918, Lorentz specifically referred to them.

> Everybody knows that in several belligerent countries organizations have been established which aim to make scientists help to meet various demands. England has the "Advisory Council of Scientific and Industrial Research," America the "US National Research Council," in France Min. Painlevé has established a similar committee, and we can be sure that in Germany science has been 'mobilized.' In our country, the question has also been raised whether the practitioners of pure and applied science should not be working, more than before, for the benefit of the national welfare and defence.[29]

---

28 Lorentz's personal notes. State Archives North-Holland, Haarlem. Archives of H.A. Lorentz, 364. Files on Lorentz Committee, 166. F.M. Jaeger to H.A. Lorentz, October 28, 1917. Ibid., Files on Subcommittee L.: Minerals and Chemical Industry 167.

29 Opening speech by Lorentz in *Mededeelingen betreffende de Wetenschappelijke Commissie van Advies*

The phrase, "national welfare and defence," was probably copied from the foreign institutes, most of which had indeed been installed for this dual purpose: to keep the economy up to strength by tackling product shortages, and to sustain military defence by catching up in the chemical arms race. But the Dutch research would concentrate mainly on problems of the first category. The subcommittees started to pursue their investigations on topics such as alternative foodstuffs, indigenous sources of fertiliser, '*Ersatz*' fibres, the production of cattle feed from straw, and other ways to replace products or production processes which the war had made impossible.[30] By their nature, many of these material substitutions demanded chemical investigations. And all of the subcommittees' chemical researches were carried out in university and polytechnic laboratories. For example, in the Utrecht organic chemistry laboratory the preparation of formaldehyde from methane was investigated; its Delft equivalent examined the possibility of turning naphthalene into benzoic acid; Amsterdam tried to extract turpentine from pine-needles, etc. etc.[31]

The research projects were often supported directly from the Lorentz Committee's budget, or through the appointment of temporary scientific collaborators. Meanwhile, however, the academicians involved also tried to establish more enduring ways of sustaining the scientific activity. Many subcommittees developed elaborate plans for the creation of special research institutes, embracing permanent laboratories and staff. For example, the Subcommittee for Chemical Industry proposed to build an "Institute for Scientific Fuel Research" modelled after German, British and American examples.[32] The Subcommittee for Foodstuffs made extensive plans for a "National Nutrition Institute" (alias vitamin laboratory), for which it even undertook a study tour to the United States and had a full architectural design completed. The Subcommittee for Ammunition, in turn, sought to establish an Institute for Ammunition, and there were at least two more initiatives to found other national research laboratories.[33]

---

*en Onderzoek in het Belang van de Volkswelvaart en Weerbaarheid*, **1** (1918), 4-10, esp. 4.

30 The Dutch scientists chose a quite original strategy for resolving the fertiliser problem. While their German colleagues were perfecting the sophisticated Haber-Bosch process for the fixation of nitrogen from the air, the Dutch sought to cover shortages by collecting the output of urinals and public toilets. Although the nitrogen subcommittee realized that there were objections "of both hygienic and aesthetic nature," they believed that the testings so far undertaken had been worthwhile.

31 First year report of the Lorentz Committee in *Mededeelingen betreffende de Wetenschappelijke Commissie van Advies en Onderzoek in het Belang van de Volkswelvaart en Weerbaarheid*, **4** (1919).

32 Those were the Kaiser Wilhelm Institut für Kohlenforschung in Mülheim, the Institute for Fuel Research in Leeds, and the Bureau of Mines in the US.

33 *Mededeelingen betreffende de Wetenschappelijke Commissie van Advies en Onderzoek in het Belang van de Volkswelvaart en Weerbaarheid*, **4** (1919). A. de Knecht-van Eekelen, "Het Vitaminen-Laboratorium," *Gewina*, **19** (1996), 43-45.

However, none of these plans was to be realized. In fact *all* of the Lorentz Committee's activities were terminated shortly after the war. In 1918 the government granted another 100,000 guilders, but from then on the subsidy was rapidly reduced, and, moreover, no more research assignments were submitted. The subcommittees became literally out of work and after two and a half years of non-activity the Lorentz Committee was finally dropped from the government budget in 1922. An important reason for this was, of course, the end of the war and Holland's relief from economic isolation, which had been the primary reason for the committee's existence in the first place. Since the committee was a 'paper' organization, embracing no permanent posts or institutions of its own (yet), its termination was relatively easy. But the most fundamental reason for the end of the Dutch research institute was that the intended connection between science and practice had still not been satisfactorily achieved.[34]

Already during the committee's period of activity, members had complained about the fact that nothing was being done with the results of their research. The recommendations based on laboratory investigations seemed to be buried in bureaucratic files. Therefore, when the minister asked in November 1919 whether the organization should continue to exist in peacetime, the academy answered in the affirmative, but added that the follow-up to its scientific advice needed to be improved. The academicians saw quite well that the "knowledge push" provided by their own researches would still be judged worthless if no connection was established with any "market pull" from industry or agriculture. Thus their advice to the minister was that the basic idea of a national research institution was still very good and valuable, even after the war, but that its over-all organization, especially on the output side, needed to be improved.[35]

The question was how to organize research so as to make it really useful to the practical sectors of society. The standpoint of the academy members had so far been quite clear, but perhaps it had also been somewhat one-sided. Their approach and proposals had invariably centred attention on their own academic type of research, and had projected the rest as a matter of subsequent 'application,' in which no inherent problems were expected or even thought of. A month after the recommendations from the academy, however, the minister was confronted with a radically different point of view by a representative of 'practice' itself: the chemical technologist C.I. van Nieuwenburg, industrial chemist at the Leerdam Glass Factory. Van Nieuwenburg's assertions as to what kind of research organization industry really demanded differed so strongly from the academy's accounts that the government decided to freeze the Lorentz Committee's

[34] Minister of Education, Arts and Sciences to Royal Academy of Sciences, October 11, 1920. State Archives North-Holland, Haarlem. Archives of the Royal Academy of Sciences, 64. Files on Applied Scientific Research, 350.

[35] Science Department of the Royal Academy of Sciences to Minister of Education, Arts and Science, November 1919. Ibid. See also the speech of the minister at the installation meeting of the Went Committee, February 16, 1924. Ibid.

activities, and first have these matters straightened out. But the straightening out proved to be a quite complicated matter in which many parties presented many views on how to connect research to practical demands. The whole process ended up taking more than five years, a freezing period which the Lorentz Committee did not survive.

Van Nieuwenburg had voiced his opinions in a published speech before the Delft Scientific Society of 19 December 1919.[36] In it he had critically reviewed the Lorentz Committee's work and tried to analyze why it had produced so little effect. His answer was in fact quite simple: the activities had been led by the wrong kind of people. One would have normally expected that research undertaken for the benefit of industry would need to be directed by representatives of industry, who knew its specific character and demands, both technically and commercially. But the opposite had taken place: the government gave control to the academy, an institute of "scholars *pur-sang*" which by its very nature "nearly *has* to be un-technical." "The curiosity of the whole affair," Van Nieuwenburg continued, "resides not even in this tactical mistake of the Government, but in the fact that the Academy has accepted the task, yes, even has taken the initiative to it." Van Nieuwenburg quite rightly noticed the double agenda of the leaders of the academy: the goal of the Lorentz Committee was not only to serve the practical sectors of society (in which it had failed anyway), but also to expand university research and enhance its importance. True, polytechnicians were also members of the academy, but apparently they had not made much of a difference. The Lorentz Committee had still produced only "dry and barren committee work" that was of value to nobody.

The basis of Van Nieuwenburg's argument was the distinction he made between different kinds of research, or different kinds of 'work,' as he called it. For Van Nieuwenburg there was an "almost unbridgeable gap" between purely scientific work and "scientific-technical work," the latter being the kind that industry really needed. Purely scientific work consisted of the curiosity-driven investigations performed in university laboratories while scientific-technical work originated in the factory and consisted of research undertaken to improve products or production processes. These two modes of research followed different methods, but above all required different mentalities of the people who undertook them. In scientific-technical work many more factors needed to be taken into account (especially financial and material restrictions) and therefore it could never be successfully led by pure scientists who had no experience with considerations of this kind. The best organization for this type of research, Van Nieuwenburg thought, was one like the Mellon Institute in Pittsburgh, since both academy and universities had been kept out of its direction. Therefore he

---

[36] C.I. van Nieuwenburg, "De Nationale Organisatie van Wetenschappelijk-Technisch Werk," *Chemisch Weekblad*, **17** (1920), 70-75. Van Nieuwenburg became professor of analytical chemistry at Delft polytechnic in 1920, after which his views on research lost some of their exclusive focus on industry. See for example his presidential speech at the 1924 meeting of the Dutch Chemical Society, *Chemisch Weekblad*, **21** (1924), 223-224.

proposed that Dutch chemical industry would found such an institution, while the government could fulfill its "technical-chemical duties" by establishing a Central State Laboratory. Both were to be led, of course, by representatives from industry.

Van Nieuwenburg's essay evoked strong but mixed reactions. The only issue which all responding parties agreed on was that the chemical industry needed research. That conviction would not fade anymore. But disagreement centred on the kind of research needed and on how to undertake its organization. The Association of the Dutch Chemical Industry found itself closest to Van Nieuwenburg's model; the idea of a Central State Laboratory had in fact already been discussed at its own meetings. A different reaction came from H.I. Waterman, a new professor of chemical technology in Delft, who agreed with the plan of a central laboratory, but thought it best to establish it *in* his own polytechnic. Delft professors could very well organize research that was valuable for industry, and the laboratory would in fact guarantee a continuous contact. Academy president Lorentz also reacted on Van Nieuwenburg's ideas, understandably disapprovingly. He did admit (once more) that the Lorentz Committee had so far produced little effect, but denied strongly that this had been due to the unfitness of its leaders. "I truly do not see the smallest ground for the opinion that the Committee would have done better if its Board had consisted of technicians," Lorentz told the minister, who had specially requested him to respond to Van Nieuwenburg's piece.[37] The variety of reactions still left the minister confused, and therefore he decided to ask yet another opinion from somebody who could be expected both to respect science and to be familiar with industry: Isaac P. de Vooys, extraordinary professor in mechanical technology at Delft, and the new director of the recently established Dutch Society for the Exploitation of Coalfields.[38]

De Vooys took a month to prepare a lengthy and detailed report for the minister, and in which he critically analyzed Van Nieuwenburg's statements, and developed alternative ideas where he thought that was necessary – which was in many places.[39] De Vooys could still agree with the objective of the argument (making research and researchers useful to industry) but he strongly disagreed with Van Nieuwenburg's conceptualization of the basic sorts of research. According to De Vooys, "scientific-technical work" was a

---

37 "Eerste Jaarverslag van de VNCI," *Chemisch Weekblad*, **17** (1920), 2-4, esp. 4. H.J. Prins, "Organisatie van Wetenschappelijk-Technisch Werk," Ibid., 125-126. H.I. Waterman, Ibid., 158. H.A. Lorentz to minister, September 11, 1920. State Archives North-Holland, Haarlem. Archives of H.A. Lorentz, 364. Files on Lorentz Committee, 166.

38 In 1930 De Vooys became director of AKU, the joint venture of Enka and HKI artificial silk producers and their German counterpart Glanzstoff. As noted, AKU was one of the first and main firms to incorporate large scale research.

39 The minister had made this request orally to De Vooys on November 2, 1920. His report is given in Is.P. de Vooys to Minister of Education, December 12, 1920. State Archives North-Holland, Haarlem. Archives of the Royal Academy of Sciences, 64. Files on Applied Scientific Research, 350.

confusing term, mixing up 'technics' and "technological science," his own preferred categories. Not only had Van Nieuwenburg incorrectly lumped those two together, he had also grossly exaggerated their contrast with pure science speaking of an "unbridgeable gap."

De Vooys projected an alternative scheme in which there were not two sorts of research divided by a gap, but three sorts of activity closely connected. The first he called *techniek* (technics), which was basically what went on in the factory; the second was "technological science," which was performed at the polytechnic; and the third was "pure science," the trademark of the university. Pure science and technological science resembled each other in method (both were science) but they differed with respect to the choice of their objects. While the former could pick whatever its curiosity led to, the latter was, by its nature, restricted to practical topics. So technological science was also related to industrial *techniek* because industry delivered its research problems. Thus, technological science took the central position among the three sorts of research, and in fact also in De Vooys's intention: over-all, his considerations emphasized the middle category.

As indicated, De Vooys's scheme of types of research was linked to a distribution over different institutions. And unlike Van Nieuwenburg, De Vooys thought that Delft was the ideal place for the crucial middle category. "Technological-scientific research" would provide good practical exercise for the students, who would then in turn become more valuable for industry. *Not* locating it in the polytechnic would come down to declaring that institution superfluous. For what would Delft be good for if not for technological science? So De Vooys advised on concentration of his central type of research in the Delft polytechnic. In order to prepare this concentration and find a suitable organizational structure, he proposed that the minister appoint a special committee to "investigate by which measures and in which forms technological-scientific research in this country can be made more serviceable to the public good."

The minister immediately forwarded De Vooys's report to the academy (which was after all still the government's principal consultant in scientific matters) whose response was generally positive: the academicians agreed with the criticisms which De Vooys had fired at Van Nieuwenburg and they also concurred with most of his personal views. They disagreed, however, with De Vooys's indication of the polytechnic as the ideal home for the new research undertaking. Instead they suggested the creation of a completely new institution which should take a position between the academy and the government, thus strengthening the former's advisory function.[40] The minister added this proposal to De Vooys's general recommendations, and forwarded those to a new committee to determine the best organizational form of the new research institution. This committee was a delegation from the academy, and again it was named after its

---

[40] Reaction by Science Department of the Royal Academy of Sciences. Ibid.

president, the successor of the deceased Lorentz, Utrecht professor of botany F.A.F.C. Went.

But before the Went Committee started its business, the academy members insisted on adopting a change in terminology: De Vooys's term "technological science" should be replaced by "applied science" while "technological-scientific research" should be substituted by "applied scientific research." The declared reason was that the latter term was "more comprehensive" than the former; but apparently there were also deeper motives. "Applied science," unlike "technological science," seemed to suggest the necessary existence of something that can *be* applied, which, of course, would be "pure science." Thus, the term "applied science" tacitly indicated the superiority of the latter, and, in fact, the academicians would expose this hidden message in many of their statements made during the subsequent work of the Went Committee. Already in the inaugural speech, on 16 February 1924, Went emphasized that the applied science which the committee was going to organize always needed the fruits of pure and autonomous science for its development. And the authors of the committee's reports repeatedly pointed at this dependency relation. It was even added that pure science, even when it did not feed its applied counterpart, was useful in itself, for it often yielded inventions that were of great practical utility. It was only the unpredictable nature of this process that precluded its formal organization, and that was why the committee needed to concentrate on the better-to-plan applied science.[41]

Thus, the members of the Went Committee moved the focus very much back to pure science, stressing the value and necessity of university research. But they could not entirely push their old point anymore, for that would come down to a return to the Lorentz Committee, the organizational form for which they were supposed to find an alternative. However, the plans for the new institution which the Went Committee presented during the subsequent years still very much reflected the structure of the old and beloved Lorentz Committee. Just as the latter had been governed by an overarching committee of scientists and representatives from industry and agriculture, the new institution was to be headed by a "Central Organization" with exactly the same composition. And just as the Lorentz Committee had presided over topical subcommittees through which the actual research work was carried out, the new Central Organization was to be the umbrella over "Special Organizations," arranged according to subject area, and taking care of the execution of the investigations. The important difference was that the new institution would no longer be a mere extension of the academy, staffed by academicians and utilizing university laboratories, but a completely new and separate organization, possessing its own laboratories and personnel.[42]

---

[41] F.A.F.C. Went, installation speech, February 16, 1924. Ibid. Draft report discussed in Went Committee meeting of October 24, 1924. Ibid.

[42] Th.J. van Kasteel, *Een Kwarteeuw TNO 1932-1957. Gedenkboek bij de voltooiing van de eerste 25 jaar werkzaamheid van de organisatie TNO op 1 mei 1957* (Den Haag: Nederlandse Organisatie voor Toegepast-

Such was the plan as it finally emerged, after five years of the Went Committee's work, intermediate proposals and political negotiation. On 30 October 1930, the Dutch parliament accepted the "Went act," as it was called, and two years later the new organization was officially installed. The government did not immediately start to build new laboratories, but began by placing existing laboratories, such as inspection services, experimental stations, etc., under the direction of the new organization, fitting them into its structure and purposes. Only at a later stage (especially after the Great Depression) would new laboratories be established and the new research body be fully "dressed up." But the framework was already in place, and from 22 April 1932, Dutch scientists could look up to a new national research institution, literally named "Applied Scientific Research" (ASR).

Overseeing its whole origination from 1917 to 1932, the "Applied Scientific Research" institution clearly appears as the fruit of a compromise. As we have seen, it emerged from at least three rather conflicting views on how to set up practically useful research. The academicians, Van Nieuwenburg and De Vooys all had a share in the final result, but each of them also had been obliged to give in. This marked the experience of the university scientists too, who had hoped during World War I that the new organization would make the results of their research an essential ingredient for industry and agriculture. Those results were certainly more often regarded as such, but the link between university laboratory and factory floor, as far as it was provided by ASR, was very much an indirect one. The personal contacts between professors and industrialists which had grown during the same period constituted a much more solid connection.

But even if ASR stood between industry and university, it did provide an important rhetorical bridge, for by its very name, ASR represented a sort of activity which was needed by industry, and which itself needed the university. Applied Scientific Research implied the necessity of pure scientific research, and as such it had given university research a broader significance and justification. No longer were academic laboratories merely used to educate students; university science was now also practiced for its results, which formed the basis of applied scientific research. That this abstract image could concretely pay off was proved after World War II, when the government created a sister organization to ASR, intended to support its complementary kind of research. This organization was to sponsor university research on an unprecedented scale and its name was directly connected to ASR: "Pure Scientific Research" (PSR). Intimately entwined, Pure Scientific Research and Applied Scientific Research would usher the Netherlands into the era of Big Science.

## 6. Conclusion

In the three episodes of debates and negotiation, we have seen how Dutch university

---

Natuurwetenschappelijk Onderzoek, 1957), chapters 1-7.

chemists struggled to defend their enterprise and carve out new roles for themselves in society. Looking back, we see a remarkable consistency in their aims: to strengthen their position as researchers by broadening the significance of their work without in the process sacrificing their autonomy. In this they succeeded remarkably well: by the end of the period university research was justified for far more reasons than as a means to educate students who would become teachers at the HBS. What had enormously changed over the episodes were the sorts of things that research was good for. While it remained as a preparation for teaching, a training in university chemistry now made students fit for industrial work as well. Moreover, the direct results of pure research were also deemed essential to the practical parts of society, or at least to ASR. The 'market' for research had substantially grown, and although the university chemists had had to face competition from the Delft polytechnic and to secure a steady demand from the chemical industry, they finally had taken a comfortable share. As indicated in the introduction, by the 1930s many university graduates made their ways to industrial laboratories and many professors sold research services to chemical companies.

At the start of this paper we asked how this success came about, and pointed at the university chemists' own promotion as one important factor in the increased demand for research and researchers. So let us now shortly look back at the plethora of rhetoric which characterized the three episodes' debates, and try to identify some main traits. If we review the wide variety of rhetorical strategies that were employed by Dutch university chemists, as well as by the polytechnicians and industrialists, we see one sort of argument that perhaps appears most often. Throughout all episodes scrutinized, various images of the science-technology relationship were repeatedly presented and defended by various groups of chemists. We have seen the university professors project a picture of chemistry as an integrated science, against their polytechnic colleagues' sharp distinction between chemistry and chemical technology (section 3). We have seen Wigersma single out factory experience, while Kruyt envisaged a useful pure science (section 4). We have seen Van Nieuwenburg sketch scientific-technical work and pure science divided by an unbridgeable gap; De Vooys replaced this image by a continuum of technics-technological science-pure science; and Went and others changed this picture into a hierarchy of pure and applied science (section 5). Most of these images were not new: Dutch chemists generally drew upon a variety of representations of the science-technology relationship that had been developed in the past and often in foreign countries. For example, the view that intellectual *and* practical concerns were both part of one integrated chemical science, as put forward by the university professors in the debates of 1903-04, was taken from a long tradition, starting perhaps with Liebig, repeated in the vision behind the British Chemical Society, and continuing beyond. And the distinction between pure and applied chemistry went back to the 18th century, although the identification of pure and applied *research* may be of more recent date.[43]

---

[43] Robert Bud and Gerrylynn K. Roberts, *Science versus Practice. Chemistry in Victorian Britain* (Manchester: Manchester University Press, 1984), pp. 69-95. Ernst Homburg, *Van Beroep "Chemiker": de opkomst van de industriële chemicus en het polytechnische onderwijs in Duitsland, 1790-1850* (Delft:

But what matters here is not where these images came from, but that they were picked up again in the Dutch context and utilized under new circumstances for specific rhetorical goals. It is significant in this respect that the research organization which took shape as a result of the debates received its very name from a particular view of the science-technology relationship. "Pure Scientific Research" and "Applied Scientific Research" literally institutionalized the shared image that had been the outcome of long-standing debates and negotiations between different professional groups of scientists. Therefore, these names suggest to us how we can understand past images of the science-technology relationship. Rather than expressions of epistemological conviction, we should regard them as tools – as rhetorical instruments employed with verve and vigour to secure a professional position.[44]

---

Delftse Universitaire Pers, 1993). Christoph Meinel, "Theory or practice? The eighteenth-century debate on the scientific status of chemistry," *Ambix*, **20** (1983), 121-133.

[44] Perhaps this could be a specification of Otto Mayr's project as outlined in "The science-technology relationship as a historiographic problem," *Technology and Culture*, **17** (1976), 663-673. For a recent follow-up to his suggestions see Ronald Kline, "Construing 'technology' as 'applied science:' Public rhetoric of scientists and engineers in the United States, 1880-1945," *Isis*, **86** (1995), 194-221.

# Part 4

# Different Routes to Competitive Advantages

# MODERNIZING INDUSTRIAL ORGANIC CHEMISTRY: GREAT BRITAIN BETWEEN TWO WORLD WARS

ANTHONY S. TRAVIS
*Sidney M. Edelstein Center for the History and Philosophy of Science, Technology and Medicine, Givat Ram Campus, The Hebrew University, Jerusalem, 91904, Israel*

## Introduction

Throughout World War I, the British chemical industry struggled with varying degrees of success to imitate the processes that it had previously neglected, but that had been mastered in Germany. Among these were the manufacture of refined aromatic intermediates and complex dyestuffs. The first British-made indigo was produced at the sequestered Ellesmere Port factory of the Hoechst dyeworks after "the task of working out the practical details was speedily solved by the united efforts of the chemical staff in the research laboratory and the engineering staff in the factory. Within six weeks large supplies of [the intermediate] phenylglycine were available. In November, 1916, the Union Jack was hoisted at Ellesmere Port, the plant was set into operation, and British synthetic indigo came on to the market in considerable amount, the quality generally admitted to be equal to that of the German product."[1]

These products were brought into strategic planning by the needs of war: intermediates for explosives and drugs, dyes for uniforms, and chlorine and its compounds for poison gas. Strategic planning focused not only on products, but also on the organization of the chemical industry.[2] The impact was to be long-lasting. Government-encouraged

1 Gilbert T. Morgan, "Synthetic indigo. A romance of the chemical industry," in "Science and Industry. The Organic Chemical Industry in England," supplement to the *Manchester Guardian*, 30 June 1917, p. 21. In 1916, Morgan, at one time editor of the *Journal of the Chemical Society*, had been appointed professor of applied chemistry, City and Guilds Finsbury Technical College, London. For an account of the first production of indigo in England see Peter Reed, "The British chemical industry and the indigo trade," *British Journal for the History of Science*, **25** (1992), 113-125.

2 See in particular Roy MacLeod's contribution to this volume. For French attempts to imitate German processes in the immediate post-war period see Lothar Meinzer's paper. Compare also Rolf Petri's description of the situation in Italy, where notable progress was made in the high pressure synthesis of ammonia. Unlike most other contributors to this volume, I place the greatest emphasis on products and



*A.S. Travis et al. (eds.), Determinants in the Evolution of the European Chemical Industry, 1900–1939*, 171–198.

mergers led to the creation of the British Dyestuffs Corporation (BDC), while the post-war threat of foreign competition brought about the formation of Imperial Chemical Industries (ICI) and the entry into agreements with international dyestuffs and nitrogen cartels. Great Britain, however, still had a special relationship with its Empire, a relationship that provided ready markets as well as resources for exploitation. Certain resources had lost their former importance, notably indigo. Indeed, it was the availability of the natural indigo product from British India that had hindered research into a synthetic alternative in Britain (where great efforts were made to protect the trade in the colonial product) but that had enhanced the search for the synthetic product in Germany at the end of the 19th century (in order to break the British monopoly). Between the two world wars, the same thing happened in the case of rubber.

However, British industrial inventiveness in chemistry was not lacking between 1914 and 1939. In many areas British industry was no longer imitative, as it had once been of the German dye companies, but sought out new directions, stimulated by awareness of new material needs. Here, partly through the eyes of contemporaries, I explore the circumstances of some of the advances, especially of those novel organic products, such as plastics, resins, and adhesives, that were to become essential strategic chemicals during World War II, and everyday products after 1945. Significantly, neither ICI nor the traditional sites of chemical manufacture were prominent at first. New innovations and new-user industries were to be found in the Midlands. The same was true for suppliers of raw materials. It was in Birmingham, Oldbury and West Bromwich, for example, that the rise of the British plastics industry was heralded.

**Chemical industry in 1939: The raw material question**

Early in 1939, an editorial in the journal *Chemistry and Industry* drew the attention of readers to two historical accounts of recent progress in British chemistry and, especially, chemical industry.

> In 1924 visitors to the Chemical Hall at the Wembley [British Empire] Exhibition were able to purchase a volume, specially prepared for their edification, entitled Chemistry in the Twentieth Century which put on record, with special reference to British investigators, the progress in chemical and physical science during the preceding thirty years. In a series of three Cantor Lectures, delivered in February of this year before the Royal Society of Arts, Sir Gilbert Morgan has described the achievements of British Chemical Industry in the last twenty-five years. There is an overlap of ten years in the two stories which taken together give an informative and authoritative

---

processes, in part because so many of them, such as amino plastics and novel adhesives, have escaped the historian's attention. Nevertheless, the stress is on the ways in which novel products and processes were created to satisfy new and different needs, partly in response to the clearer recognition that the manufacture of organic chemicals had become an important strategic industry.

> account of modern British chemical developments such as the future chronicler of our days will find invaluable .... The recovery since the war has been striking; it has been brought about by an awakening of the chemical conscience, by two Acts of Parliament, by co-operation, in particular that effected by the Association of British Chemical Manufacturers, and by a variety of other causes.[3]

The editor's reference to 'co-operation' in Britain hardly included the type of close industry-academic links that had existed in Germany since the 1880s. He was referring to the Association of British Chemical Manufacturers' encouragement of co-operation between chemical producers, and of close links with government departments, which to some extent influenced parliament. In common with many innovations in the British chemical industry, the association was itself a product of enforced wartime co-operation, having been founded in 1916. After World War I, a number of chemical trade and scientific research associations had been formed in Britain. The latter had been set up under the aegis of the Department of Scientific and Industrial Research, itself a wartime creation (1916), and had close links with universities. The Chemical Research Laboratory, renamed the National Chemical Laboratory in 1928, was also an outcome of these initiatives.

Certainly there was cause for some celebration in the months leading up to September 1939. British chemical industry had during the preceding decade successfully pioneered polythene and 'Beetle' (urea-thiourea-formaldehyde) plastics, while complete outsiders whose interests were stimulated by the need for novel textile fabrics and insulating materials and improved methods of aircraft construction had made significant advances in fibres, resins, and adhesives. ICI's high pressure process for ammonia, which had operated at Billingham since 1923, was adapted to hydrogenation of coal, that is conversion to fuel oil.

Certain of these successes in organic chemistry represented a move away from the traditional raw material, coal (what the British called King Coal), though this was not always readily apparent. It was Herbert Levinstein, formerly of BDC, but now involved in the manufacture of viscose rayon (artificial silk), who, when promoting industrial planning, reminded readers of *Chemistry and Industry* of how things had changed since before 1914:

> There are three sources of raw materials for the manufacture of organic

3 "Achievement," *Chemistry and Industry*, **58** (8 April 1939), 297. Sir Gilbert T. Morgan, who in 1917 wrote the account on synthetic indigo (note 1), had recently retired as director of the National Chemical Laboratory, Teddington, and was coauthor, with David Doig Pratt, of *British Chemical Industry. Its Rise and Development* (London: Longmans, Green, 1938), an historical account that emphasized the need for self-sufficiency. Relevant here also is Morgan's, "The trend of researches in plastic materials," *Chemistry and Industry*, **57** (1 January 1938), 3-11, based on his Presidential Address to the Institute of the Plastics Industry.

> industrial products and three sources only: (1) The large group of naturally growing things [for methanol and ethanol]; (2) Coal; (3) Petroleum.

Levinstein was here referring to the raw materials used to produce macromolecules, the synthetic polymers that followed the introduction of Bakelite and the related Novolak (phenol-formaldehyde products):

> Plastic substances serve many different purposes. They are, however, for the most part derived from very simple and very similar raw materials. Thus the three sources of raw material overlap in the most remarkable fashion. This is not the case with the older organic industries, based on coal tar, which present a close analogy in some respects. In these a simple substance like benzene may be converted into an astonishing variety of dyes or maybe pharmaceuticals, perfumes, or photographic chemicals, but they cannot be made from things that grow or from petroleum. *In the plastics industry alcohol may be said to play the part of benzene.* Alcohol is easily converted into acetic acid, acetic anhydride, acetone, or even ethylene. Ethylene is used not only as a raw material for plastics, but for making mustard gas. So too, ether and acetone are used in the manufacture of propellants. [Emphasis added.] [4]

Ethylene was the building block of ICI's newly commercialized plastic known as polythene. Levinstein went on to remind British readers that even the source of ethylene had changed: "All the ethylene used during the war in this country was made from alcohol obtained by fermentation. Ethylene is, however, obtainable in enormous quantities from cracked (vapour phase) petroleum or from natural gas." Another source, and one that would permit self-sufficiency, was suggested by Levinstein. This was the large-scale production of calcium carbide from coal:

> You will see on the walls the well-known acetylene tree diagram. This shows the wide range of products that can be made from calcium carbide and are made elsewhere – alas, not here – alcohol, acetaldehyde, acetic acid, ether, acetone, ethylene, the vinyl products, solvents, plastics, artificial silk, resins, rubbers ... Calcium carbide is the only substance which provides these indispensable commodities without requiring one ton of shipping. It is the only route by which they can be made without adding a penny to our already unfavourable balance of trade. It is the only material which costs neither blood nor treasure in war time ... Our imports of calcium carbide were last year about 70,000 tons. They are increasing.[5]

---

[4] Herbert Levinstein, "The impact of plastics on industry: The need for national planning," *Chemistry and Industry*, **58** (8 February 1939), 189-191.

[5] Ibid., p. 191. Manufacture of calcium carbide was originally carried our for lighting purposes, and was

At that time, calcium carbide was imported from Norway, a source that would soon be cut off. From 1935, there had been attempts by several British firms to gain government support for carbide production. The principal proposal was for a plant with an output of 20,000 tons of calcium carbide at Llannerch-Y-Mor, near Mostyn, Flintshire, to be constructed by the North Wales Carbide Corporation Ltd. Despite Levinstein's warning that "[y]ou can shirk for four years putting down a carbide plant; you may not be able to shirk the consequences," this scheme met with no success. The outcome was that during World War II the government was forced to install a calcium carbide plant in South Wales, close to sources of coal and limestone.

The most important products required in the manufacture of the big molecules until the 1930s were phenol and formaldehyde. In the former case, coal tar phenol alone no longer sufficed. The price had risen so much as a result of expansion in Bakelite manufacture that it was found worthwhile to manufacture phenol from benzene (synthetic phenol), as had been employed in World War I. In March 1939, H.V. Potter, managing director of Bakelite Ltd., explained to the Plastics Group of the Liverpool Section of the Society of Chemical Industry:

> Before the war the most important uses for phenol were for the dye, pharmaceutical, and propellant industries, and ample supplies were available from natural [that is, coal] sources ... For the first time, in 1937, the demand for phenol exceeded the natural supply, and a synthetic plant was established which produced 1,500 tons, the quantity available from natural sources being 4,000 tons. The plastics industry alone absorbed 6,000 tons in that year, and phenol had, as a consequence, to be imported. Increased production of phenolic resinoids in this country caused still further demands to be made for phenol, with the result that prices fluctuated to an alarming extent. The position was stabilized by the increased output of the synthetic plant, and the increased supplies of phenol available today can be attributed to the four following reasons:
> (1) Increase in coke ovens;
> (2) Increase in synthetic plant;
> (3) New sources from hydrogenation of coal; and
> (4) Better extraction from waste liquor.

Formaldehyde was essential not only for the production of Bakelite, but also for some of the important plastics introduced, in part, because of the "difficulty of producing water-white sheet material from phenolic resinoids [Bakelite]." This had led to the "development of urea plastics in 1924 and later vinyl, styrol, and acrylic resins." In

---

widely used by railways and cyclists. It remained an important source of lighting in countries where the construction of coal gas pipelines and electricity networks were not economically viable. This, of course, was not the case in Britain.

Britain, amino-plastics and Bakelite were responsible for the major part of consumption, with the result that "[t]he plastics industry is one of the largest users of formaldehyde." The source of formaldehyde was methanol: "This was of course formerly obtained from the destructive distillation of wood, but is now produced synthetically from [the high pressure reaction between] carbon monoxide and hydrogen."[6] In 1939, the main producers of formaldehyde for the plastics industry were Synthite Ltd., of West Bromwich, and B.N.R. Ltd., of Barking, Essex (whose selling agent was Victor Blagden & Co. Ltd.).

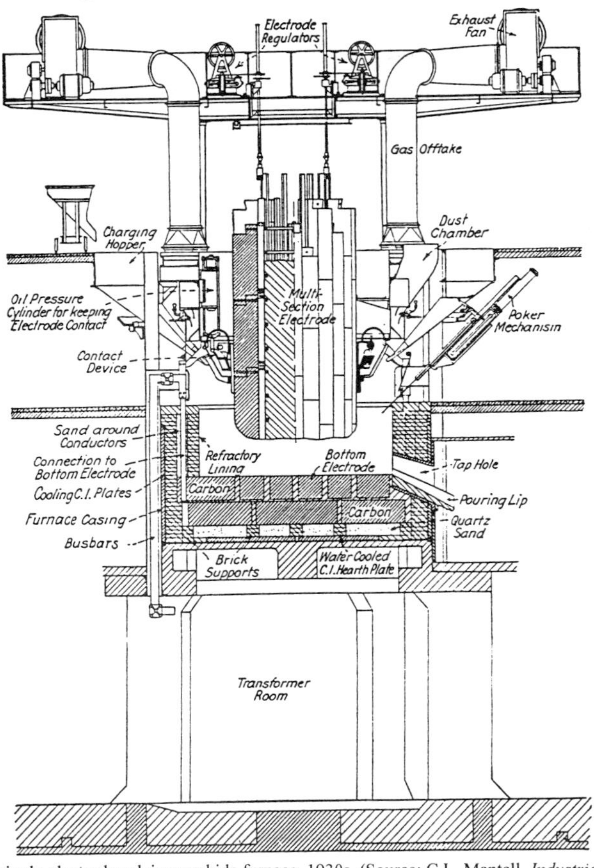


Single-phase, single-electrode calcium carbide furnace, 1930s. (Source: C.L. Mantell, *Industrial Electrochemistry* [New York: McGraw-Hill, 2nd edition, 1940], p. 513.)

6 "Plastics group and Liverpool section," *Chemistry and Industry*, **58** (8 April 1939), 309-310, on p. 309.

**Resins, Plastics and Adhesives from Calcium Carbide and Formaldehyde**

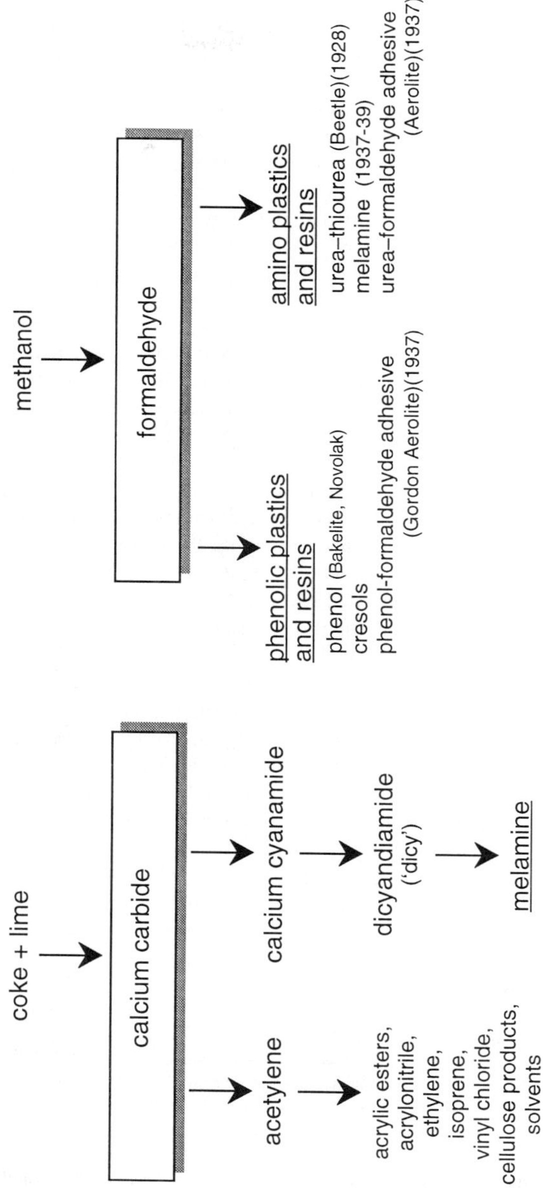

**New needs**

It is self evident that novel products must satisfy human needs if they are to become the basis of commercial manufacturing processes. The range of organic products changed as a result of World War I, particularly with the increased use of electrical equipment and motorized transport. Further, the absence of previously abundant raw materials, as well as the fear of shortages, stimulated research and development in both Allied and German academic and industrial laboratories. Work on the production of nitric acid from synthetic ammonia, and on synthetic petroleum and synthetic rubber, underscored the new needs, particularly the growing reliance on new industries. It mattered little that the basic chemistry was not understood. It was more important that the new products could be easily processed by users.

The German chemical industry understood this well; indeed through its attention to the needs of dyers and textile printers it could draw on over half a century of experience in fashioning products to the needs of users. This paved the way to economic success, especially for the synthetic polymeric materials that entered the market place from around 1910, and in the two decades after 1918.[7]

**Phenol: From explosives to plastics**

Phenol was a versatile aromatic compound that found widespread use in dyestuff manufacture and as an antiseptic. From the end of the 19th century it was employed extensively in the manufacture of explosives. The original source was coal tar. Coal tar benzene was also converted into phenol, via benzene sulphonic acid, at Read Holliday, of Huddersfield, during the Boer War and World War I. The main use of phenol from 1914 was for picric acid or lyddite. From 1915 to 1919, British synthetic plants operated at Ruabon, North Wales, and Northwich, Cheshire. More significant was the route to "synthetic phenol" via chlorobenzene, adopted in Germany before World War I. In 1915 a substantial manufacturing facility for synthetic phenol was opened in the United States. This partially freed America from dependence on British phenol and, along with the manufacture of aniline and other aromatic compounds, represented the first stage of the synthetic organic chemicals industry in the United States.[8]

With the end of war, the enormous surplus of phenol caused the price to fall from "one shilling and threepence per pound to something below sixpence per pound." Within a short time this was followed by an increase in demand for Bakelite production. By 1922, the surplus in the United States had become a shortage. In Britain the price fluctuated "violently over the years 1920 to 1927." Japan was the largest buyer of

---

7 For a historical survey of plastics, see S.T.I. Mossman and Peter J.T. Morris, eds, *The Development of Plastics* (Cambridge: Royal Society of Chemistry, special publication no. 141, 1994).

8 Kathryn Steen, "Wartime Catalyst and Postwar Reaction: The Making of the United States Synthetic Organic Chemicals Industry, 1910-1930," University of Delaware Ph.D. thesis, 1995. UMI Microfilm no. 9610494 (1996).

British phenol, used for sulphur black dyes, and later for plastics. The price was not to be stabilized until the early summer of 1927 when there began discussions among British manufacturers that ended with the formation of the Phenol Convention (the Phenol Producers' Association) on 1 October 1927, "[t]he basic schedule price being agreed at sixpence-farthing per pound." This remained in force until 1930, and in the meantime tar distillers produced "phenol and cresols [homologues, that is chemical relatives, of phenol] to close-cut specifications, with control and elimination of impurities to a degree as yet undreamed of." As the demand from Japan fell, following the commencement of phenol manufacture there in 1930, demand from British plastics manufacturers increased. Phenol production in 1927 had been around 3,000 tons; it had more than doubled by 1939.[9] The increasing use of continuous vertical retorts in British gas works did not favour the extraction of phenol (nor of naphthalene and anthracene), and there was a reliance on coke ovens and synthetic phenol.[10] Major phenol producers were Yorkshire Tar Distillers, Ltd. (founded in 1906), and Monsanto, at Ruabon.

**Phenol-formaldehyde resins: Bakelite**

Leo Hendrik Baekeland's process for the phenol-formaldehyde polycondensation resin that he called Bakelite was patented in 1907 and commercialized by his General Bakelite Company, established in Perth Amboy, New Jersey, in 1910. A British factory, a subsidiary of Bakelite's German holding company, was opened at Cowley, Middlesex, in 1914. Baekeland also coined the name Novolac, which at one time was used as the trade mark for Novolak phenol-formaldehyde resins, following his observation that they behaved like lac resins. Phenol-formaldehyde resins were found to be thermosetting (hard and unmouldable when heated), and formed by condensation:

Excess phenol → dioxydiphenylmethane → Novolak

Excess formaldehyde → Bakelite A (liquid while hot and soluble) → Bakelite B (viscous and partly soluble) → Bakelite C (insoluble and fusible)

Baekeland's success derived from the fact that he had managed to perfect a process which had been much studied after 1900. By careful control of the proportions of phenol and formaldehyde, and recognition of the catalytic action of acids and bases, he produced a homogenous product.

The story of manufacture in England began at the turn of the century when the

---

9 It is of interest that during World War II, British benzene for conversion into phenol was sent to the United States under the Lend-Lease arrangement, which made Britain somewhat dependent on external supplies after 1945. Whereas in World War I the main use of phenol was in explosives manufacture, in World War II it was in plastics manufacture.

10 H.W. Graesser-Thomas, "The story of phenol: From small beginnings to a thriving industry," *British Plastics*, (October 1948), 456-460, 470.

consulting engineer Sir James Swinburne investigated improved insulation materials for electrical cables. He acquired rights to Adolf Luft's 1902 patent for phenol-formaldehyde resins, and formed the Fireproof Celluloid Syndicate in London during 1904. Swinburne improved on Luft's work, but delayed his patent application until 1907, only to find that his invention was covered in Baekeland's patent of the same year. Swinburne's syndicate was dissolved in 1910, from which year he traded as the Damard Laquer Co. Ltd., which made phenol-formaldehyde laquers in Birmingham. In 1916 Damard took over the Bakelite Cowley factory, which had closed down at the outbreak of the war. Until 1918, the entire output served the electrical needs of the British government. New uses after the war included cigarette holders, tea pot and umbrella handles, and cabinets for radio sets, all strategic products in most British homes. Phenol-formaldehyde laminated sheets were manufactured from 1922.

Bakelite was the first fully synthetic plastic. It was the main plastic of the inter-war period, finding enormous application as an insulator in the new electrical and car industries, and also in a range of components for cars, household and other purposes. The use of Bakelite in the rapidly emerging British car industry led to the construction of a new factory near Birmingham during 1920-21 (from which time the Cowley and Birmingham plants were run down). From 1924, Damard specialized in the new fast-curing phenol-formaldehyde moulding powders, perhaps the most important advance in the history of Bakelite and similar products.[11] In 1926, the British Bakelite Company was formed to bring together the firms in the phenol-formaldehyde sector, as had happened in the United States, following the end of patent suits (1922).[12] These were Damard, Mouldensite Ltd., of Darley Dale, Derbyshire (founded in 1921 as a subsidiary of the USA-based Condensite Co.), and Redmanol Ltd., of London (subsidiary of the US Redmanol Chemical Products Company). Swinburne was chairman of the new company, which wisely became involved in application as well as in production. From 1928 a new works was erected at Tyseley, Birmingham. Bakelite enabled tremendous advances in the electrical engineering industry, and despite the reliance on Malayan rubber for use in telephone and electrical cables, did encourage research into similar

---

11 The demand for Bakelite is discussed in Wiebe E. Bijker, "The social construction of Bakelite: Toward a theory of invention," in Wiebe E. Bijker, Thomas P. Hughes and Trevor J. Pinch, eds, *The Social Construction of Technological Systems* (Cambridge, Mass.: MIT Press, 1987), pp. 159-187. A full account of the early history of Bakelite will be found in Wiebe E. Bijker, *Of Bicycles, Bakelite, and Bulbs* (Cambridge, Mass.: MIT Press, 1995), pp. 101-197. An excellent survey of the impact of plastics during the interwar period, though almost exclusively related to the United States, is Jeffrey L. Miekle, *American Plastics: A Cultural History* (New Brunswick: Rutgers University Press, 1996).

12 Percy Reboul, "Britain and the Bakelite revolution," in Mossman and Morris, op. cit. (7), pp. 26-37, and Leo Hendrik Baekeland, "Impress of chemistry upon industry: Bakelite, an example," *Industrial and Engineering Chemistry*, **27** (May 1935), 538-543, esp. p. 539. Baekeland pointed out that United States production of phenol, "most of which is now made synthetically, either from benzene-sulfonic acid or by the hydrolysis of chlorobenzene," had grown from 3,310,991 pounds in 1923 to 24,177,000 in 1929, "and the average price per pound has decreased from 34 cents in 1923 to 14.25 in 1933" (p. 539), which makes an interesting comparison with the figures presented by Graesser-Thomas.

novel compounds suited to the electrical industry. At the end of the 1930s, Bakelite was applied to railway carriage and ship fittings, including on the London Midland & Scottish Railway's 'Coronation Scot' train that toured the United States in 1939, and on the ss 'Mauretania.' By this time, injection moulding of resins was in widespread use, a process hardly used a decade earlier.

**Amino plastics: urea-, thiourea- and melamine-formaldehyde products**

In the early 1920s, the British Cyanides Company Ltd. of Oldbury faced bankruptcy from its highly unprofitable cyanide manufacture. Desperate, it turned to production of new chemicals. One of these was thiourea, made from ammonium thiocyanate, as used in the company's cyanide process. Edmund Charles Rossiter, chief chemist at the firm, undertook research into the condensation of thiourea with formaldehyde, and produced an organic glass-like condensate of urea. The reaction of formaldehyde with thiourea was found to differ considerably from that with urea and though no satisfactory resin was produced, some "fascinating applications in tiles and laminates showed the possibilities and attracted considerable attention at the British Empire Exhibition at Wembley in 1924." Eventually, in 1925, Rossiter found that a promising material resulted from simultaneous condensation of both urea and thiourea with formaldehyde. The complex condensate was used to produce a successful moulding powder. The new product was displayed before the public for the first time in December 1926 at Harrods store in Knightsbridge as items of moulded tableware for the Christmas season. It was marketed in 1928 as Beetle.[13] After Bakelite, this was the second most important primary development in the field of thermosetting plastics. Bakelite plastics, however, had one big disadvantage in that they were brown or black only. The urea "condensates were translucent but not *transparent*, and with their commercial success as moulding powders the search for glass-like materials in the urea-formaldehyde field was finally abandoned."[14] Laminated sheets made up from urea-formaldehyde resins became available in 1931. Two years later, some 400-500 tons of these new aminoplastics were produced. Since they were thermosetting, rapid hot moulding was possible. Demand came from the electrical and car industries, as well as from manufacturers of household goods who could now offer a range of coloured plastics.

The international success of Beetle plastics was reflected in the negotiations for licenses with firms in France, Spain, Holland, Germany, Switzerland and the USA, and the considerable amount of litigation over licensing and patent rights, especially during 1933-34, in the United States and elsewhere. Thus in Germany during 1934, the holders

---

[13] For the early history of Beetle see lecture given by Kenneth M. Chance, managing director of British Cyanides, at the Science Museum, London, on the occasion of the 1933 plastics exhibition: "The Plastics Exhibition, 1933. Twelfth lecture. Plastics: Urea-formaldehyde types and their uses," *British Plastics and Moulded Products Trader*, **5** (1934), 313-318. This journal was founded in 1929, as the official organ of the British Plastic Moulding Trade Association (from 1933, the British Plastics Federation).

[14] D.W.F. Hardie and J. Davidson Pratt, *A History of the Modern British Chemical Industry* (Oxford: Pergamon, 1966), p. 199.

of the Beetle patent rights were engaged in a lawsuit with CIBA of Basel over the latter's Cibanoid urea-type moulding powders. In the same year many of the disputes were resolved, including with ICI. British Cyanides' Beetle process was licensed to American Cyanamid, which, from 1930, was the fourth largest US chemical corporation, and anxious to diversify away from agricultural products, especially during the Depression years.[15] "American technicians visited Oldbury to see the process, plant was quickly erected at Bound Brook, New Jersey, and production started within the space of a year."[16] In 1935, Societa Italiana Resine began the manufacture of 'Beetle' in Italy. Success was also reflected in the change of British Cyanides' name, which in 1936 became British Industrial Plastics Ltd. (which traded with the associated Beetle Products Co. Ltd.). Furthermore an English invention had opened up the new field of commercial aminoplastics, which soon became an international endeavour. British Industrial Plastics' American associate, American Cyanamid, was to become synonymous with the next major advance, the melamine plastics and resins.

The triazine melamine (2,4,6-triamino-1,3,5-triazine), was first isolated, and named, by Liebig in 1834. The triazine formula was adopted in 1902, following the work of Diels, and Chattaway and Wadmore. The industrial reaction for preparing melamine was perfected at the end of the 1930s by American Cyanamid, and was, significantly, in view of British lobbying for carbide plant as described by Herbert Levinstein, based on calcium carbide.[17] Since melamine and urea contained highly reactive amido/amino groups the mode of condensation with formaldehyde was the same. Further, while the melamine polymers were similar in many respects to urea-formaldehyde condensates, they displayed greater resistance to moisture and heat. Melamine-formaldehyde resins and butylated melamine resins were developed in Britain and elsewhere during 1938. I.G. Farben, Henkel and CIBA also used melamine as the basis of new resins formed by condensations with formaldehyde.[18] The potential offered by melamine resins and

---

[15] In the United States, Beetle was made by the Synthetic Plastics Co., in which British Cyanides was originally a shareholder, but which by 1933 was mainly owned and controlled by American Cyanamid.

[16] W. Blakey, "The history of aminoplastics: The sixth Chance Memorial Lecture of the Society of Chemical Industry," *Chemistry and Industry*, (25 July 1964), 1349-1357, on p. 1352. Blakey received a D.Phil. in organic chemistry from Kings College, Cambridge, in 1928, and shortly after joined British Cyanides to undertake research into aminoplastic moulding compounds. Blakey was appointed chairman of British Industrial Plastics in 1962, one year after the firm was merged with Turner & Newall Ltd. This lecture is an extremely useful source of information on aminoplastic developments. See also C.P. Vale and W.G.K. Taylor *Aminoplastics* (London: Iliffe Books, 1964), which includes a historical introduction, and a portrait of Hanns John who in 1918 patented the first industrial use of urea-formaldehyde resins.

[17] Other processes had been patented by Henkel in Germany (1934), by CIBA, in Europe, the United States and Canada (from 1937), and by I.G. Farben, in France (1938).

[18] CIBA had developed Cibanoid urea-formaldehyde, Cibanit aniline-formaldehyde, and melamine-formaldehyde resins. See Hans Stäger, "Modern organic materials at the Swiss National Exhibition Zurich, 1939," *British Plastics and Moulded Products Trader,* **11** (1939), 254-257.

melamine laminates aroused great interest during 1939 in Britain, where it was noted that novel products were obtained by undertaking the condensations in the presence of phenol and aniline.[19] Melamine plastics represented a tremendous advance in the discovery of aminoplastics, and, because of their superior moulding and mixing properties, for certain purposes they superceded those based on urea-thiourea.

**Adhesives and Aero Research Ltd.**

In 1934, British Industrial Plastics introduced a urea-formaldehyde adhesive based on I.G. Farben technology ('Kaurit W'). It could be employed under milder conditions than phenolic adhesives then used in plywood manufacture. In the same year, Dr. Norman A. de Bruyne, physicist and Fellow of Trinity College, Cambridge, founded Aero Research, Ltd., Duxford, Cambridgeshire. As part of a programme of novel aircraft construction, a low-wing cantilever monoplane was assembled at Duxford in 1935. It was called the Snark, and later purchased by the Air Ministry. De Bruyne's research as a physicist suggested that synthetic resins might be used in aircraft manufacture. De Bruyne engaged Cambridge University chemist Dr. R.E. Clark to investigate novel adhesives for aircraft structures. The outcome was the 'Aerolite' urea-formaldehyde adhesive, which, unlike conventional glues, resisted water and microorganisms. Claude Rayner, later research manager at de Bruyne's firm, took charge of further developments, and found that gap bridging (GB) hardeners, incorporating formic acid, enabled Aerolite to be employed as an assembly glue.[20] Aerolite was the first adhesive of its type invented and manufactured in Britain, and was used in resin-bonded plywood.[21] In May 1937, de Bruyne was awarded the Royal Aeronautical Society's Simms Gold Medal for his lecture on "Plastic materials for aircraft construction," which described the use of phenolic resins as the basis for aircraft structural components. Later in the year Aero Research's 'Gordon Aerolite' adhesive, based on a phenolic resin, was introduced.[22] This was followed in 1942 by the complex

[19] See reviews, "Melamine and its potentialities in plastics manufacture," *British Plastics and Moulded Products Trader*, **10** (1939), 27-28, and "Melamine resins," *British Plastics and Moulded Products Trader*, **11** (1939), 326. For an early account of production of the monomer, see P.P. McClellan, "Melamine preparation," *Industrial and Engineering Chemistry*, **32** (September 1940), 1181-1186.

[20] *Fifty Years at Duxford: Ciba-Geigy Plastics* (Duxford: Ciba-Geigy, 1984).

[21] Norman A. de Bruyne, "The establishment and development of Aero Research Limited," in Ronald S. Edwards and Harry Townsend, *Business Enterprise: Its Growth and Organisation* (London: Macmillan, 1964), Appendix I, pp. 103-111; and Elisabeth Garnsey, "An early academic enterprise: A study of technology transfer," *Business History*, **34** (1992), 79-98.

[22] See for example, Norman A. de Bruyne, "Plastics materials for aircraft construction," *British Plastics and Moulded Products Trader*, **10** (1939), 515-519, summary of a lecture given to the Royal Aeronautical Society on 4 January 1939. By 1938, De Havilland Aircraft Co., Ltd. was using Bakelite on a large scale in aircraft construction, and in 1939, Du Pont, Bendix Aviation Corporation and Atlas Corporation formed a joint venture to manufacture plastic parts for aircraft in the United States. A useful phenol-formaldehyde plywood adhesive was Tego, the invention of Theodore Goldschmidt of Essen, which appeared in England and the United States during the 1930s. An account of Tego will be found in Sheldon Hochheiser, *Rohm and Haas: A History of a Chemical Company* (Philadelphia: University of Pennsylvania Press, 1986), pp. 49-50,

bonding agent 'Redux.'

With these products reliable metal-metal and metal-wood bonding became possible. 'Gordon Aerolite' saw application in the Spitfire, and Redux in the de Havilland Hornet and the Royal Navy's version of the Mosquito. In 1938, the company had devised a honeycomb panel, which subsequently became a useful lightweight core material once strong metal adhesives for aluminium became available. In 1938, also, various investigators, particularly at Bakelite Ltd., were considering the use of plastics for the manufacture of aircraft parts, such as airscrews and spars, windscreens, and fuselages. Within a few years observers of the aircraft industry predicted that the time would come when the "all-metal, streamlined" plane would incorporate plastic components.[23] British Industrial Plastics developed a phenolic gap bridging hardener early in World War II; it was kept on the secret list at the demand of the Air Ministry, and used in the manufacture of aircraft, gliders, tank-landing craft, pontoons, and motor launches.

After the war, Aero Research manufactured CIBA's Araldite and melamine products (licensed by American Cyanamid to CIBA) at Duxford, and in 1948 was taken over by CIBA.[24] This had been a case of the successful introduction of chemical technology into an enterprise that had no prior experience in chemical manufacture. Through its expertise as user and producer of new-technology resins and adhesives, Aero Research eventually directed all its efforts towards their production.

**A new corporation**

Britain's ICI (Imperial Chemical Industries) was formed in 1926 by the merger of Brunner, Mond & Co. Ltd., United Alkali Co., BDC (British Dyestuffs Corporation), and Nobel Industries Ltd. Both BDC and Nobels were the outcomes of mergers and acquisitions encouraged by wartime conditions. BDC was created in 1919 by the merger of British Dyes Ltd. (formed, with government help, by Read Holliday of Huddersfield in 1915), and Levinstein Ltd., of Blackley, Manchester (this fusion was also government-inspired). Nobels was the outcome of a 1918 merger of several companies, originally known as Explosives Trades Ltd. The new name was adopted in 1920. Brunners and BDC had embarked on programmes of diversification and further acquisitions.[25] In 1925, BDC purchased Scottish Dyes Ltd., created in 1917 with

---

72-73.

23 Vernon J. Clancey, *Chemistry and the Aeroplane* (London: Thomas Nelson, 1942)

24 "Progress in plastics. British adhesives: Exhibition and demonstration at Aero Research, Ltd.," *British Plastics*, (October 1948), 500-501; C.J. Moss, "Araldite: A new adhesive, coating and casting resin," *British Plastics*, (November 1949), 521-527; Norman de Bruyne, *Norman de Bruyne: My Life* (Cambridge: Midsummer Books, 1996).

25 Thus in 1917, Brunners acquired a controlling interest in Chance & Hunt; in 1919 a controlling interest in Buxton Lime Firms Ltd., and full ownership of Ammonia Soda Co.; in 1920 Castner-Kellner became a

government assistance to ensure the future of Solway Dyes Ltd. (the dye-making side of Morton Sundour Fabrics).[26]

This new corporation, a response to the rapid growth of the United States' Du Pont, Union Carbide & Carbon Corporation, and Allied Chemical & Dye Corporation, and the creation of Germany's I.G. Farben, increasingly dominated the British chemical industry through cartel arrangements with foreign firms, preferential trade agreements with the Empire, and government protection, which included subsidies. The latter arose from the perceived need for a strong national chemical industry, one that could maintain Britain's strength as a trading nation and that in time of war could be mobilized rapidly. Nevertheless, the mergers were not without problems, both organizational and technical. These were aggravated by the absence of long term vision and technically qualified and experienced men in the boardroom.

In this context, it is not without interest to note that Herbert Levinstein, trained as a chemist and sole or joint holder of some 25 chemical patents, was voted off the board of BDC in 1922, mainly because of his criticisms of the Huddersfield division (the former works of Read Holliday). During World War I, Levinstein Ltd. had successfully manufactured indigo, novocain and acriflavine at the sequestered Ellesmere Port factory of Hoechst, and mustard gas, as well as dyes and intermediates, at the original Levinstein Ltd. Blackley factory. Herbert Levinstein also engaged the services of the academic chemist Arthur G. Green, who built up a research team at Blackley that had close links with several British university chemistry departments. (In 1916, Green moved from the University of Leeds, where he had been professor of tinctorial chemistry, to the University of Manchester, where he was appointed associate professor of chemical technology.) However, the type of research organization fostered by wartime conditions did not appeal to Levinstein's partners. ICI's successes of the mid 1930s, such as phthalocyanine dyes and polythene took place more by accident than by design, mainly as a result of the insight and awareness of research staff. Such insight and awareness was, as demonstrated by British developments in plastics and adhesives, not the sole preserve of ICI. Nevertheless, through strategic national and international alliances, diversification of research, control of patents, and its substantial resources, ICI did continue to hold sway over many markets in organic chemicals. This was demonstrated by strong interests in dyes and, as elsewhere, in fibres and insulating and moulding materials. These interests merged, quite by accident, with the discovery of polythene, which was also influenced by ICI's research into high pressure processes. Changes in demand and production problems did, however, pose considerable risks.

---

subsidiary, Electro-Bleach & By-Products Co. was purchased, and Synthetic Ammonia & Nitrates Ltd. was formed.

[26] For a full account of the prehistory, and early development, of ICI, see William J. Reader, *Imperial Chemical Industries. A History.* 2 volumes (London: Oxford University Press, 1970).

Thus, ICI's foray into synthetic fibres was not successful until after 1945.

**Synthetic fibres: Celanese, Ardil and Nylon**

Among the surplus products available in abundance after 1918 was cellulose acetate, used during the war as a dope for aircraft wings, for transparent panels inserted in the upper wings of aircraft (so that pilots had a clear view above), and as a film for covering the gaps left by shattered windows. This had been manufactured by the Safety Celluloid Company of Park Royal, close to Ealing, West London. From 1916, there was a second manufacturer, British Cellulose and Chemical Manufacturing Company, at Spondon, set up by the Swiss brothers Henri and Camille Dreyfus.[27] Using the surplus cellulose acetate, British Cellulose embarked on the manufacture of what was known as Celanese fibre. Later the firm became known as British Celanese Ltd. The success of cellulose acetate fibres depended on the discovery of suitable dyestuffs for their coloration. The outcomes were the first disperse dyes, the Ionamines, based on aminoazo compounds, discovered in England at BDC during 1922-23, and applied as finely divided aqueous suspensions. In 1923, BDC and British Celanese together developed anthraquinone-based disperse dyes, which extended the range of available colours considerably. British Alizarine Company, Scottish Dyes, and, from the mid 1930s, Yorkshire Dyeware & Chemical Co., Ltd., also undertook research into disperse dyes. Du Pont purchased rights to the British Celanese technology in 1926.[28]

In addition to acetate 'rayon,' there was considerable expansion in the manufacture of viscose 'rayon,' also based on cellulose, and originally developed in Britain during 1891 by Charles F. Cross, Edward J. Bevan and Clayton Beadle. Herbert Levinstein, following his enforced resignation from BDC, became involved in the manufacture of this artificial silk, as founder-director in December 1925 of the Neura Art Silk Co. Ltd., and from 1928 as technical director of Lansil Ltd. Courtaulds Ltd., a major manufacturer of viscose rayon, began manufacture of acetate fibre in 1928, but two years later was engaged in litigation over patents with British Celanese. Novel uses for cellulose acetate were in the production of gramophone records (instead of shellac composition), which was well established in the United States by 1933.[29]

By this time there was considerable understanding of the chemical constitution of polymers following the work of Hermann Staudinger, who as a consultant to BASF around 1910 studied isoprene, the building block of rubber. His interest in the constitution of the natural product led him on a path that in May 1922 culminated in the

---

27 Hardie and Davidson Pratt, op. cit. (14), pp. 79-80.

28 The fibre from cellulose acetate was also known as acetate silk. In general, it was called Celanese in Britain and Lustron in the USA. The term 'rayon' was introduced in 1924.

29 British Celanese was acquired by Courtaulds in 1957, which during the ensuing years bought up smaller manufacturers and merchants.

announcement that the "Macromolekül" was formed by linkages of molecules through normal chemical bonds. Despite much initial skepticism from academics, Staudinger's theory was generally accepted after extensive studies by methods of physical organic chemistry.[30] The implications for industrial research were considerable. Thus in 1935, Professor W.T. Astbury, director of the Textile Physics Laboratory, Leeds, and Professor A.C. Chibnall, of Cambridge University, suggested to ICI that textile fibres might be produced by dissolving vegetable protein in urea and extruding the solution into a suitable curing bath. At that period ICI was involved in research connected with the British Empire's natural resources, in particular the government sponsored groundnut scheme. Some eight million tons of groundnuts were annually imported into Britain for use in margarine manufacture. The residue, after removal of oil, contained a high proportion of vegetable protein, which, it was thought, would provide a good source of starting material for production of the proposed fibre. The two inventors and ICI signed an agreement for development of the process and experimental work was undertaken by ICI, under the direction of Dr. David Traill, at its Nobel Division, Ardeer, from which place the name of the new fibre, Ardil, was taken. In 1938, some of the fibre, produced in a newly constructed pilot plant, was woven into suits, but the properties were not at that time attractive. Work was stopped until after World War II.[31]

More significant was the formation early in 1939 of a joint company by ICI and Courtaulds Ltd. "to exploit du Pont de Nemours' nylon ... and it was stated that nylon yarns and fibers manufactured by the joint company will, in due course, be made available to existing silk manufacturers and other textile industries."[32] .

**Phthalocyanine dyes: Serendipity at ICI and academic research**

During the early 1920s, dyestuffs research at BDC was in decline, partly as a result of board room differences. Herbert Levinstein, whose wartime successes included technical assistance to Du Pont, had hoped to gain the upper hand and create a dye-manufacturing industry that would compete on equal terms with the dye industry of Germany. Flushed with success, over-confident, and highly critical of anything that was not of Levinstein Ltd. origins, he generated so much antagonism that at the earliest opportunity he was ejected from his position as co-chairman of BDC (in 1922). Harmony may have returned to the executive suites and dining rooms, but not to the research laboratories. In protest, Arthur Green resigned as director of research in 1923.

---

30 A brief but highly readable account of polymer history is Peter J.T. Morris, *Polymer Pioneers* (Philadelphia: Center for the History of Chemistry, 1986).

31 *Industrial Chemist*, (August 1957), pp. 401-407. See also David Traill, "Industrial fibres and their possible production from synthetic resins," *British Plastics and Moulded Products Trader*, **10** (1939), 488-494.

32 "I.C.I.-Courtaulds combine to produce Nylon in England," *Industrial and Engineering Chemistry. News Edition*, **17** (10 May 1939), 320

For the next decade, completely new innovations in dye-making, apart from disperse dyes, were confined mainly to Scottish Dyes Ltd., over which ICI gained full control, through BDC, in 1928. Around 1932-33, Arthur Dandridge, working at the Grangemouth works with three colleagues, Dunworth, Drescher and Thomas, found that a blue impurity was obtained when phthalimide was prepared from phthalic anhydride and ammonia. They "realized that, despite a dull and unattractive colour, the general stability and novelty of the substance under examination was highly interesting."

The product was found to contain iron, and became the focus of academic study at Imperial College, London, where in 1933-34, R.P. Linstead and co-workers proposed that the constitution was similar to those of the pigments in chlorophyll and blood. It was in fact a phthalocyanine compound, of the type first prepared in 1907 (by A. Braun and J. Tcherniac). The structure was established in 1935 by J.M. Robertson's x-ray analysis, which showed "quite clearly the position of every atom in the molecule and confirms Linstead's structure with absolute certainty."[33] The four industrial chemists at Grangemouth were indeed "distinguished from earlier workers who undoubtedly had the phthalocyanines in their hands, but failed to realize their potential importance." ICI (Dyestuffs Group) chemists had already adapted the chance observation to the manufacture of the copper analogue, known as Monastral fast blue. Introduced in 1934, it was, over half a century later, hailed as the most important colorant discovered in the 20th century.[34] In 1934, also, I.G. Farben improved on the process and introduced the rival Heliogen Blue B. A telling comment from ICI was that the I.G. Farben product was "a revelation of the resources they are able to throw into any opening for research, whether provided by their own initiative or by that of their competitors."[35] The German process was far superior, to the extent that ICI acquired a license for its use.[36]

**Polyvinylchloride (PVC), methyl methacrylate (Perspex), and neoprene**

From 1930, ICI undertook research into polyvinylchloride, but despite its popularity in other countries did not embark on manufacture. There was, understandably, no reason to do so, since natural rubber for electrical insulating applications was available in abundance, and calcium carbide, the source of starting acetylene, was, as Herbert

---

33 N.H. Haddock, "Phthalocyanine colouring matters - Their chemistry and uses," *Journal of the Society of Dyers and Colourists*, **61** (March 1945), 68-73.

34 Heinrich Zollinger, "Some thoughts on innovative impulses in dyeing research and development," *Textile Chemist and Colorist*, **23**, no. 12 (December 1991), 19-24.

35 Reader, volume 2, *The First Quarter Century*, op. cit. (26), p. 332.

36 For other developments in Britain see Maurice R. Fox, *Dye-makers of Great Britain, 1856-1976: A History of Chemists, Companies, Products and Changes* (Manchester: ICI, 1987). For an international overview see Peter J.T. Morris and Anthony S. Travis, "A history of the international dyestuff industry," *American Dyestuff Reporter*, **81**, no. 11 (November 1992), 59-100, 192-195.

Levinstein pointed out in 1939, imported.

The acrylate plastic story was rather different. Acrylate polymerization had been studied at McGill University, Montreal, by William Chalmers who realized that the products offered the possibilities of an organic glass. Rohm & Haas in the United States and ICI in England took great interest in the processes and products. ICI researchers found that methyl methacrylate possessed a higher softening temperature than the corresponding ethyl compound. A patent for moulded articles made from polymerized methyl methacrylate was filed in 1931, and the process became viable when in 1932 ICI's John W.C. Crawford discovered an economical way of manufacturing the monomer from acetone, hydrocyanic acid, methyl alcohol and sulphuric acid. Production began in 1934, and two years later ICI licensed Crawford's process to Du Pont, in accord with the terms of a 1929 exchange of knowledge agreement between the two firms. The ICI technology was also licensed to Röhm of Darmstadt, whose product was known as Plexiglas from 1936, in return for a license to manufacture cast sheet in Britain.[37] Shortly before World War II, ICI managed to produce sheets of the polymeric material, by then known as Perspex.[38] The bulk of production until 1945 was employed in aircraft construction. Some was used to prepare false teeth, as a substitute for vulcanized rubber.

In 1933, ICI purchased an interest in urea-formaldehyde manufacturer Croydon Mouldrite Limited, forerunner of ICI (Plastics) Ltd. (1938). ICI took up research into urea-formaldehyde polymers at Billingham (and phenol-formaldehyde products at its dyestuffs division).

American polymer technology was employed in Britain as a result of the 1929 agreement between Du Pont and ICI. Notable, in view of Britain's reliance on natural rubber, was the production, from 1937, by ICI of Du Pont's synthetic rubber, neoprene: "The 'rubber' was rather solemnly declared to be incompetitive as regards uses, with natural rubber. The writer has not yet encountered many cases where it would not be extremely competitive. In many instances the price question is no longer important."[39]

---

37 Hochheiser, *Rohm and Haas*, op. cit. (22), p. 58. Hochheiser provides a good account of the history of Plexiglas in Germany and the United States.

38 "Alfred Caress and the growth of ICI," *Chemical and Engineering News*, (24 January 1966), p. 86. Caress was one of two men who first made methyl methacrylate corrugated sheet at ICI. Later he headed ICI's operations in the United States. See also, J.P. Tilley, "Versatility of acrylics," in Mossman and Morris, op. cit. (7), pp. 95-104.

39 "Neoprene enters the British market," *Industrial and Engineering Chemistry. News Edition*, **15** (20 July 1937), 312.

**Polyethylene (Polythene)**

According to Martin Sherwood, the writer on chemical matters who in the early 1980's interviewed the leading participants involved in the discovery of ICI's polythene, this plastic "had no 'logical' place in the historical trend of plastics developments in the 1930 – its appearance could have been delayed by a decade or longer."[40] The story of this achievement combined dyestuff research and high pressure chemistry, ICI's alkali division, the new contract with Du Pont, academic research in Holland, and a change in the focus of research away from high-pressure coal-to-oil conversion based on the German Bergius process. The main participants were the organic chemist E.W. Fawcett, ICI Winnington research leader F.A. Freeth, and physical chemists R.O. Gibson and M.W. Perrin. They came together at Winnington at a time when novel high pressure research topics were being conducted. Sherwood described how:

> Some years before the formation of ICI in 1926, Brunner, Mond's chief chemist, F.A. Freeth, had also argued the necessity for abstract and theoretical work as a basis for new products. This research philosophy led to close contacts between the company and academic institutions, with scientists being sent around the world to gain expertise in new techniques.
>
> Freeth's inclination was strongly towards physical chemistry and both Gibson and Perrin were physical chemists. According to Eric Fawcett, he went to Winnington because 'they wanted one organic chemist and they hadn't got any.' Fawcett joined the company in 1929, when research was being undertaken into [high pressure] hydrogenation of coal as a potential source of oil. In 1931 he was sent to the National Bureau of Standards in the USA to work on the constitution of petroleum. However, he was also very interested in polymers at the time. ICI had given him introductions to some of DuPont's technical staff - the two companies had recently [1929] formed an alliance, which involved exchange of knowledge - and he became friendly with Wallace Carothers, the discoverer of nylon.
>
> When Fawcett returned to Winnington late in 1932, 'it was in disintegration as a result of the recession.' The coal-oil project was being run down and he was transferred to BDC [by this time part of ICI]. However, it was only a few months before he was seconded back to Winnington to work with Gibson on high-pressure chemistry.

An important outsider was Professor Anton Michels, of the thermodynamics laboratory, University of Amsterdam, who undertook work related to the production of dye intermediates under conditions of high pressure. R.O. Gibson had spent several years with Michels.

---

[40] Martin Sherwood, "Polythene and its origins," *Chemistry and Industry*, (21 March 1983), 237-242, p. 237.

> In early 1931 also, Dr. Michels had been invited to talk to members of the [ICI] Dyestuffs Group Research Committee, which included several eminent chemists, among them Robert Robinson. Michels talked to them about the possibility of using pressures in excess of 1000 atmospheres to influence chemical reactions. Later that year, Robinson submitted a list of reactions which might be influenced in this way and lead to useful products.
>
> Because Perrin and Swallow knew of the Dyestuffs Committee's interest in Michels' ideas, they also wrote that 'if work of an essentially organic character be embarked upon which is considered to be of value to BDC in any connection, then an organic chemist from that company should collaborate with those engaged on the work in the Winnington laboratories.' The empirical approach won the day and Fawcett was seconded back to Winnington as the organic chemist who would work on the project with Gibson (Perrin was still in Amsterdam at the time).[41]

This is the background to how Fawcett and Gibson began to study the effects of high pressures on chemical reactions in an attempt to make organic compounds of interest to the dyestuffs division.[42] In one experiment involving ethylene and benzaldehyde, a white, waxy solid was formed that lined the wall of the reaction vessel. The date was 27 March 1933. The product was not then of special interest. The experiment was repeated in December 1935. Unknown to the experimenters, a leak in the apparatus allowed oxygen to enter, with the result that 8 g of the waxy product appeared. Its properties were established, and it was found to be a polymer of ethylene with excellent electrical insulation. A pilot plant was constructed, and by the end of 1938 a ton of the new material had been produced.[43] An undersea cable between the Isle of Wight and the mainland confirmed the suitability of the new polymer as a versatile insulating material. Large scale production began just as World War II broke out, and the insulating properties of 'polythene' proved to be particularly invaluable for use in radar equipment. The original source of ethylene was dehydrated ethanol, based on an ICI process used since 1934 at Huddersfield, where ethylene oxide and ethylene glycol were prepared.[44]

---

41 All quotes from Ibid.

42 Note in this case the British commitment to the application of academic chemistry. Compare with the situation in Germany, where there was some encouragement of pure research in industry, as described by Morris and Reinhardt, this volume.

43 For accounts of the discovery of polythene by participants see R.O. Gibson, *The Discovery of Polythene* (London: Royal Institute of Chemistry, 1964); and M.W. Perrin, "The story of polythene," *Research*, **6** (1953), 111-118.

44 A useful summary of manufacturing developments will be in found in the chapter on polythene in J.A. Allen, *Studies in Innovation in the Steel and Chemical Industries* (Manchester: Manchester University Press,

In 1937 a delegation of I.G. Farben chemists had visited Winnington, interested in ICI's latest technology. Among them was polymer specialist Otto Ambros, who later declared of polythene: "I don't know how we missed it." A year later I.G. Farben was unable to come to an agreement with ICI over sharing polymer technology. "Thus, almost casually, in a boardroom memorandum, a significant military advantage was preserved for Britain. As Bill Reader, the ICI historian, has remarked: 'There cannot have been many business propositions in history which have broken down with such advantage to Great Britain.'"[45] ICI's process was, however, licensed to Du Pont, which undertook large-scale manufacture from 1941. For his services, Fawcett was discharged by ICI in 1938. Following a period of unemployment he joined the Anglo-Iranian Oil Co., where he spent the rest of his career.

**Sulphonamides**

World War I encouraged the increased production of existing and new pharmaceutical products by British chemical, drug and foodstuffs firms, including Allen & Hanburys, Beecham, Boots, Joseph Nathan & Co. Ltd. (of Glaxo Baby Food fame), and the Wellcome Company. The antiseptic value of acriflavine, proflavine, and brilliant green were exploited in Britain, and these compounds were manufactured on a large scale at Levinstein's Blackley works. At the former Hoechst factory situated in Ellesmere Port, Levinstein manufactured Paul Ehrlich's recently discovered "poisoned arrow," the first chemotherapeutic compound, known as Salvarsan. The outcome was that all these firms began to participate in similar types of organic manufacture and research.[46] However, there were no more major advances until 1936, when the drug known as Prontosil was introduced by Bayer. Prontosil was a red azo dye that was biologically active, it appeared, against certain bacterial infections. This finding aroused enormous interest. French workers soon established that sulphanilamide (*p*-aminobenzenesulphonamide), formed by breakdown of the dye, was the curative agent. The preparation of sulphanilamide had been described in 1908, and, therefore, could not be patented. This enabled several companies to undertake its manufacture and engage in research into what were soon known as sulpha or sulphonamide drugs. Outside of Germany, the most successful firm was May & Baker, of Dagenham, Essex, and incorporated in 1890.

---

1967), pp. 7-52.

[45] *Carol Kennedy, ICI. The Company that Changed our Lives* (London: Hutchinson, 1986), pp. 66-67. Chapter 4 covers the development of polythene until 1940. See also G.D. Wilson, "Polythene: the early years," Mossman and Morris, op. cit. (7), pp. 70-86.

[46] Useful corporate histories are: R.T. Davenport and Judy Slinn, *Glaxo: A History to 1962* (Cambridge: Cambridge University Press, 1992); Geoffrey Tweedale, *At the Sign of the Plough: 275 Years of Allen & Hanburys and the British Pharmaceutical Industry, 1715-1990* (London: John Murray, 1990). Comparisons and connections with the Norwegian pharmaceutical industry are explored in Rolv Petter Amdam, "Professional networks and the introduction of research in the British and Norwegian Pharmaceutical industry in the inter-war years," *History and Technology*, **13** (1996): 101-114.

During 1937-38, Arthur James Ewins and coworkers at May & Baker prepared over six hundred compounds related to sulphanilamide and its derivatives. In the spring of 1938, Ewins and M.A. Phillips synthesized sulphapyridine, soon to be known as M&B 693 (since it was the 693rd compound made at Dagenham), a drug highly effective against pneumonia, and sulphathiazol. Sulphapyridine was licensed to the Calco Division of American Cyanamid (Bound Brook, New Jersey), which became the leading producer of sulphonamides in the United States. In all, 3,000 derivatives were prepared at Dagenham, of which five were therapeutically active.

Interest in the new drugs was considerable at ICI, in keeping with its new-found interest in diversified research. In 1936, ICI had created a Medicinals Chemicals Division as a subsection of its dyestuffs research department. The division drew on expertise in coal tar dye chemistry, as did sulphonamide studies, since the reactions used to produce dye intermediates were almost the same as those used to prepare many pharmaceutical products. This enabled ICI to make substantial contributions during World War II, again drawing on earlier German discoveries, notably the antimalarials Atebrin (Mepacrine), and Plasmoquin.[47]

**Solvents: Distillers and ICI**

The British chemical industry was occasionally assisted by legislation based on lessons of the pre-1918 period. Thus the "Safeguarding Industries Act" of 1932 placed a customs duty of 33⅓ per cent on imported synthetic organics. This was known as the Key Industry Duty. Originally designed to protect the synthetic dye industry, it had the effect of encouraging manufacture of other organic chemicals in Britain. This had a considerable impact on the raw material question, especially for novel solvents.

The use of alcohol, which had been duty free since early in the century, again to assist the dye industry, was further encouraged by an "inconvenience allowance," a form of compensation for the bureaucracy encountered by manufacturers. This ensured that ethanol continued to be the principal source of aliphatic products. Industrial ethanol production was dominated by the Distillers Company Ltd., which embarked on large-scale fermentation of molasses (instead of grain) following the formation in 1928 of its subsidiary British Industrial Solvents, at Hull. There, the ethanol was converted into acetone (for which demand had grown due to its use as a solvent in the manufacture of cellulose acetate silk), acetaldehyde, butaldehyde, butanol, and acetic acid. At Carshalton, Surrey, the Distillers' subsidiary Methylating Co. Ltd. converted products made at Hull into ethyl and butyl acetates, used as solvents and plasticisers. In 1929, Distillers undertook research into the conversion of petroleum hydrocarbons into ethanol and other products. Similar work was undertaken by British Petroleum during the 1930's.

---

[47] Hardie and Davidson Pratt, op. cit.(14), pp. 169-172. For M&B 693 see July Slinn, *A History of May and Baker, 1834 - 1984* (Cambridge: Hobsons, 1984), pp. 122-126.

Apart from methanol, produced by the high pressure process at Billingham since 1925, ICI did not enter the solvent business. No doubt if feared that Distillers might well respond by manufacturing chemicals that rivalled ICI's products. Neverless, Distillers entered the plastics business in 1937, through its ownership of British Resin Products Ltd., which in 1938 was the first British company to produce styrene (styrol, vinylbenzene, phenylethylene), starting from ethanol. This was then converted into polystyrene, the Distillers' Distrene. Manufacture of plastics such as polystyrene in England was favoured by the abundance of coal-tar derived aromatic compounds. Products made from polystyrene resins were appreciated for their "glass-like clarity, low water absorption, [low] solubility, and unique insulating characteristics."[48]

This no doubt influenced the decision by Distillers and ICI to come to an agreement over their respective spheres of influence in 1938. One year later, in September 1939, Distillers announced its purchase of 50 per cent of BX Plastics Ltd., the manufacturing subsidiary of British Xylonite Ltd., the long-established (in 1877), and highly science-based, firm in the celluloid, cellulose acetate, and casein fields.[49]

Between 1914 and 1918, solvent and general chemical production had been geared to the needs of war. At war's end, the chemical surpluses included, in addition to phenol and cellulose acetate, the acetone and butanol from the Weizmann fermentation process and chlorine from poison gas production. The latter, made by electrolysis of sodium chloride solution, became the basis of the British chlorinated solvents industry. Here the outstanding firm was Albright & Wilson, which began manufacture of carbon tetrachloride in 1925. Cleaning and degreasing processes led to production of trichloroethylene, for which demand increased dramatically between 1928 and 1936. The principal manufacturer was Weston Co., later taken over by ICI.

**ICI's coal-to-oil process**

The high pressure coal-to-oil process of Friedrich Bergius was patented in 1913, and developed in Germany during the following years. Bergius realized that the difference between coal and crude oil is the carbon to hydrogen ratio. Since coal has a higher carbon content it is solid rather than liquid. Bergius reasoned that coal could be hydrogenated to oil under conditions of high temperatures and high pressures. A pilot plant was opened in 1922 at Rheinau, near Mannheim, operating at 100 atmospheres

---

48 H.M. Stanley, "Styrene and its polymers," *Chemistry and Industry*, **57** (20 January 1938), 93-98. By 1938, polystyrene was manufactured on a large scale in Germany and the United States.

49 Foster Sproxton was both director and chief chemist at British Xylonite. See for example his "The rise of the plastics industry," *Chemistry and Industry*, **57** (25 June 1938), 607-616. Research into physical properties of plastics at the same firm was headed by W.G. Wearmouth, author of "Physics and the plastics industry," *Chemistry and Industry*, **57** (17 December 1938), 1176-1182. The studies of Wearmouth and de Bruyne represented the high levels of understanding in industry of the physics underlying the behaviour of plastic materials. Many industrial chemists in the plastics industry demonstrated considerable mastery, through their trade and other publications, of the basic principles behind the processes of polymerization.

pressure and elevated temperature. The product mixture contained 25 per cent petrol and 40 per cent oil. In the mid 1920s the process was investigated on a large-scale by I.G. Farben which developed improved catalysts. In 1927, ICI bought rights to Bergius's process from I.G. Farben, and eight years later, on 15 October 1935, opened a plant at Billingham designed to afford over 100,000 tons a year of motor fuel from coal and coal tar. Soon after, planned capacity was increased to over 150,000 tons of oil, two-thirds from coal. The opening of the hydrogenation plant by Ramsay MacDonald was considered "one of the most important industrial events that has taken place in the United Kingdom for many years." Hopes ran high that Britain would become self sufficient in oil. The engineering work was no less impressive: "The plant took about 18 months to build, and except for a few instruments the whole was made in the United Kingdom. Special materials, never before employed in this country, were used. The steel industry deserves great credit for the successful production of plant from these special steels, particularly when the size and weight of the high pressure portions are considered."[50]

However, despite the British government's earlier tax on imported oil (1928), and the Hydrocarbon Oil Act (1934), the project only proved its worth during World War II, when Billingham fuel, made from creosote oil, was used by the Royal Air Force. Then, as in Germany, normal market forces that would have made the process uneconomic no longer applied.

**Coal-to-oil at Low Temperature Carbonisation**

ICI was not alone in undertaking the development of coal-to-oil processes that would serve the growing motor industry, much as the development of plastics had done. Nor were expensive German high pressure processes required. Relatively simple chemical technology, compared with Bergius-type processes, was used in Britain to convert coal to liquid fuel. This was the domain of Low Temperature Carbonisation Ltd., formed in 1917 through the merger of two companies to develop processes of low temperature conversion of coal into smokeless fuel and fuel oils. The processes were based on Thomas Parker's 1906 patent, but had taken another twelve years to become viable. The main product was 'Coalite' smokeless fuel. The enterprise owed much of its success to the efforts of Colonel W.A. Bristow, chairman and managing director, Sir Arthur G. Hazlerigg, Bart., and Commander C. Buist, directors, who since 1928 had supported "the continuous attention of the Company's engineers and chemists" in their quest for "solutions of hundreds of technical problems." On 13 May 1939, Low Temperature Carbonisation's new Coalite works and coal refinery at Bolsover, near Chesterfield, was opened by Mr. Geoffrey Lloyd, the Minister of Mines: "This is the only plant in the country where the coal is taken from the mine and converted into finished motor spirit and Diesel oil, and other refined chemical products... If required a very large proportion

---

[50] C.H.S. Tupholme, "Gasoline from coal: Formal opening of the Billingham hydrogenation plant," *Chemical and Engineering News. News Edition*, **13** (10 November 1935), 414-415. The mention of foreign (German) instruments is of interest; see Stuart Bennett's contribution to this volume.

of the oil could be prepared as a fuel oil for the navy, or for other steam-raising purposes." The company's petrol "has been used by the Air Force for the past five years, and in the words of the Secretary of State [Kingsley Wood] ....' - is now considered part of their normal supply.'" In common with other processes and products then in use, the potential for war use weighed as heavily as consumer consumption.[51]

Apart from considerations of self sufficiency in time of war, there was a great deal of optimism that the fuel oils would be profitable under peace-time conditions, since in 1939 known reserves of petroleum were expected to be exhausted within two decades. This optimism was shared by *Chemistry and Industry*: "The present success of the recovery of petrol from low temperature carbonisation depends upon the fact that the majority of motorists are young or middle-aged people who insist upon motoring whenever they can make the time and that when the price of petrol rises they can always economize by having two drinks at a road-house instead of three." This prospect no doubt infuriated the drink and drive lobby of the time. And would anyone have become alarmed at the environmental impact of a nationwide chain of Low Temperature Carbonisation refineries serving these eager motorists? Certainly not *Chemistry and Industry:* "We hope that Low Temperature Carbonisation will continue to flourish; that it will open new refineries in all kinds of charming villages in England, Scotland, and Wales."[52]

Abundant byproducts of the Coalite process were also low temperature tars, containing cresylic acids (cresols; the products were referred to as meta cresol and cresylic acid). With the sudden increase in the price of phenol around 1937, and further rises in 1938, both as a result of demand from the plastics industry (despite a general downturn in trade), it was suggested that Low Temperature Carbonisation Ltd. investigate catalysts suited to condensations of cresylic acids (as an alternative to phenol) with formaldehyde. Certainly there was a greater dependence in Britain on cresylics than was the case in the United States, because of their availability from coal-based processes. In the meantime Monsanto, which manufactured coal tar cresols (as did Yorkshire Tar Distillers), was considering the erection of a second British synthetic phenol plant.

**Conclusion**

After 1918, many developments in British industrial organic chemistry were determined by awareness of changing needs. This provided opportunities for new firms to enter the industry, as well as diversification for the well established companies. Though the industry was dominated by ICI, this behemoth was perhaps at first so preoccupied with perpetuating traditional lines and remoulding its disparate entities that prior to the mid 1930s it was unable to make major contributions towards the modernization of organic chemistry. Martin Sherwood, in his account of the discovery of polythene, remarked

---

51 "Low Temperature Carbonisation, Ltd.," *Chemistry and Industry*, **58** (20 May 1939), 492-494, p. 494.

52 "Editorial," *Chemistry and Industry*, **58** (20 May 1939), 477-478, on p. 478.

that "One gains the impression that research at that time [the 1930's] in ICI was carried out with little understanding of how innovation could be managed. ICI had only recently been formed and philosophical differences between its constituent parts were not yet settled. At the same time, Britain was deep in economic recession, which presumably focused management attention almost exclusively on short-term survival goals. The idea of the importance of research as a basis for wealth production ... was still a matter of faith rather than irrefutable argument."[53] That same lack of management control over innovation, effectively the control of research and development, had been carried over from past practice, based on the same notions that drove the 19th century British chemical industry.

This is not to say that German and American corporate strategies were in every way superior, for they were also confronted with severe difficulties. Thus the eventual success of high pressure processes did not occur under normal market conditions. The large-scale German synthesis of ammonia from its elements was originally justified by the need for explosives, and only to a lesser extent by fertiliser production, during World War I. Coal-to-oil conversion, as well as synthetic rubber, almost crippled I.G. Farben before they were adopted as major components of Hitler's autarky programme. For similar reasons plastics were adopted on a large scale in Germany from the mid 1930s.[54] As for petroleum reserves, they turned out to be far greater than anyone anticipated in 1939. Consequently, synthetic oil could never have effectively competed with petroleum, at least under peace time conditions.[55] Britain still relied on coal as the basis for its organic chemical manufactures, including for the new plastics, notwithstanding Herbert Levinstein's pertinent remarks regarding alternative sources in 1939. This reliance on coal was just as well as far as the next six years were concerned.

Why, we might ask, did ICI, despite the apparent absence of planning and organization, and the late start in high pressure processes, manage to come up with a number of successes? Perhaps it was the lack of planning, and even of adequate chemical theory, that permitted certain types of unexpected discoveries? Serendipity, chance, and recognition that novel products might be useful, were still important components in the innovation process, and could not be imposed or dictated by management.

Most important, however, was the availability of staff in industry, at ICI and elsewhere, with both the appropriate chemical training and the breadth of experience that enabled recognition of potential new products, of entrepreneurs interested in problem-solving

---

53 Martin Sherwood, "Polythene and its origins," op. cit. (40), p. 237.

54 T.H. Grant, "Plastics and National Socialism," *Chemistry and Industry*, **57** (19 March 1938), 263-266.

55 It was only in 1960 that the British government decided that coal to oil conversion had no future. See "Oil from coal 'not a commercial proposition,'" *Chemistry and Industry*, 27 August 1960, 1100. There was a short-lived revival of interest in the late 1970s.

related to new industries, and the need for strategic materials suited to the electrical industry, aviation, motorized transport and the fight against bacterial infections. This led to the development of an acute degree of awareness that was sometimes well in advance of approaches in Germany and the United States. The awareness was enhanced by various elements, including an unlikely mix of research interests, and even desperation, when a company's very survival was at stake. There were also signs of a change of attitude at the end of the 1930s. Thus in 1938 alone, the Sir Robert Hadfield Metallurgical Laboratories were opened at the University of Sheffield, Pilkington's of St. Helens, which jointly investigated synthetic glass-like materials with ICI, opened the largest glass research laboratory in Britain, the British Coal Utilisation Research Association decided to establish a research laboratory in London, the Medical Research Council announced a substantial government grant for studies on chemotherapy, and ICI had on its payroll over one thousand people allocated to research stations that were operational or authorized.[56] And, at the end of the 1930s, with war on the horizon, British chemists and leaders of chemical industry surveyed progress over the previous two decades, sought lessons from the past, and began to lobby for national planning.

Finally, quite apart from the level of organization, it should be emphasized that whereas during 1914-18 British chemical manufacturers struggled to catch up with Germany, just over two decades later they entered World War II sharing novel chemical technologies with corporations in the United States. This was an outcome of the largely independent application of organic chemistry to new and immediate needs.

**Acknowledgments**
Elisabeth Garnsey, University of Cambridge, Ernst Homburg, University of Maastricht, and Peter J.T. Morris, Science Museum, London, are thanked for assistance with information used in this paper, and also for critical comments.

---

[56] See also the editorials "New laboratory at Blackley," and "The society of glass technology," *Chemistry and Industry*, **58** (15 January 1938), 93-94.

# SCALING UP: THE EVOLUTION OF INTELLECTUAL APPARATUS ASSOCIATED WITH THE MANUFACTURE OF HEAVY CHEMICALS IN BRITAIN, 1900-1939

COLIN DIVALL AND SEAN JOHNSTON
*Department of History*
*University of York*
*Heslington*
*York, YO1 5DD, UK*

**Introduction**

At the turn of the century, men trained in chemistry but not in any of the engineering disciplines held sway over the technical side of much of the business of chemical manufacturing in Britain. However, the growing economic importance of bulk chemicals after around 1880 had been accompanied by the emergence of new specialists in the design and operation of chemical plants. This became particularly significant with the widespread introduction of high pressure equipment from around 1920.[1] The outcome was that by the outbreak of World War II chemists were under some pressure to cede certain of these responsibilities to a new breed of technical worker, the chemical engineer.

All would-be professions have to claim and maintain jurisdiction over a class of occupational practices if they are to prosper, and the evolution of chemical engineering serves to illustrate the importance of intellectual apparatus for this process.[2] Although only loosely defined and poorly organized before World War I, chemical engineers promised a different way of organizing the production of heavy – or bulk – chemicals. In tandem with these new industrial practices, practitioners and educators developed a new way of conceptualizing chemical manufacturing. With the establishment in 1922 of a professional organization, the Institution of Chemical Engineers, great effort was put into defining and promoting chemical engineering as a subject that came to be taught

[1] For the technical problems, organization and personnel involved in such work see, for example, Victor E. Parke, *Billingham - The First Ten Years* (Billingham: Imperial Chemical Industries, 1957).

[2] Andrew Abbott, *The System of the Professions* (Chicago: Univ. Chicago Press, 1986), pp. 52-58.



*A.S. Travis et al. (eds.),*
*Determinants in the Evolution of the European Chemical Industry, 1900–1939,* 199–214.

quite widely as a separate discipline at the university level. If by 1939 chemical engineers could claim some standing within industry, it was at least in part because of a growing acceptance that the intellectual framework of chemical engineering amounted to more than just a synthesis of earlier technical disciplines.

Most historical studies of chemical engineering focus on the emergence of the profession in the United States, and they tend to assume that the intellectual substance of chemical engineering in Britain was almost entirely the result of American influences.[3] Some scholars have highlighted certain national differences in the cognitive scope of the discipline in European countries other than Britain.[4] Only quite recently have historians started to acknowledge the indigenous roots of British chemical engineering.[5] In this chapter we expand on the work of James F. Donnelly and of one of us (Divall), by looking at how the particular circumstances under which chemical engineering emerged as a profession in Britain shaped the associated intellectual discipline.[6]

**Conceptual foundations of the chemical engineering profession**

Much of the historiography of chemical engineering as a kind of industrial practice and

---

3 Many of these accounts were written by former practitioners; the two principal collections of such material are, both William F. Furter, ed., *History of Chemical Engineering* (Washington, DC: American Chemical Society, 1980), and *A Century of Chemical Engineering* (New York: Plenum Press, 1982). On the US experience see, for example, Martha M. Trescott, "Unit operations in the chemical industry: an American innovation in modern chemical engineering," in *A Century of Chemical Engineering*, pp. 1-18; F.J. Van Antwerpen, "The origins of chemical engineering," in *History of Chemical Engineering*, pp. 1-14; David F. Noble, *America by Design: Science, Technology and the Rise of Corporate Capitalism* (New York: Alfred A. Knopf, 1977), pp. 26-27, 38, 79, 192-195; John W. Servos, "The industrial relations of science: chemical engineering at MIT, 1900-1939," *Isis*, **71** (1980), 531-549; T.S. Reynolds, *75 Years of Progress: A History of the American Institute of Chemical Engineers, 1908-1983* (New York: American Institute of Chemical Engineers, 1983), *passim*; Reynolds, "Defining professional boundaries: chemical engineering in the early 20th century," *Technology and Culture*, **27** (1986), 694-716.

4 E.g. Klaus Buchholz, "Verfahrenstechnik (Chemical Engineering) - its development, present state and structures," *Social Studies of Science*, **9** (1979), 33-62 and, less convincingly, Karl Schoenemann, "The separate development of chemical engineering in Germany," in *History of Chemical Engineering*, op. cit. (3), pp. 249-272.

5 J.F. Donnelly, "Chemical engineering in England, 1880-1922," *Annals of Science*, **45** (1988), 555-590; Colin Divall, "Education for design and production: professional organization, employers and the study of chemical engineering in British universities, 1922-76," *Technology and Culture*, **35** (1994), 258-288; Divall, "Professional organization, employers and the education of engineers for management: a comparison of mechanical, electrical and chemical engineers in Britain, 1897-1977," *Minerva*, **32** (1994), 241-266; Clive Cohen, "The early history of chemical engineering: a reassessment," *British Journal for the History of Science*, **29** (1996), 171-194.

6 Divall, op. cit. (5), and "A measure of agreement: employers and engineering studies in the universities of England and Wales, 1897-1939," *Social Studies of Science*, **20** (1990), 65-112. Donnelly, op. cit. (5), and "Representations of applied science: academics and chemical industry in late nineteenth-century England," *Social Studies of Science*, **16** (1986), 195-234.

a profession has been concerned with its early academic connections. Chemical engineering courses were important because they gave a sense of identity to the would-be profession and made explicit its cognitive scope. Moreover, by providing practitioners with a unique intellectual foundation for the changing techniques of chemical manufacturing and the design of chemical plant, chemical engineers were given the means to exclude competing occupations from these and certain other related kinds of industrial work. But if we are to understand the evolution of the intellectual apparatus of chemical engineering in Britain, we need to be clear about the peculiarities of other countries, and particularly of the United States.

Both Britain and the United States, in contrast to many of the countries of continental Europe, were characterized before World War II by political systems that laid great emphasis on the voluntary, professional regulation of those scientific or technical occupations that commanded social and economic prestige. Similarly, the universities were not subject to any significant measure of control by the state. Thus in both countries intellectual developments relating to the chemical industries were largely a matter of tacit negotiation between academics and certain groups of business officials, consultants and other would-be professionals. But the British chemical industry was by no means identical to the American industry in terms of the markets it served or the typical structure of firms, and advocacy of educational programmes of chemical engineering in the two countries took place under rather different circumstances.

Generally speaking, British chemists and engineers were more firmly entrenched in the industrial division of labour, and enjoyed older and more effective forms of professional organization than did their counterparts in the United States. Chemistry, in particular, was well established as a university discipline. It was therefore more difficult for chemical engineers in Britain to press their claim to professional status, and to establish a novel intellectual underpinning for their expertise. But while the scale of their achievement was perhaps not as great as that in the United States, British chemical engineers nevetherless proved remarkably inventive when it came to defining the intellectual content of their discipline.

Scholars agree that from the 1880s certain groups of technical experts dissatisfied with their standing in the industrial hierarchy made repeated attempts in Britain and the United States to wrest control of certain industrial practices from chemists; in the intellectual realm, this struggle was paralleled by disputes between advocates of, on the one hand, chemistry and, on the other, "chemical engineering." This process was first analyzed by historians working on the United States. Terry Reynolds has shown that by the first decade of the 20th century, groups of senior chemists employed in chemical manufacturing were seeking to distinguish themselves from analytical chemists, who had a low status, by claiming as their own the task of scaling up chemical processes from the laboratory to the industrial level. They were joined by some consultants to the chemical manufacturers, and a number of businessmen.[7] These nascent chemical engineers were opposed by the existing professional society for chemists, the American Chemical Society, the leaders of which argued that the development of new manufacturing processes could be understood within the conceptual framework of industrial chemistry. Mindful of the need to establish its claim to a distinctive specialism, the American Institute of Chemical Engineers (AIChE), founded in 1908, chose to restrict full membership to persons over 30 years of age who possessed several years of practical experience in both chemistry and another branch of engineering. Would-be members also had to be actively involved with the application of chemical principles to manufacturing.[8] Thus early members of the AIChE used their knowledge of theoretical chemistry as a way of distinguishing themselves both from self-trained factory hands and from mechanical engineers who had become involved with chemical plant design.

These measures had some limited success, but, according to Reynolds, the key to legitimating American chemical engineering as a profession was the evolution of a new cognitive base – the unit operation. This conceptual framework was not formally employed in the struggle for professional status until 1922, when the AIChE's Education Committee chaired by Arthur D. Little, head of one of the largest American consulting engineering firms, recommended that it be employed to define the scientific core of academic educational programmes. However the unit operations had already been evolving over a period of twenty years or more in several American universities, chief among which was MIT.[9] Hence Reynolds sees the American profession as being shaped by at least three factors: first, the desire of certain groups of technical workers to differentiate themselves from the more heterogeneous and less prestigious category of 'chemists;' secondly, their appropriation of a cognitive realm (first physical chemistry, and then unit operations) as a way of underpinning claims to technical expertise; and thirdly, a gain in legitimacy in academic, engineering and industrial circles following the successful application of this intellectual apparatus to the design and operation of large chemical plants.

Tracing the origins of concepts is difficult when the inchoate professions that are trying to appropriate them are in a state of flux, and, reflecting this, the intellectual features of chemical engineering up to 1939 have been described in various and inconsistent ways. Olaf Hougen and Franklin J. Van Antwerpen, for example, identify the unit operations

---

7 Reynolds, "Defining professional boundaries," op. cit. (3), pp. 697-708.

8 Ibid., pp. 697, 706.

9 By 1905 the concept of unit operations implicitly informed the curriculum of a new chemical engineering course at what was to become MIT. From 1900, William K. Walker, the course instructor, had a close association with the Arthur D. Little engineering consultancy. John Servos has discussed the currency of the notion of unit operations in the US in this early period, showing that Little was linked closely with such concepts in the curricula of American institutions before 1915.

as being crucial from 1916 to 1925, although they also note the increasing importance of material and energy balances to 1935, and of applied thermodynamics and process control from then until the end of World War II, and beyond.[10] However, Gianni Astarita's work on Italy distinguishes unit operations as the sole organizing concept.[11] Clive Cohen largely concurs with this simple genealogy for the United States and Britain, while moving the influence of unit operations back to about 1910, and suggesting that the profession in both countries could have developed on the basis of other – but largely unspecified – intellectual foundations.[12]

All sources agree, however, that the unit operations were first described as such in 1915 by Arthur D. Little. He defined them as discrete physical processes employed in chemical manufacturing, such as distillation, roasting, filtering and condensation. Each described a particular way in which material could be transformed physically, as, for example, by the reduction in size of solid matter, or by the mixing or separation of solids, liquids or gases. These basic processes would be performed in sequence to obtain a final product; while the number and order might vary from one product to another, any manufacturing process could be understood in terms of the same set of building blocks. The steps could either be carried out on batches of material, or performed as a continuous series of processes in which the material is transformed at different locations in a plant.

Accounts of the origins of the unit operations before 1915, however, diverge considerably. Martha Trescott identifies them as almost exclusively an American innovation. While acknowledging that a British consultant, George E. Davis (discussed at greater length shortly) may have described unit operations in his 1901 *Handbook of Chemical Engineering*, she claims that the concept was popularized and integrated with physical chemistry exclusively by American practitioners. Trescott suggests that three national characteristics explain this American 'lead:' first, the American predilection for rational mechanical design, leading to the careful design of equipment intended for specific uses; secondly, the influence of Frederick W. Taylor's system of "scientific management" in American engineering, which stressed the technically and financially efficient division of manufacturing tasks into unitary operations arranged sequentially; and finally, an American willingness to innovate, to scale up industrial processes, and to engage in the production of goods for mass markets.[13]

---

10 Van Antwerpen, op. cit. (3); Olaf A. Hougen, "Seven decades of chemical engineering," *Chemical Engineering Progress*, **73** (1977), 89-104.

11 G. Astarita, "L Evoluzione dei fondamenti teorici dell ingegneria chimica," *Chimica e Industria (Quaderni dell Ingegnere Chimico Italiano)* (supplement), **8** (1972), 112-114.

12 Cohen, op. cit. (5), pp. 175-186.

13 Trescott, op. cit. (3).

Chemical engineering in the United States was certainly associated with the manufacture of chemicals on a large scale. But beyond this, not one of Trescott's observations is particularly convincing. In the first place, the distinctive American system of manufacturing, involving long runs of relatively small items of machinery and consumer goods of a mechanical or electrical nature, was of limited relevance to the design of large chemical manufacturing plants or the bulk production of chemicals, many of which were intermediates destined for further processing.[14] Moreover, as Donnelly argues, and as was often the case from around 1860 (for example William Perkin's mauve process, introduced in 1858, and his, and BASF's, alizarin process of 1869-70) chemical manufacturing on a considerable scale arose from the scaling up of laboratory procedures long before unit operations were conceived, and efforts to rationalize chemical production – particularly by the introduction of labour-saving processes – pre-dated Taylor's activities.[15]

The close connection between unit operations and mass production is, however, remarked upon by other advocates of the pioneering role of the United States. Klaus Buchholz postulates that the practical problems of exploiting the country's vast oil fields played a substantial part in the American genesis of chemical engineering. He puts forward a particular direction of evolution: from market forces and existing chemical technologies, to intellectual concepts and an emergent profession. Attributing the occupation's growth to the "need for mass-production prompted by the expanding market" in petrochemicals from the turn of the century, Buchholz asserts that this pressure promoted the adoption of economical production methods and encouraged the rational design of industrial plants.[16]

More importantly, he claims that the particular technical requirements of petrochemical processing – namely physical operations such as distillation – explain the readiness of US manufacturers to accept the conceptual framework of the unit operations.[17] By contrast to the American case, a unique occupation combining mechanical and chemical expertise failed to coalesce in Germany; the main reasons were that the petrochemical industry there was negligible, and much more complex chemical syntheses dominated the dyestuffs and pharmaceuticals industries. The occupational specialization involved

---

[14] David A. Hounshell, *From the American System to Mass Production, 1800-1932: the Development of Manufacturing Technology in the United States* (Baltimore: Johns Hopkins University Press, 1984), *passim*; John K. Brown, *The Baldwin Locomotive Works, 1831-1915: A Study in American Industrial Practice* (Baltimore: Johns Hopkins University Press, 1995). Brown finds no unique features in the American context of the capital goods sector.

[15] Taylor's first paper on scientific management appeared in 1895; he died in 1915, but Taylorist analysis and management were widely applied in American industry through the inter-war period.

[16] Buchholz, op. cit. (4), 42.

[17] Ibid., p. 37.

in *Verfahrenstechnik* was organized around specific industrial products and their manufacturing processes, and chemical plants were designed and maintained by a combination of chemists and mechanical engineers with a strict division of labour.[18] In a similar vein, Jean-Claude Guédon argues that US manufacturers, particularly in the petrochemical industry, were much more welcoming towards unit operations than the British, and that the American system of higher education was more receptive to the new conceptual framework.[19]

Yet Buchholz says that neither the American experience nor the concept of unit operations were essential for the solution of the problems of expanding industrial production in petrochemicals.[20] Neither, it should be pointed out, was the establishment in the United States of the profession of chemical engineering and its associated intellectual apparatus the inevitable outcome of the growth of the manufacture of bulk petrochemicals. This particular market certainly provided fertile ground for the emerging profession in the United States. However, gaining a measure of acceptance for a novel conceptual framework also required agreement between academics, chemists and engineers working in the chemical industries, and their business leaders, that this was to their mutual advantage. We must therefore examine just how successfully such an alliance was constructed in Britain.

**The origins of chemical engineering in Britain, 1900-1914**

In their efforts to organize as a profession, British chemical engineers faced a higher degree of organized opposition from other groups of technical workers than their counterparts in the United States. Also, wider developments in the chemical industries that were largely outside their control proved less than favourable. Thus although there were striking similarities between the evolution of key intellectual concepts in the two countries, it is arguable that by 1939 the academic definition of chemical engineering was not as firmly established in Britain as it was in the United States.

In the late 19th and early 20th centuries, the most significant portion of the British chemical industry was, like that in the United States, involved with the bulk manufacture of mainly inorganic products such as bleaching powder, alkali and sulphuric acid. Organic chemicals, particularly coal tar products, and fertilisers began to

---

18 Ibid., pp. 38, 44-51.

19 Jean-Claude Guédon, "Conceptual and institutional obstacles to the emergence of unit operations in Europe," *History of Chemical Engineering*, op. cit. (3), pp. 45, 56.

20 Buchholz , op. cit. (3), p. 37. Other historians reverse Buchholz's arrow of causality. Van Antwerpen, for example, concludes that the explosive development of large-scale chemical plants had to await the development of chemical engineering as a distinct engineering discipline [Van Antwerpen, op. cit. (3), 11]. We suggest that the co-evolution of professional organizations, academic concepts and industrial practice is a more realistic evaluation.

assume a growing importance by the turn of the century.[21] Firms manufacturing these products in the English industrial regions of Lancashire, Merseyside and Tyneside, and in Scotland on Clydeside, began to employ growing numbers of specialist workers from the 1860s. But until the last quarter of the 19th century, technical innovation relied more upon empirical developments than on organized scientific research by trained investigators. Production processes were dominated by mechanical plant, and improvements in these, such as a better roasting device for sulphur extraction, were introduced by mechanical engineers assisted, rather than directed, by chemists.[22]

This empirical tradition began to falter in the last decades of the century partly because of the more complicated developmental work and manufacturing techniques associated with the rise of synthetic dyes, itself a response to rising international competition, especially from Germany.[23] The British alkali industry, still reliant on the LeBlanc soda process, saw its exports captured by overseas manufacturers employing the newer and more efficient Solvay process.[24] Meeting these challenges required a more specialized approach to the design and operation of chemical plants; by the end of the century, a rough and ready division of technical labour was fairly well established.[25]

Chemists and engineers had largely proved capable, however, of claiming jurisdiction of the slowly evolving industrial order. Works chemists usually were responsible for the technical supervision of manufacturing processes and managing the labour force; this gave them a status far superior to the analytical chemists that made up the greater proportion of the skilled workforce. Chemists also helped to scale up manufacturing processes from the laboratory to the industrial level in the few firms, such as those manufacturing organic products, in which the development of new chemicals was

---

21 On the British heavy chemicals industry, see, for example, L.F. Haber, *The Chemical Industry During the Nineteenth Century: A Study of the Economic Aspect of Applied Chemistry in Europe and North America* (Oxford: Clarendon Press, 1958); Haber, *The Chemical Industry (1900-1930): International Growth and Technological Change* (Oxford: Clarendon Press, 1971); M. Boas Hall, "La croissance de l'industrie chimique en Grande-Bretagne au XIXe siècle," *Revue d'Histoire des Sciences et Leurs Applications*, **26** (1973), 49-68. On the alkali industry, see K. Warren, *Chemical Foundations: The Alkali Industry in Britain to 1926* (Oxford, 1980) and Frank Morton, "A short history of chemical engineering in the North-West of England," *A Century of Chemical Engineering*, op. cit. (3), pp. 19-28.

22 Guédon, op. cit. (19), pp. 45-75.

23 Ibid., esp. 47-52.

24 Haber, *The Chemical Industry During The Nineteenth Century*, op. cit. (21), pp. 10-11.

25 Robert F. Bud and Gerrylynn K. Roberts, *Science Versus Practice: Chemistry in Victorian Britain*, (Manchester: Manchester University Press, 1984), pp. 111-117; J.F. Donnelly, "Consultants, managers, testing slaves: changing roles for chemists in the British alkali industry, 1850-1920," *Technology & Culture*, **35** (1994), 100-128. See also Colin A. Russell, with Noel G. Coley and Gerrylynn K. Roberts, *Chemists by Profession: The Origins and Rise of the Royal Institute of Chemistry* (Milton Keynes: Open University Press, 1977).

comparatively commonplace; they were often assisted in this task by engineers. But within all of these neat designations, actual responsibilities could vary widely, particularly in the myriad small firms that provided much of the country's chemical output.[26] Manufacturing plants were increasingly designed, and sometimes erected, by consultants trained either in engineering or in chemistry, or even elements of both. Small firms, lacking the resources to develop their own expertise, were particularly reliant on consultants.[27]

The strength of the professional organizations for chemists and engineers partly explains why, compared to the United States, there was less of a sense of urgency in Britain of the need to establish an autonomous organization and identity for chemical engineers. The Institute of Chemistry (founded in 1877) and the Institution of Mechanical Engineers (1847) catered for many of the more senior workers in the chemical industries. The term "chemical engineer" was used by a handful of specialists in plant design and the management of chemical works, but their attempts in the 1880s to establish an autonomous organization had foundered in the face of opposition from some industrialists and the Institute of Chemistry. The Society of Chemical Industry (1881), a learned society which lacked the constitutional powers to engage with matters of professional concern regarding education, was the rather unsatisfactory alternative open to those manufacturers, chemists and consultants who were unhappy with the leading societies.[28]

The professional societies enjoyed a large measure of influence with academics. Of all the scientific subjects, chemistry was the best established in the universities, and academic chemists successfully guarded their claim to train men destined for industrial employment. From the 1880s, variants such as "industrial chemistry," "applied chemistry," "chemical technology" and even – at the Central Institution, London – one called "chemical engineering" were developed in anticipation of a demand from manufacturers for graduates with skills relating specifically to industry. Despite the increasing availability of these courses, for the most part firms preferred their workers to acquire such knowledge through practical experience.[29] Engineering was less well established as a university subject, but as a source of trained personnel it was rapidly growing in importance by 1900.[30]

---

26 For an anecdotal but revealing practitioner's account of this period, see Norman Swindin, *Engineering Without Wheels: A Personal History* (London: Weidenfeld & Nicolson, 1962), *passim*.

27 See, for example, Sydney Gregory, "Hugh Griffiths and consultancy," *Chemical Engineer*, (October, 1986), 3.

28 Donnelly, op. cit. (5), pp. 581-587; Russell, Coley and Roberts, op. cit. (25), pp. 186ff.

29 Bud and Roberts, op. cit. (25), pp. 111-117; Donnelly, op. cit. (5), pp. 557-561.

30 Divall, "A measure of agreement," op. cit. (6), pp. 68-70.

Lacking a distinctive conceptual framework for their work, would-be chemical engineers found it hard to resist the argument that they were no more than a hybrid of chemist and engineer. But this gradually began to change in the early 1900s with the development of new intellectual principles – the unit operations – based initially on the ideas of George E. Davis.

As we have seen, even historians who are keen to propound the American origins of chemical engineering usually cite the work of Davis. A Briton and sometime chemical worker, government Alkali Inspector and consultant, Davis is widely credited with having presented one of the first chemical engineering courses, at the Manchester Technical School in 1887, and publishing the first textbook of chemical engineering.[31] Yet there is some disagreement about the significance of his work, even among historians of British chemical engineering. James Donnelly says that Davis's lecture course "resembled a plant manufacturer's catalogue," and that consultants like him "frequently did not design machinery so much as select and combine that of specialist equipment manufacturers."[32] Following on from this description, and his own examination of manufacturers' catalogues, Clive Cohen has concluded that attributions of the concept of unit operations to Davis are misguided. They confuse, he suggests, a logical ordering by equipment manufacturers of certain standardized kinds of apparatus with an intellectual categorization by an allegedly seminal chemical engineer.[33]

And yet, near-contemporaries cited Davis's text as both novel and useful, and equipment manufacturers found a ready market for standardized items such as boilers and heat exchangers. This suggests that Davis was enunciating a strategy of plant design that was in fairly common – if not widely recognized – use by the consultants of his time. Donnelly, in fact, argues that Davis was constructing a conceptual framework for chemical engineering in the context of certain social and occupational tensions. He sought to define a new intellectual specialism while avoiding two dangers: first, appropriation of the nascent subject by the disciplines of chemistry and engineering (particularly, but not exclusively, of the mechanical kind), and secondly, a possible inadequacy of his lectures and writings owing to manufacturers' demands for confidentiality with regard to the industrial processes to which, as a consultant, Davis had privileged access. Because standard forms of plant equipment were available, and not subject to the constraints of commercial secrecy, Donnelly suggests that machinery "became an important medium for the public conceptualization of industrial chemical activity."[34]

---

[31] George E. Davis, *A Handbook of Chemical Engineering* (Manchester: Davis Bros., 1901).

[32] Donnelly, op. cit. (5), pp. 563, 566.

[33] Cohen,, op. cit. (5), pp. 176, 183-184, 189.

[34] Donnelly, op. cit. (5), p. 567.

Donnelly also suggests that the construction of the subject was perceived as a resolution of a problem of higher technical education. If chemical engineering were to exist as a subject of instruction it had to fit in with the prevailing views of British educators, industrialists and leaders of the professional societies concerning the nature of technical education; such courses should emphasize the 'principles' that informed practice rather than seek to develop the detailed competencies required in particular industries.[35] Thus we find a specifically British motive for adoption of the concept of unit operations: pedagogical generality and facility of instruction.

Davis was therefore careful to distinguish chemical engineering from chemistry and its derivatives such as industrial chemistry and chemical technology. Chemical engineering was, he argued, a branch of applied science founded on a specialist blend of commercial and technical knowledge rather than any existing scientific or engineering discipline.[36] He also emphasized the professional consultant's tasks of scaling up technical experiments, dividing chemical manufacturing processes into manageable sub-tasks, and systematizing equipment. Hence the notion of unit operations, in practice if not name, was very likely familiar to those active in the field, and was given a crude but recognizable focus by Davis.

Davis's work was taken up in the academic sphere before World War I by John W. Hinchley, a consultant and specialist in plant design who was later to play a leading role in the founding of the Institution of Chemical Engineers, or IChemE. Hinchley established two courses of instruction in London, one, in 1909, at Battersea Polytechnic for part-time students, the other, a year later, in the chemistry department at the Imperial College of Science and Technology. These appealed largely to students who wished to supplement their training in chemistry. Although the term unit operations was not used at first, there is no doubt that unit operations heavily informed Hinchley's teaching.[37]

Like Davis, Hinchley was acutely sensitive to the need to distinguish chemical engineering from chemistry and engineering. He also insisted that academic courses should inculcate in students the ability to apply theoretical knowledge to the kinds of commercial and technical problems they would face in industry. Both considerations were satisfied by the requirement that students had to undertake the design of "commercial plant." Costing was an important consideration in this exercise, reflecting the recommendations of another consulting engineer from Manchester, Jacob Grossmann, who had urged in a 1906 textbook that the key dimension for the design of

---

35 Divall, "A measure of agreement," op. cit. (6), pp. 78-80.

36 Donnelly, op. cit (5), p. 564.

37 Ibid., p. 576; Cohen, op. cit. (5), pp. 178-180.

any chemical plant was its financial viability.[38] Thus Hinchley's students were asked to factor manufacturing processes into discrete operations and optimize their economic return. Employing the results of their laboratory investigations into a particular process, they identified the most economic method of undertaking each unit operation under given conditions. By determining quantitatively the flow of materials through a complete process ("material balance"), and by calculating the requirements for both heating and cooling it ("energy balance"), they arrived by a process of trial and error at that combination of the unit operations which offered the most economic way of producing any given chemical. This method became known as "process design." Hinchley's students, however, had to go much further than this. By drawing on training in civil, mechanical and electrical engineering, they had to complete designs of the manufacturing plant, and even the complete factory ("plant design").

These design projects brought together in a unique fashion the conceptual foci of unit operations, plant-centred design, and economic viability that, on the face of it, had only weak intellectual links. Hinchley's teaching distanced the academic speciality of chemical engineering from those of chemistry and mechanical engineering, and, by demonstrating a coherent approach to academic instruction, helped to make the case for further departments of chemical engineering. By establishing a new language through which consultants and their clients could plan, design and evaluate chemical plant, the new intellectual apparatus also helped to provide would-be chemical engineers with a formalized knowledge with which to underpin their technical expertise, and thus eventually to lay claim to professional status. But success in this lay in the future: there were not many chemical engineers of the kind trained by Hinchley in the British chemical industry before 1914.

**Professional organization, plant design and the unit operations**

Despite strong opposition from the professional societies for chemists and engineers, a formal organization of chemical engineers, the IChemE, was established in 1922. The institution forged an alliance between a handful of prominent business leaders, educators, consultants and manufacturing chemists of some seniority who were all, for quite different reasons, dissatisfied with existing ways of training chemists and others for industrial posts. It drew up a definition of an intellectual framework of chemical engineering based heavily on Hinchley's work, and was successful in getting this accepted in a handful of universities and some parts of the chemical industry. But progress prior to World War II was less rapid than the institution would have liked, largely because many of the largest chemical firms, and in particular Imperial Chemical Industries (ICI), could see no good reason to abandon the by-then traditional division of labour between chemists and engineers.

---

[38] Jacob Grossmann, *The Elements of Chemical Engineering* (London: Charles Griffen, 1906), pp. 123-125; Edith M. Hinchley, *John William Hinchley: Chemical Engineer* (London: Lamley, 1935), pp. 27-30, 40-44.

Consultants dominated the leadership of the IChemE in its early days, and they were particularly influential in defining the intellectual basis of chemical engineering. Their views were shaped by their experiences of designing chemical plants and, in many cases, that of constructing and operating manufacturing facilities for explosives, chemical weapons, dyestuffs and intermediate products that had been sponsored by the state during World War I.[39] This shared background was important for securing agreement within the IChemE over the details of the intellectual framework for chemical engineering, and it proved critical as well for defining the links between the IChemE, industry, government and academe.

The IChemE settled on a set of intellectual principles by 1925; in its fundamentals, this served the profession until well after World War II. The influence of North American ideas cannot be denied. The highly successful enlargement and reconstruction of the explosives industry during 1914-18 had been guided by the design techniques of an American chemical engineer, K.B. Quinan. Close links thereafter between the IChemE and the AIChE ensured a ready awareness in Britain of the American technical literature, especially textbooks. More particularly, the successful programme of chemical engineering at MIT was well enough known in Britain to be presented to the IChemE in 1922 as an ideal to emulate.[40] Not surprisingly, there were striking parallels between the list of subjects judged to be distinctive of chemical engineering that the IChemE published in 1925, and the recommendations of Arthur D. Little's report of 1922 for the AIChE. Both lists placed unit operations at the core of the new academic discipline.

There were, however, marked differences in emphasis. Reflecting the dominance of plant manufacturers and consultants among its leadership, the IChemE made the capacity to design manufacturing plant on a commercial basis central to its definition of the intellectual foundations of chemical engineering. This embodiment of separable chemical stages in discrete equipment was conceptually and practically distinct from the American version of unit operations, and the British also proved quick to adopt new technologies such as control equipment.[41] The continued focus on cost-effective chemical staging of processes also helped to maintain educational programmes for chemical engineers in a more intimate connection with business ideals than did courses for chemists and engineers.[42]

---

39 See, for example, Roy MacLeod, "Chemistry for King and Kaiser: revisiting chemical enterprise and the European war," this volume and *Idem*, "The chemists go to war: the mobilization of civilian chemists and the British war effort, 1914-1918," *Annals of Science*, **50** (1993), 455-481.

40 Robert Edgeworth Johnstone, "The dark days," *Chemical Engineer* (January 1987), 38-39.

41 See Stuart Bennett, "The use of measuring and controlling instruments in the chemical industry in Great Britain and the USA during the period 1900-1939," this volume.

42 IChemE, *The Training of the Chemical Engineer* (London, 1925), 5-6; Russell, Coley and Roberts, op.

There was one other important respect in which British conceptual definitions were more extensive than those current in the United States: not until the mid-1930s was there any significant attempt in the latter country to define a conceptual framework that would function in the domain of *chemistry* as unit operations did for the *physical* analysis of chemical manufacturing. In Britain, by contrast, the 1925 IChemE enumeration listed various "reaction treatments of materials," which were to become more commonly known as unit *processes*, and later still as chemical reaction engineering.[43] The list of treatments included calcining, electrolysis, hydrolysis, oxidation and fermentation. The earlier British emphasis on analyzing the chemistry of manufacturing in a manner analogous to that of unit operations can be attributed partly to the experiences of the earliest educators of chemical engineering. Some, such as E.C. Williams, the first professor of chemical engineering at University College, London and Herbert W. Cremer at King's College in the same city, were trained as chemists and had quite extensive experience of chemical manufacturing. Their educational programmes consequently paid considerably more attention to manufacturing techniques than did, for example, Hinchley's course at Imperial College.[44]

By 1939, six institutions of university rank taught chemical engineering broadly in line with the IChemE's definition; together these bodies had turned out nearly 400 trained men. Indeed, accredited chemical engineers were proportionally more numerous in Britain than in the USA.[45] But it is arguable that the idea of the unit operations and associated concepts was not as convincingly accepted in Britain as in the USA. The links between British educational programmes, professional organization and industrial practice were, on the whole, fairly weak, although there were certain firms and parts of the chemical industry where the connections were strong. Oil companies and coal-gas

---

cit. (25), pp. 158-185, 264-283.

43 Nikolaos A. Peppas and R.S. Harland, "Unit processes against unit operations: the educational fights of the thirties," in N.A. Peppas, ed., *One Hundred Years of Chemical Engineering*, (Dordrecht: Kluwer Academic Press, 1989), pp. 125-142.

44 The role of manufacturers' interests was also influential in the curriculum. At UCL, the department owed its very existence wholly to the generosity of industrialists, who funded the establishment of a chemical engineering laboratory in 1923. A further appeal four years later to secure the financial future of the department called upon the services of eight leading industrialists from companies such as ICI, Shell, Lever Brothers, Dunlop, and the Gas, Light & Coke Co.

45 By 1940, there were 761 accredited chemical engineers in Britain (IChemE 'corporate' members) compared with 1349 in the US (AIChE 'active' members). However, the number of student members in the two institutions (for example, 50 in the IChemE versus 1400 in the AIChE in 1932) suggests a significantly higher rate of production in America. This is supported by academic records: by 1940 some 40 American colleges offered degrees in chemical engineering, MIT alone averaging 58 graduates per year between 1920 and 1934. In Britain, the combined output of diploma, certificate and undergraduate degree programmes was some 15-20 per year. For Ph.D.s in chemical engineering, the US rate by 1940 was over fifty per year; the British output was less than one-tenth this figure.

manufacturers proved particularly enthusiastic employers of chemical engineers. But elsewhere, chemical plant was still likely to be designed by mechanical engineers and chemists, and to be operated and managed by chemists alone.

This was undoubtedly partly due to conservative business practices; many plant designers and managers had been trained in Germany before World War I, and rose to positions of authority thereafter. In the 1930s, a still common reaction of many industrialists among larger firms was that chemical engineers were a hybrid breed fit only for smaller manufacturing units which could not afford better.[46] ICI was a major disappointment. After its government-inspired formation in the mid 1920s, ICI dominated the highly rationalized business of chemical manufacturing in Britain; it took over a number of firms, such as the British Dyestuffs Corporation, that had shown signs of recognizing the potential of the chemical engineer.[47] Throughout the 1930s, ICI felt little need to employ academically-trained chemical engineers. Its technically important high pressure plant at Billingham was designed, built and operated by using a typically German division of labour between chemists and engineers. In the period 1928-38, the company employed only 16 per cent of the 159 chemical engineers turned out by University College, London, which ran the largest programme in Britain between the world wars. Only in the late 1940s did the company welcome chemical engineers in some of its divisions; and until the late 1950s, few chemical engineers were employed at Billingham.[48]

Nor were all educators convinced of the value of a training in chemical engineering. Professors of chemistry were in a powerful position within many British universities, and they were understandably reluctant to allow the founding of courses that might reduce recruitment to their own programmes. Even among those chemists and other academics who thought that more should be done to provide a training tailored to industrial skills, there was continued unease over whether the unit operations provided a scientifically rigorous enough set of general principles upon which instruction could be based.[49] But we need to relate this point more closely to the kind of tasks to which the

---

46 Lord McGowan, *Proceedings of the Chemical Engineering Group (Society of Chemical Industry)*, **25** (1943), 27.

47 W.J. Reader, *Imperial Chemical Industries: A History* (London: Oxford University Press, 1970), vol. 2, pp. 249-257, 268-270, 354-363; "Firms employing former students. . . between July 1928 and July 1937," archives of University College, London (UCL) (1937), 32/3/3; "Report on the Department of Chemical Engineering, 1937-38," UCL (1938-1939), 32.

48 John A. Oriel, "The chemical engineer in plant operation and management," *Chemical Engineer* (April 1955), xxxviii-xlii; R.H. Simpson, "The chemical engineer in a medium sized chemical company," *Chemical Engineer* (October 1960), A45-A49; Ronald Holroyd, "The acceleration of progress in the chemical industry due to activities during the second world war," *Chemistry and Industry* (June 14, 1969), 766-770; P.W. Reynolds, "Chemical engineering with ICI in Billingham," *Chemical Engineer* (March 1965), CE55-CE56.

49 Donnelly, op. cit. (5), p. 588; Cohen, op. cit. (5), pp. 175-186.

majority of chemical engineers could expect to be assigned in industry. Most were probably destined for jobs in the management or operation of chemical plant, not its design or construction. Although the IChemE said that a training in plant and process design was of great value for would-be manufacturing chemists, the various derivatives of chemistry, such as applied chemistry, industrial chemistry and chemical technology, provided alternative ways of conceptualizing these tasks. For much of the period between the wars, Imperial College was split by these two competing perspectives on how best to train chemists for industrial jobs; either through the medium of the unit operations, or as a body of knowledge best imparted as separate industrial specialities. The University of Leeds also opted for the latter interpretation.[50]

**Conclusions**

Intellectual abstractions evolved along with a professional identity for chemical engineering and the expansion of the heavy chemical industry in Britain. The defining concept for the emerging profession, at least in Britain and the USA, was a set of largely empirical categorizations collectively known as "unit operations." The origins of the concept appear to have had distinct national origins. In the USA, unit operations have been postulated as an outcome of the Taylorist programme in manufacturing, or as a consequence of American enthusiasm for mechanical innovation and large-scale manufacturing. The unit operation, as first coherently articulated by A.D. Little, was defined as an isolable physico-chemical stage in a manufacturing process.

In Britain, by contrast, unit operations initially were popularized as plant-based chemical operations rather than the manufacturing processes per se; that is, the notion of chemical stages was reified in mechanical equipment. The adoption of this intellectual principle was initiated by professional consultants such as G.E. Davis from the 1880s and eventually taken up by academics and professional societies, particularly after World War I. The reasons for the adoption are various: thus for consultants, the concept provided a technical vocabulary for discussing plant design in operational terms without divulging company secrets; while for academics, unit operations generalized and organized the particularities of chemical plant design into a manageable curriculum.

The appropriation of the unit operation and other intellectual constructs as unique features of chemical engineering helped to consolidate the profession and to explicate its works to its clients, to the wider public, and to practising engineers themselves.

**Acknowledgement**

Financial support for some of the research for this chapter was provided by the Institution of Chemical Engineers.

---

50 Donnelly, op. cit. (5), p. 580.

# THE USE OF MEASURING AND CONTROLLING INSTRUMENTS IN THE CHEMICAL INDUSTRY IN GREAT BRITAIN AND THE USA DURING THE PERIOD 1900-1939

STUART BENNETT
*Department of Automatic Control & Systems Engineering*
*The University of Sheffield*
*Mappin Street*
*Sheffield S1 3JD, UK*

**Introduction**

In 1938, Corrington Gill, Assistant Administrator of the US Works Progress Administration (WPA), in a letter accompanying the report "Industrial Instruments and Changing Technology," wrote:

> These industrial instruments increase the productivity of labor in industry by permitting changes in processes, by increasing the productive capacity of installed machinery, by reducing fuel consumption, spoilage of materials, or machinery stoppages, or by providing records on the basis of which management may improve the utilization of its labor force and the flow of work in process.
>
> Skilled workers are eliminated or the number of skilled workers is reduced when instruments serve to standardize or eliminate specialized operations. The judgment of a trained worker is frequently made unnecessary by an instrument which indicates a condition so precisely that an unskilled operative can give the process the necessary attention. Where a new skill is introduced it is likely to be of a different sort, for instance, one requiring an ability to read meters and perform calculations or an engineering training giving familiarity with the entire production process.
>
> When instrumentation renders an entire process highly automatic or continuous, jobs in all categories of skill may be eliminated. The types of instruments sold, as well as the new instruments placed on the market, indicate a shift from manual to automatic control.



*A.S. Travis et al. (eds.),*
*Determinants in the Evolution of the European Chemical Industry, 1900–1939*, 215–237.

> Older industries, whose production techniques have long been established, have been slow in adopting industrial instruments. But even in many of these industries instrumentation has had an important function in improving the quality of the product, reducing waste, and cutting labor costs. On the other hand, many new industries and older industries whose basic production methods have undergone a major change depend upon instruments for their existence and smooth operation. Some of the chemical industries, for example, are founded on processes which could not be used on a commercial basis without instruments adapted to the accurate measurement and maintenance of operating conditions.[1]

Gill succinctly summarized the arguments and justifications being used by the manufacturers and suppliers of instruments and by advocates of their use, such as Major Behar, editor of the journal *Instruments*, and suggested that the consequences of the increased use of instruments were increased productivity, reduction in the need for skilled labour, and ability to use new processes.

Some of the questions which underlie Gill's conclusions are: Was the use of instruments part of a systematic attempt by managers to organize scientific, technical and craft knowledge? Were quantitative measurements supplied by instruments used to transfer knowledge gained in the laboratory to the production process, or were quantitative methods adopted as part of an attempt to apply the "scientific method" in order to gain a better understanding of production processes? Were the qualitative, craft-based measurements replaced because this reduced the power of the craft workers and gave managers greater control over the manufacturing processes? Was automatic control introduced because the process technology demanded it or simply as a way of reducing the amount of labour required? What drove these changes? Was it the need or desire for standard products? Was it de-localization of production and the growth of conglomerates? Were managers motivated by managerial theories and practices, for example, the scientific management movement? Did these changes lead to increased productivity? And, in relation to these issues, are there differences in the extent to which change took place in Britain and the USA, and if so how can these differences be explained?

## Growth in use of industrial instruments

There was sporadic use of instruments in the process industries (including the chemical industry) during the 19th century, and a gradual increase during the early years of the

---

1 The letter, addressed to Harry L. Hopkins, Administrator of the Works Progress Administration (WPA), is included in the report, G. Perazich, H. Schimmel and B. Rosenberg, *Industrial Instruments and Changing Technology*, Work Progress Administration, National Research Project, Report No. M-1 (Philadelphia, PA.:GPO, 1938), reprinted in I. Bernard Cohen, ed., *Research and Technology* (New York: Arno Press, 1980). The Work Progress Administration was the government body which administered the projects set up under the various "New Deal" acts. After 1939, it was called the Works Projects Administration.

20th century, followed in the USA by a rapid increase during the 1920s (see figure 1).[2] Data from the sales of instruments, based on value, does not *per se* demonstrate growth in use. The data shown have not been corrected for general price inflation or for any sectorial price changes related specifically to instruments. However, post 1919 evidence from manufacturers' trade catalogues suggests that there was very little change in the dollar price of recorders and controllers between 1920 and 1935 in spite of the large changes in the complexity of the instruments; hence increased sales in dollar terms can be interpreted as an increase in number of instruments sold. The period 1921 to 1929 was a time of sustained economic growth, and any increase in instrument sales did not necessarily represent a broader, more widespread use. Of greater significance is the ratio of instrument sales to general machinery sales (see figure 2). These show a significant increase in the proportion of money spent on instruments relative to general machinery and this ratio increased during the Depression although the total value of sales of instruments declined.[3]

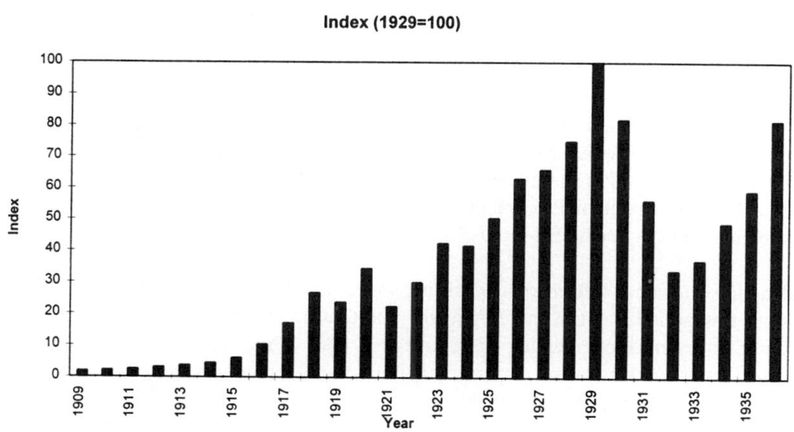


Figure 1. Index of instrument sales in the USA, 1909 to 1935 (Data from WPA Survey, 1938).

[2] Although the data given are based on a sample of seven major companies, and sales data for all the companies was not available for the whole period, the sales figures given are believed to be representative of the industry as a whole; see Ibid., Appendix C.

[3] This data, from the WPA survey, needs interpreting with care since the prices for general machinery may have fallen relative to instrument prices.

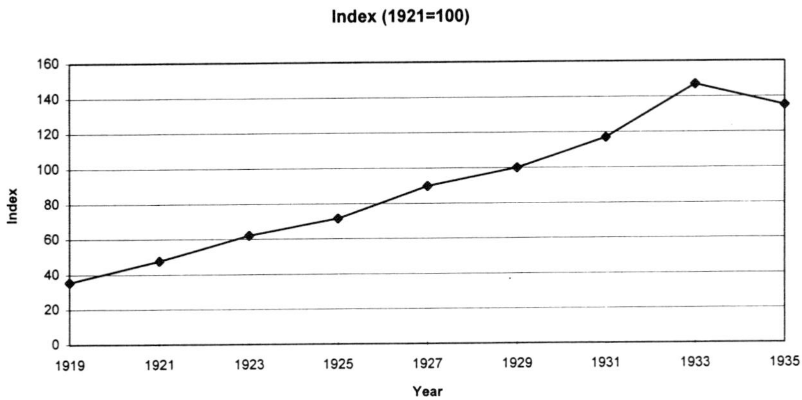


Figure 2. Ratio of instrument sales to machinery sales in the USA, 1919 to 1935 (Data from WPA Survey, 1938).

Other evidence which suggests that the growth shown by the WPA data was real includes growth in the number of companies manufacturing and selling industrial instruments; by the mid 1930s there were over 600 such companies in the USA alone. In 1936, *Instruments* listed over one thousand different measuring instruments.[4] The data given in the WPA survey shows that by 1935 the chemical industry in the USA was a large purchaser of instruments, accounting for 5 per cent of the total sales (the petroleum industry accounted for 16.3 per cent of sales).

Obtaining statistical information regarding instrument sales and use in Great Britain during this period is much more difficult. The Census of Production surveys did not include instrument engineering as a category until 1968. In the earlier censuses, industrial instruments were dispersed within several broad categories and are not included in the re-constructed data given under Order VIII, Instrument Engineering, in the *Historical Record of the Census of Production 1907 to 1970*.[5]

Data compiled by S. N. Broadberry shows that productivity in the British chemical industry measured in terms of output per employee declined relative to that in the USA

---

4 *Instruments*, **11** (1936), supplement "Index of Instruments." This journal, devoted to industrial instruments, was founded in 1929.

5 *Historical Record of the Census of Production 1907 to 1970* (London: HMSO, 1978), p. 88.

over the period 1907 to 1950. Table 1 shows the changes. Productivity is obviously dependent on other factors but they do suggest that the chemical industry in Britain did not adopt the same strategy as that of the USA and, we might suppose, this implies that it made less use of instruments.

Table 1. US/UK manufacturing output per employee (UK=100), from Broadberry.[6]

| | 1909/1907 | 1937/1938 | 1950 | 1967/1968 | 1975 | 1987 |
|---|---|---|---|---|---|---|
| Chemicals | 143.1 | 226.9 | 356.4 | 281 | 226.8 | 152.4 |
| All sectors | 208.6 | 217.9 | 273.4 | 276 | 224.7 | 186.6 |

There was much less difference between productivity in Germany and Britain, with a Germany/Britain ratio of 122.9 in 1935 and 124.0 in 1967/1968.

**Types of instruments.**

Throughout the period under consideration the majority of the instruments in use were concerned with the measurement and control of flow, temperature, pressure, liquid level and speed, and all were based on a small range of physical sensors which converted a change in the quantity being measured to either a voltage change or a change in pressure.[7]

Voltage changes can be converted to mechanical movements by a range of mechanisms, the most common being the permanent-magnet, moving-coil based instruments. These take two forms: indicators and direct acting recorders such as Horace Darwin's, and the Bristol Company recorder shown in figure 3; and recorders which used sensitive galvanometers with electro-mechanical follow-up mechanisms such as the Callendar recorder (Cambridge Scientific Instrument Company) and the Leeds & Northrup recorder.

Pressure changes can be converted into mechanical movement to operate both indicators and recorders by the use of bellows or helically-wound tubes. From the early years of the century circular chart recorders of the type shown in figure 4 became a familiar sight in process plants: gradually control elements were added. These controllers were of two types: an on-off type which makes and breaks electrical contacts and thus acts as a relay (figure 5) and a pneumatic device (figure 6). Controllers of these types were widely used until well into the 1950s. However, they have low sensitivity thus requiring a large change in temperature before there is any movement of

---

6 S. N. Broadberry, "Comparative productivity levels in manufacturing since the Industrial Revolution: Lessons from Britain, America, Germany and Japan," *Structural Change and Economic Dynamics*, **6** (1995), 71-95.

7 In addition to the above quantities, John J. Grebe reported the use of instruments for the measurement of time, specific gravity, volume, humidity, gas composition and ion concentration. See John J. Grebe, Ray H. Boundy, and Robert W. Cermak, "The control of chemical processes," *Transactions of the American Institute of Chemical Engineers*, **29** (1933), 211-256.

the device controlling the heat input; this led to continuous fluctuation in the temperature about a mean value.

Figure 3. Bristol Company Instruments, *circa* 1912 (from Bristol Company Catalogue).

During the 1920s, a pneumatic amplification mechanism, the flapper-nozzle amplifier, that greatly increased the sensitivity of the pneumatic system, was developed. The principle of the flapper-nozzle based controller is shown in figure 7, and figure 8 shows a Foxboro controller *circa* 1922.

The use of the flapper-nozzle amplifier increased the sensitivity of the controller but at the expense of its proportional action. The flapper-nozzle was so sensitive that the controller had a proportional range of about 1 per cent of full-scale movement and hence became the equivalent of an on-off controller. Developments during the 1920s concentrated on trying to modify the flapper-nozzle valve action so as to increase its proportional range, and by 1930 several instrument companies offered controllers with a proportional band of 5 to 7 per cent of full range.

In 1931, the Foxboro company announced its *Stabilog* controller which had a wide throttling (proportional) band and also included "reset action." The advantages of increasing the throttling range and adding reset action are shown in figure 9.[8] The crucial features of the *Stabilog* arose from the discoveries of Clesson E Mason, of the Foxboro Company, and involved using negative feedback around the flapper-nozzle amplifier (this was the pneumatic equivalent of the electronic negative feedback amplifier of Harold Stephen Black, details of which were published in 1934).[9]

Figure 4. External view of a typical controller used from around 1925 onwards (from Foxboro Company catalogue).

[8] The figure is taken from Thomas J. Rhodes, *Industrial Instruments for Measurement and Control* (New York: McGraw-Hill, 1941), figure 175, p. 422.

[9] For further details see Stuart Bennett, *History of Control Engineering: 1930-1955* (Stevenage: Peter Peregrinus, 1993), particularly chapter 2. The date of Black's invention was 1927, a patent application was filed in 1929 and a patent issued in 1937.

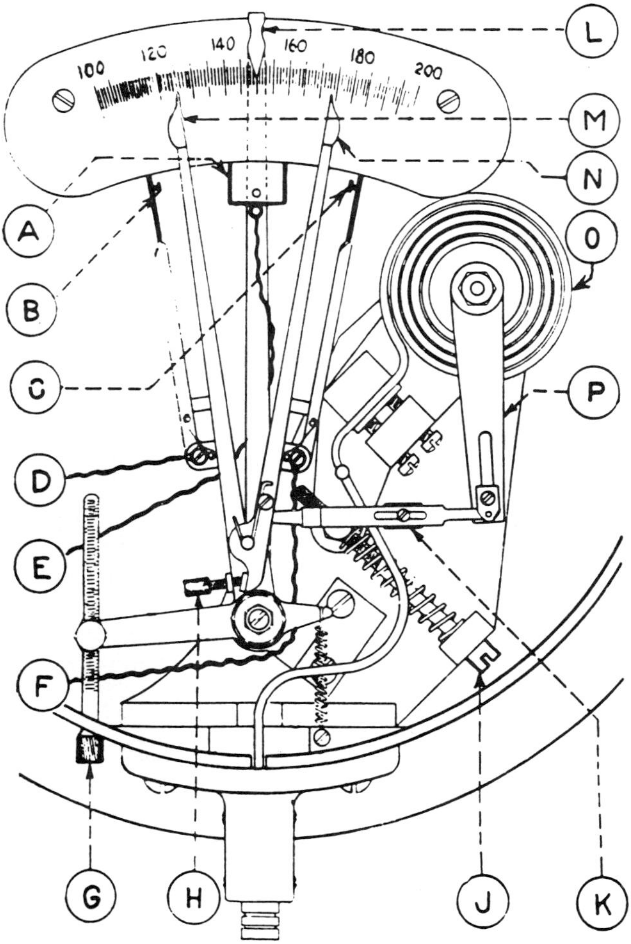


Figure 5. Schematic diagram illustrating an on-off controller, *circa* 1925 (from C. J. Tagliabue Company catalogue 900, 1928).

In the on-off device, the sensing element is connected to the spirally wound tube (O) and changes in pressure cause the tube to move such that a lever (P) moves the indicator (L), which indicates the temperature of the system being measured (about 150°F in the figure). Attached to indicator (L) is a contact (A). Placed either side of (L) are moveable arms, (M) and (N), which indicate the minimum and maximum temperature deviation before some corrective action is taken. These two arms are connected to (B) and (C) which also carry electrical contacts. Thus if the temperature drops to 125°F then the contact (A) (on L) makes contact with (B) and presumably results in some form of heat being supplied to the system. On-off systems result in limit cycling, in which the controlled quantity oscillates between the upper and lower limits (many room temperature control systems still work on this principle, and if the upper and lower limit are too far apart one is alternately too hot and then too cold).

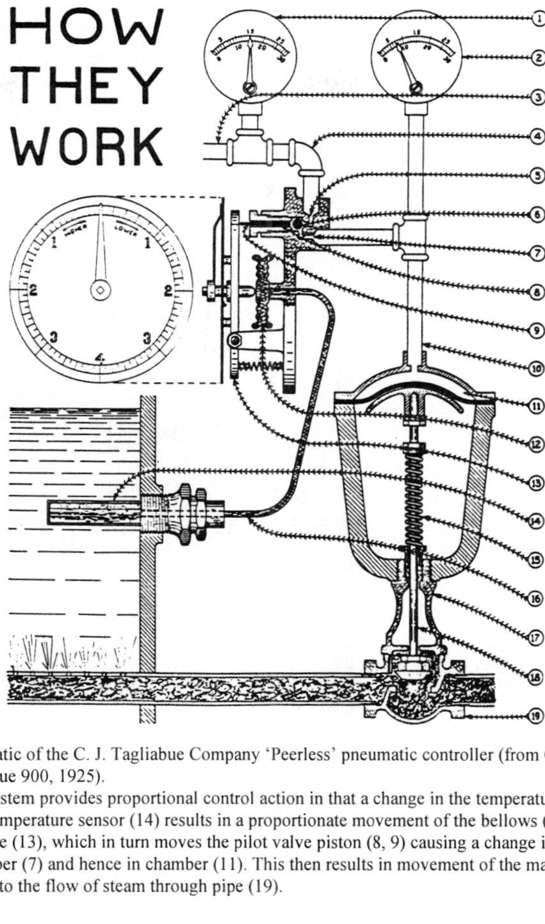


Figure 6. Schematic of the C. J. Tagliabue Company 'Peerless' pneumatic controller (from C. J. Tagliabue Company catalogue 900, 1925).
The pneumatic system provides proportional control action in that a change in the temperature of the bath detected by the temperature sensor (14) results in a proportionate movement of the bellows (12), causing the movement of plate (13), which in turn moves the pilot valve piston (8, 9) causing a change in the air pressure in chamber (7) and hence in chamber (11). This then results in movement of the main valve thus making a change to the flow of steam through pipe (19).

By 1939 all the major industrial instrument companies (Foxboro, Bristol, Brown, Leeds & Northrup, Mason Neilan, Tagliabue, Bailey, Taylor, George Kent) were offering controllers that provided proportional plus reset action. In 1939, Taylor added 'pre-act' and in the same year Foxboro offered a re-designed *Stabilog* with reset and hyper-reset actions. Pre-act and hyper-reset are what is now referred to as derivative action, that is the control signal is made proportional to the rate of change of the error. Process control engineers often still use the term reset but it is now more commonly referred to as integral action, that is the control signal is made proportional to the integral of the error. Proportional plus integral plus derivative control (PID) can be used effectively to control a wide range of processes (often only PI control is needed) and through the work of N. B. Nichols and J. G. Ziegler of Taylor Instruments, details of a simple, empirical method of choosing the optimum proportions of P and I or P, I and D were published in 1942.

AUTOMATIC TEMPERATURE RECORDER-CONTROLLER

**Operating Principle**

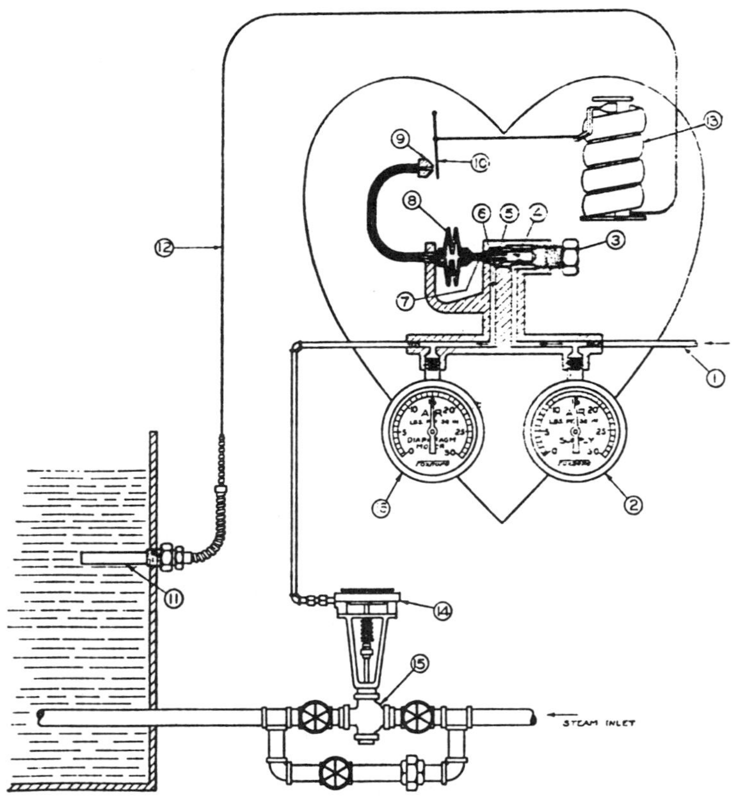


Figure 7. Diagram illustrating the principle of the Foxboro company pneumatic controller (Foxboro Company Bulletin 177-1, circa 1925)
A temperature sensor (11) is connected by a capillary tube (12) to the helical element (13). A change in temperature produces a change of pressure inside the helical element causing it to rotate and move the flapper arm (10) and hence change the gap between the arm and the nozzle (9). Compressed air is supplied through a pressure reduction valve (7) to the nozzle and the bellows (8). The amount of air escaping through the nozzle, and hence the pressure in the pipe connecting the nozzle to the bellows, is determined by the gap between the nozzle and the flapper. Reducing the gap increases the pressure and vice-versa. The pressure changes cause the bellows to expand or contract thus moving the pilot valve (15) which controls the pressure of the air supplied to the diaphragm motor (14). This in turn controls the position of the steam flow valve (15) and hence changes the amount of steam supplied to heat the vessel.

**Use of instruments in the chemical industry**

The WPA survey referred to earlier showed that the major growth in the use of instruments came after World War I and this is borne out by the style of the papers and reports published in the professional and trade journals. For example, accounts in the British *Journal of the Society of Chemical Industry* suggest there was little change in the types of devices used until after 1918; the pattern is similar in the USA in respect of the professional journals. However, in the USA, the more general management journal *Engineering Magazine* from about 1910 carried advertisements and short articles promoting the use of instruments as well as articles on the work of Frederick W. Taylor and explanations of the ideas of scientific management.

Figure 8. Details of internal structure of a Foxboro controller, *circa* 1922 (Foxboro Bulletin 112-2, 1929)

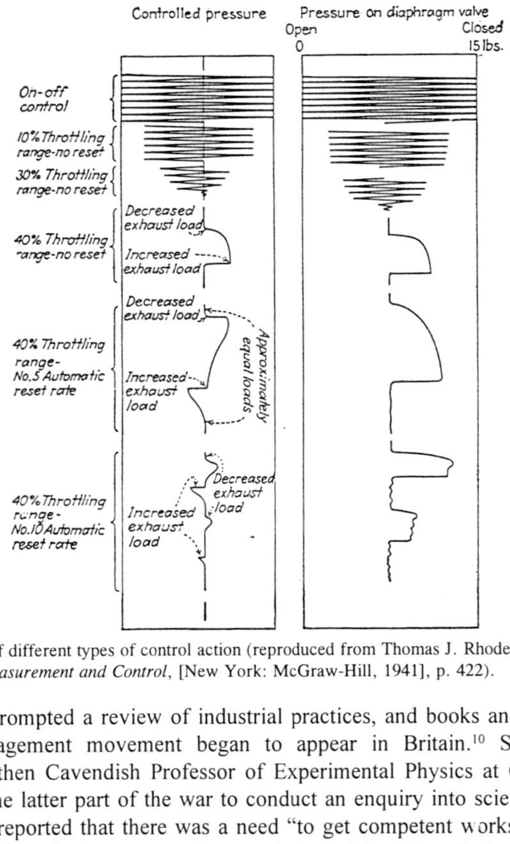


Figure 9. Effect of different types of control action (reproduced from Thomas J. Rhodes, *Industrial Instruments for Measurement and Control*, [New York: McGraw-Hill, 1941], p. 422).

World War I prompted a review of industrial practices, and books and reports on the scientific management movement began to appear in Britain.[10] Sir Joseph John Thomson, the then Cavendish Professor of Experimental Physics at Cambridge, was asked during the latter part of the war to conduct an enquiry into scientific education; his committee reported that there was a need "to get competent works managers with enough knowledge of science, enough appreciation of the point of view of the engineer and enough power of thinking quantitatively on questions of energy."[11] Of course to think quantitatively required measurements.

The then recently formed Chemical Engineering Group of the Society of Chemical

[10] M. A. McKillop and A. D. McKillop, *Efficiency Methods: an Introduction to Scientific Management* (London: George Routledge, 1917), and Sidney Webb, *The Works Manager Today* (London: Longmans, Green and Co., 1917).

[11] Quoted from L. F. Haber, *The Chemical Industry 1900-1930* (Oxford: Clarendon Press, 1971), p. 363. From the "Report of the Committee...on the Position of Natural Science in the Educational System of Great Britain" (London: HMSO, 1918), Cd 9011.

Industry held conferences on pumping machinery in 1919, and on labour saving devices in 1920.[12] The subject matter of these two conferences illustrate some of the changes which were taking place in the British chemical industry: these relate to improving the efficiency of batch processes through the use of pumps to move liquids and slurries, and a range of mechanical handling equipment, cranes, aerial tramways, etc., to move large quantities of other materials. At the 1920 conference the W. & T. Avery Company promoted its range of automatic and self-indicating weighing machinery.[13] Later, in 1926, W. A. Benton, Head of Research of the Avery Company extolled the benefits of self-indicating weighers commenting that these

> are by no means limited to the increased speed of operation, or saving of labour and its resulting fatigue. Careful study of the mistakes made by skilled or unskilled weighmen engaged in technical weighing on steel yard machines shows that while the latter—the unskilled—make mistakes of all kinds, the former generally correctly balance the varying loads, but more or less often make mistakes in reading the weights, or the steelyard indications, or both. That is a very important point, and one which as a result of personal investigation, I have often called [to] the attention of factory and laboratory superintendents. In the case on self-indicating machines, it is very seldom that a weighman reads or records the dial readings incorrectly, and in the case of printing machines the human element is practically eliminated.[14]

Benton was, in an aside to the article, disparaging about the use being made of carbon dioxide and temperature readings for boiler control, arguing that the best method was to record coal and water consumption on charts using Avery's coal and closed water weighers.

The extent to which steam boilers in power plants were both fitted with instruments and automated during the first thirty years of the century makes it reasonable to suppose that steam generating plant used in chemical processing was similarly automated.[15]

---

[12] The second conference of the Chemical Engineering Group, held in Newcastle-upon-Tyne, on 15 December 1919, was on pumping machinery, and the third conference, held in Birmingham on 23 April 1920, was on labour saving machinery.

[13] S. H. Johnson, "The development of automatic weighing machinery and its application to the chemical industry," paper presented at conference on "Labour Saving Devices in Chemical Works," Birmingham, 1920, published in *Proceedings of the Chemical Engineering Group (Society of Chemical Industry)*, **2** (1920), 29-32. S. H. Johnson worked for Messrs. W. and T. Avery, the weighing machine manufacturers.

[14] W. A. Benton, "Measuring and weighing as applied to chemical works," *Proceedings of the Chemical Engineering Group (Society of Chemical Industry)*, **8** (1926), 83-84.

[15] F. P. Stainthorp, writing about process control, commented that there had been extensive use of automatic control since 1920 and that "industry adopted controllers which had been developed particularly for boiler control." He further commented that only since 1945 had control been seen as "an essential and integral part of design." See volume 9 of Herbert W. Cremer and Sidney B. Watkins, eds, *Chemical*

Instruments used on steam boilers included pressure and flow indicators and controllers, carbon dioxide recorders, speed controls for mechanical stokers and liquid level indicators and controllers. As chemical manufacturers began to abandon batch processing in favour of continuous processing plants and also paid greater attention to fuel efficiency the possible uses of the type of instruments employed in steam raising plant must have become apparent.[16]

Evidence for the use of instruments began to appear in the publications of the Chemical Engineering Group: in the 1921 volume of the *Proceedings of the Chemical Engineering Group, Society of Chemical Industry*, a picture of M. Georges Claude and his synthetic ammonia apparatus clearly showed pressure gauges. This is not surprising since safe and efficient operation of high pressure equipment requires knowledge of pressures and temperatures. The industrial use of high pressure and high temperature equipment began in 1913 when BASF used the Haber-Bosch process at Oppau[17] but I have not been able to find much comment on the use of measuring equipment until this later period. In the same volume, a paper on the design of ammoniacal liquor stills showed the chart from a temperature recorder; and there is evidence of the use of control systems in a de-gassing system for boiler feed-water together with the illustration of a flow indicator.[18] However, papers on costings for chemical works published in 1924-25 assumed that all record keeping in chemical works was manual. A paper in the 1926 volume described a control system which appears to have been developed by the Dunlop Rubber Company for controlling both temperature and the timing of operations in the moulding of rubber.[19] Further papers relating to measuring

---

*Engineering Practice*, (London: Butterworths, 1965) p. 316.

16 Haber op. cit. (11), p.3. Papers on boiler plants appeared from time to time in the publications of the Society of Chemical Industry. For example, a Mr. B. A. Oldham read a paper on "Carbon dioxide recorders and their application in boiler efficiency control" at a meeting in Manchester (*Society of Chemical Industry. Review*, **III** (1920), 144). Llewellyn Jones and Frederic I. Scard, in the second edition of their book *The Manufacture of Sugar Cane* (London: Duckworth & Co., 1921, first edition 1909), included a chapter on the "Scientific control of the factory" which was largely concerned with techniques for minimizing the energy used in the refining process, and paid close attention to the efficiency of the boilers.

17 See Meinzer, and Travis "High pressure industrial chemistry," this volume.

18 J. H. West, "The Claude synthetic ammonia process and plant," *Proceedings of the Chemical Engineering Group (Society of Chemical Industry)*, **3** (1921), 33-40. Paul Kestner, "De-gassing and purification of boiler feed water," *Proceedings of the Chemical Engineering Group (Society of Chemical Industry)*, **3** (1921), 41-65 (Kestner was Président of the Société de Chimie Industrielle de France); and P. Parrish, "Observations on the design of ammoniacal liquor stills," *Proceedings of the Chemical Engineering Group (Society of Chemical Industry)*, **3** (1921), 80-109.

19 H. Willshaw, "Moulding and vulcanisation of rubber articles," *Proceedings of the Chemical Engineering Group (Society of Chemical Industry)*, **8** (1926), 111-121. The Bristol Instrument Company claimed to have introduced "process cycle controllers" in 1923, specifically with tyre moulding machines in mind (Bristol Company catalogue 1603, 1932). The illustration in Willshaw makes no mention of the Bristol

systems or the need for such systems appeared in 1926 and 1927.[20] From these passing references to control systems it is difficult to establish whether or not it was simply common practice being mentioned or whether the comments were made because the systems represent a departure from common practice.

Clearly by the 1930s automatic regulators were attracting attention and in 1936 the Chemical Engineering Group held a two day conference on "Automatic Regulators" at which ten papers were read. There were two theoretical papers.[21] Of the practical papers some came from representatives of instrument manufacturers (Douglas Harrison of Negretti & Zambra and W. J. A. Copeland of the Brown Instrument Company) and some from users (W. J. Clark of ICI Billingham, A. Callendar and A. B. Stevenson from ICI Alkali Research Laboratory, and A. D. Elmsly Lauchlan).

Elmsly Lauchlan began his paper with a somewhat trenchant statement: "The modern manufacturer, frequently faced with opposing problems of economy of production and the competitive necessity of a consistently high standard in the finished product, must be interested in methods which give accurate information concerning the various processes which lead up to the final product. For it is this information which enables him to control...both the quantity of the constituents and the quality of the product."[22] He went on to argue that automatic recording and control was far less complicated than many thought, that the cost was offset by the saving in wages, and that it had the advantage that the records could be stored and "irregularities in the finished product

---

Company and the illustration given in the paper appears to be of a different system than the Bristol system. The rubber industry was quick to adopt the mass production and automation practices of the automobile industry. See George Oenslager, "Chemical and engineering advances in the rubber industry," in Sidney D. Kirkpatrick, ed., *Twenty Five Years of Chemical Engineering Progress*, Silver Anniversary Volume, American Institute of Chemical Engineers, (New York: D. Van Nostrand, 1933), p. 187. Oenslager worked for the B. F. Goodrich Company, Akron, Ohio.

[20] See J. W. Hinchley, "The measurement of temperature in technical practice," *Proceedings of the Chemical Engineering Group (Society of Chemical Industry)*, **8** (1926), 122 (Hinchley was a professor at Imperial College, London); J. A. Hall, "Thermo-electric and resistance pyrometry in industry," *Proceedings of the Chemical Engineering Group (Society of Chemical Industry)*, **9** (1927), 9-17 (J. A. Hall worked at the National Physical Laboratory, Teddington); J. L. MacDonald, "Some aspects of the manufacture of fibrous cellulose," *Proceedings of the Chemical Engineering Group (Society of Chemical Industry)*, **9** (1927), 73-84 (MacDonald worked for the Inversesk Paper Company).

[21] A. Callender and A. B. Stevenson, "The application of automatic control to a typical problem in chemical industry," *Proceedings of the Chemical Engineering Group (Society of Chemical Industry)*, **18** (1936), 108-116; A. Ivanoff, "The influence of the characteristics of a plant on the performance of an automatic regulator," *Proceedings of the Chemical Engineering Group (Society of Chemical Industry)*, **18** (1936), 138-151.

[22] A. D. Elmsly Lauchlan, "Electrical control of chemical processes," *Proceedings of the Chemical Engineering Group (Society of Chemical Industry)*, **18** (1936), 87.

can then be more easily discovered at some future date." He also noted that the recorder could be located remote from the plant in the "control laboratory or the foreman's office."[23] Summing up the discussion on this paper, the chairman, Stanley Robson, was more cautious about the value of automatic recording and controlling devices, being particularly concerned that the men might come to rely on them and hence not check the plant as frequently as otherwise.[24] During the discussion of a paper by W. J. A. Copeland there was a plea from a Mr. L. R. West for "uniformity in terminology" a problem which was also exercising both academics and practitioners in America.[25]

The paper by Dr. W. J. Clark of ICI Billingham provides a useful history in the use of instruments at the new plants then being built at Billingham. Brunner Mond had built a plant for ammonia synthesis (the so called No. 2 plant) which started operation in 1923 and a larger plant was built between 1925 and 1927.[26] According to a report written by Dr. R. E. Slade, in 1925, (and quoted by Clark) the No. 2 plant at Billingham appeared to have been designed for manual operation. Slade wrote,

> [u]ntil quite recently, chemical plant was controlled by skilled men who had been specially trained on the particular plant and who had learned what was the right thing to do under given conditions. Their estimate of these conditions was formerly made by the use of their senses – trained by long experience. For example, temperatures were estimated by eye or by touch. More recently thermometers and pyrometers have been introduced, and chemical analyses are often made at short intervals. But usually a great deal more than a few weeks' training is necessary to enable a man to take charge of a plant.[27]

Clark uses that report to argue that at the early stages of development of a process a plant is best run under manual control – though with a large number of instruments, as was done with the ammonia synthesis plant – until the behaviour of the plant becomes understood and then automatic control can be used.

---

23 Ibid., p. 90.

24 Ibid., p. 94.

25 Ibid., p. 123. One of the tasks taken on by the Industrial Instruments and Regulators Committee of the Process Industries Division of the American Society of Mechanical Engineers on its formation in 1936 was to produce a standard nomenclature.

26 See Haber op. cit. (11), p. 292. The ICI Company, which incorporated Brunner Mond, was formed in December 1926.

27 W. J. Clark, "The automatic control of chemical processes," *Proceedings of the Chemical Engineering Group (Society of Chemical Industry)*, **18** (1936), 125.

The American C. J. Tagliabue Company claimed to have been the first industrial instrument company to install controllers in industrial plant when they installed three pressure controllers in a New York chemical works in the early 1900s (various dates are given, ranging from 1900 to 1907).[28] Chemical manufacturing companies are listed among the users of Bristol Company instruments in 1910, although the types of instruments used is not given. During the 1920s there was a very rapid growth in the sales of instruments to the chemical sector, followed by a collapse of sales in 1932 and 1933, the change being much more pronounced than in other sectors of industry surveyed by the WPA. The interpretation put on this data by the WPA researchers was that the introduction of continuous processing placed greater demands on the control required and made the plants difficult to control by manual means, particularly because of the need to synchronize the unit processes "for regulating the flow of materials, or for ga[u]ging temperatures and pressures." They continued: "it is questionable whether the judgement of operators, unaided by measuring devices, could be sufficiently accurate to determine the proper operating conditions, or whether manual operation alone could provide the necessary speed and sensitivity."[29] In the USA, around 1933, Arthur D. Little, an influential consultant chemical engineer, highlighted the importance of high pressure and high temperature chemistry, citing as an example a plant to separate helium from natural gas at Amarillo, Texas. He noted that this plant was "controlled by thermocouples, orifice meters (flow), differential meters for level, pressure gages and automatic helium recorders which give purity of the helium produced throughout the whole 24-hour period and which depend for their operation upon thermal capacity."[30]

The changes taking place in America during the 1920s are perhaps best summarized in the words of Herbert H. Dow, President of the Dow Chemical Company, writing in 1929:

> The most conspicuous labor-savers in a chemical plant are: (1) larger equipment, to cut down the expense of chemical control and operating labor per unit of product: (2) automatic analysis, to save the labor expense of chemists making routine analytical tests at regular intervals and to secure results more quickly and more exactly, with the chance of error due to the personal factor reduced; (3) automatic operation of the equipment, which is governed by the automatic analyzer, all leading to the final desideratum; (4) a

---

28 *Industrial Instruments and Changing Technology*, op. cit. (1), p. 15; see also Tagliabue Catalogue 1920.

29 Ibid., p. 63.

30 Arthur D. Little, "Chemical engineering research," in *Twenty Five years of Chemical Engineering Progress*, op. cit., (19), p. 10.

continuous process which is both automatically controlled and operated.[31]

The Dow Company had found that the available commercial instruments could not meet their needs and between 1924 and 1929 the control problems which they faced led them to develop their own automatic control equipment. They used what they termed "electrometric control," that is, control based on measurements of oxidation potentials, conductivities and ion concentration measurements. At first they used on-off control, which was found to be worse than manual control, and later proportional action, which resulted in success in controlling a neutralization reaction they had found impossible to control manually. They experienced difficulties in 1928 when they tried to control a plant which had not been designed for easy control (control engineers today still complain that plant designers do not take controllability into account at the design stage) and this led them to modify their controller to one which provided a form of a proportional plus derivative plus integral control action. They also argued for the use of so-called ratio control to offset transport lags.[32]

In 1936, John J. Grebe provided information to *Instruments* which illustrates the extent of the Dow Chemical Company's use of instruments. He reported the use of 486 temperature indicators and recorders, 64 automatic temperature controllers (not counting thermostats), 72 electrometric recorders and automatic controllers (for hydrogen ion, oxidation potential, gas analysis, conductivity, and specific gravity, all these of special Dow design), 15 flow indicators, recorders and automatic controllers, 5 time recorders, 147 pressure recorders and automatic controllers, and 20 level recorders and sundry other devices.[33] The number of temperature and pressure instruments in the above list is significant since it was in these areas that the commercial suppliers had the greatest experience and the most reliable sensors. Such devices also found use in both batch and continuous processes. Some instrument companies were aware of the need for instruments which could measure quantities such as hydrogen ion concentration, but they were concerned about the potential size of the market.

The chemical and petro-chemical industries were the leaders in the adoption of continuous processing and hence were the first industries to confront the particular measurement and control problems which arise from continuous processing. The overall control problems of such processes can often be simplified if some variables can be tightly controlled, that is held at some pre-determined value. The emphasis thus began to shift from simple controllers which held variables within reasonable values to

---

[31] Herbert H. Dow, "Economic trend in the chemical industry," *Industrial and Engineering Chemistry*, **22** (1930), 113-116.

[32] John J. Grebe, Ray H. Boundy, and Robert W. Cermak, "The control of chemical processes," *Transactions of the American Institute of Chemical Engineers*, **29** (1933), 211-256.

[33] *Instruments*, **9** (1936), 58.

more accurate and faster operating controllers.

By 1938, M. G. Mostin, Chief Chemist of Westvaco Chlorine Products, Newark, California, commented that "use of indicating, recording and controlling instruments...is so common that for a chemical plant to be without an instrument department is the exception rather that the rule."[34] There were also some voices of caution. E. F. Hanford of Lever Brothers Company, Hammond, Indiana, commenting on Sergei D. Mitereff's paper on "Principles underlying the rational solution of automatic-control problems," published in 1935, remarked that having installed controllers an engineer may return later to find that the operators have disconnected the devices because they have found that they could do better using manual control.[35] And, in 1939, Donald J. Bergman of the Universal Oil Products Company stressed that if operators were to learn to trust indicators, recorders and automatic controllers, then they should be reliable; he also pointed out that in some plants there had been a minimal of automatic control on the grounds that the operator might come to place too much trust in the controller and, in the event of a major disturbance, seek to adjust the controller rather than seek out the cause of the disturbance.

S. N. Broadberry has assembled evidence on productivity levels and labour and energy costs in a number of industries (including the chemical industry) for the USA, Britain and Germany. For example, over the first half of the 20th century productivity levels (measured in terms of output/worker-hour) were much higher (two to three times) in the USA than Britain (and over this period there was little difference between productivity levels in Britain and Germany). Comparing the relative costs of labour and energy in the USA and Britain he shows that over the period 1924-50 the ratio of labour cost/energy cost was between 2 to 4.5 times greater in the USA when compared to Britain. The difference arose partly from the adoption in the USA of large-scale continuous plant which resulted in lower unit energy costs, and partly from the higher labour costs in the USA. Figures quoted by Haber suggest that in the mid-1920s workers in chemical plants in the USA were paid between 2.5 and 4 times more than those in Britain.[36] This implies that any job eliminated through automation and use of instruments produced higher savings in America than in Britain. Given the higher cost of energy in Britain (relative to America) one would have expected British manufacturers to make more use of instruments and automatic control both in applications which reduced use of energy and those which increased efficiency of

[34] *Instruments*, **11** (1938), 11.

[35] E. F. Hanford's discussion of the paper "Principles underlying the rational solution of automatic-control problems," by Sergei D. Mitereff, *Transactions of the American Society of Mechanical Engineers*, **57** (1935), 159-163, in *idem.*, **58** (1936), 63. Donald J. Bergman, "Application of automatic control in the oil industry," *Transactions of the American Society of Mechanical Engineers*, **60** (1938), 651-656.

[36] Haber, op. cit. (11), pp. 389-390.

energy production. However, there is little evidence of this: W. J. Clark's paper makes no mention of this and simply states that the important factor is "improved steady running" of the plant.[37]

Broadberry argues, on the basis of a model which assumes a common technological knowledge base, that divergence in technological practice within countries or within particular industries arises from differing cultural and economic factors.[38] His evidence supports the argument that at any given time there is not "one best technological way" to operate an industry and that several different ways can lead to equal competitiveness in terms of price and quality.

Once a technology begins to follow a particular trajectory, then innovation becomes local, that is it seeks improvements which do not change the fundamental direction of the path. Similarly external changes produce a response which seeks to maintain the existing technological trajectory. The chemical industry in Britain grew on the basis of small units, with a flexible approach to production, and with great reliance on the skills of a craft-based labour force. Its response to the rationalization movement of the 1920s was to preserve the long tradition of craft work and supply of products to order, a tradition inimical to the ideas of standardization and mass production, and to seek protection through the formation of cartels. In this they were supported by a political climate that favoured the protection of employment. In such a climate there was no perceived need for instruments and automatic control.[39] The response in the USA was to seek to improve production processes and to move towards greater standardization of products, thus enabling the use of mass production techniques. The use of measuring and controlling instruments was central to these changes in that their use gave managers greater knowledge of, and control over, the whole of the production processes.

While the Broadberry model is useful, I do have concerns about the assumption of a common base of technical knowledge. Manufacturers are fully aware of the commercial value of technical knowledge and "know how" and are reluctant to publish details. One can perhaps assume a shared, common, scientific knowledge.[40] This does not entirely

[37] W. J. Clark, op. cit., (27), p. 126.

[38] S. N. Broadberry, op. cit., (6); see also S. N. Broadberry, "Manufacturing and the convergence hypothesis: what the long run data show," *Journal of Economic History*, **53**(4) (1993), 772-795.

[39] 'Rationalization' was defined in 1927 as "the methods of technique and organisation designed to secure the minimum waste of either effort or material." It was a potent idea during the 1920s and early 1930s and from it flowed ideas about standardization of both products and production processes, as well as mergers and closures of companies and the elimination of production units. This definition was agreed at the World Economic Conference, Geneva, May 1927; see J. Davidson Pratt, "Rationalization - its meaning and application with special reference to the chemical industry," *Proceedings of the Chemical Engineering Group, (Society of Chemical Industry)*, **11** (1929), 48-57.

[40] I do not wish to enter into arguments about the differences between 'technical' and 'scientific' knowledge here. I believe my argument still holds if one replaces scientific by 'open' and technical by

invalidate the model since it can work on the basis of how the available common technical knowledge is utilized. However, the possibly more interesting questions are: How, in the two traditions, is new technical knowledge generated? By whom is knowledge promulgated within and outside the organization? And who 'owns' the knowledge? The British approach, craft based, left ownership of knowledge in the hands of the skilled production workers. Inevitably this form of knowledge was less 'open' and more slowly transmitted – through apprenticeship training and through the movement of skilled workers.

The ownership of technical knowledge is important. In the British system managers and capitalists found it much more difficult to control both in terms of protecting knowledge from competitors but also in terms of codifying it and transmitting it to other employees. The apprenticeship schemes relied on skilled workers to pass on their knowledge. Also, because managers and engineers were not fully aware of how the plant was actually operated, it was more difficult to use knowledge gleaned from other plants, or from scientific developments, and also to systematically and routinely generate patents that might protect new developments. The strength of the system, however, was the detailed and widespread experience and knowledge of the process which operatives developed, and an ability to cope with slow changes in plant operation due to ageing of the plant and changing requirements.

Following from Taylor and the scientific management movement, the emphasis in the USA was for the manager and supervisor to instruct labour as to the "one best way," and this necessitated managers taking control of knowledge.[41] For the American model to succeed the manager needed to know more and more about the plant and to keep up-to-date with changes in plant performance. To do so required information taken either by manual recording, or through the use of recording instruments. Thus Dan Gutleben, Chief Engineer of the Pennsylvania Sugar Company, reported that in 1927 his firm installed $6000 worth of flowmeters in the pan steam lines to gather data for the design of a new plant.[42] Gutleben also reported the surreptitious installation of recording

---

'compact' or closed. See also Helge Kragh's contribution to this volume.

41 The issue of labour efficiency and scientific management appears to have been taken up much more quickly in America than in Britain. See for example, Haber, op. cit., (11), p. 377, who notes that Du Pont formed a "Labor Efficiency Department" in 1911 and the staff of this department published a booklet on *Scientific Management in Powder Making* in 1912. The Hoover Committee reported that the commentators from outside the USA considered that the idea of "scientific management" and the search for the 'optimum' result were major American contributions to economic developments. *Recent Economic Changes in the United States*. Report of the Committee on Recent Economic Changes of the President's Conference on Unemployment, Chairman, Herbert Hoover, (New York: McGraw-Hill, 1929), Vol. 1, p. 5. The same report commented that the outstanding theoretical contribution to production in industry was 'standardization' (see p. 80).

42 Dan Gutleben, "Instrument savings? Peek into my diary," *Instruments*, **13** (1940), 6.

meters to check on the operators. While the surreptitious installation is, perhaps, unusual, the theme of installing recorders to check on operators' performance is one that runs through the American literature from the 1890s until the 1940s: from Herbert Dow in 1893, installing an iron wire across a bromine outlet so that if the operator did not exercise careful control the wire would be corroded and a weight suspended from the wire would fall; to the Thwing Pyrometer Company which in a 1938 brochure declared, "the moral effect of Recording Pyrometers upon workmen cannot be overestimated ...men are most efficient when tab is kept on their actions."[43] The arguments were not confined to the issue of being able to rely on workers doing their jobs, but also extended to the ability to recruit and retain workers. This was put at its most crudest in the *Engineering Magazine* of 1913 in which, referring to boiler firing, the writer argued that "intelligent systematic hand firing" was superior to the use of mechanical stokers but it was difficult to secure and retain "intelligent men" at the "prevailing firemen's wage," with the "result that most boiler plants are now manned by foreigners of insufficient capacity to master the first principles of efficient firing."[44] This view, with different degrees of subtlety, is found repeatedly in the engineering and management literature of the period.

**Conclusion**

In both the USA and Britain there were many common factors which promoted the growth in the use of measuring and controlling instruments in the chemical industry. For example, from around 1920 professionally trained chemical engineers and chemists with some engineering training (as opposed to chemical analysts) began to take up employment in the industry. This group had both the scientific and quantitative skills identified by J. J. Thomson in his report and formed a natural link between the laboratory (both academic and industrial) and the plant. They had been educated in an environment which was placing greater emphasis on measurement and one in which the "scientific method" was increasingly seen as a valuable intellectual tool. Inevitably they became interested in obtaining measurements from the plant, both to apply knowledge gained from laboratory experiments and to study the plant scientifically in order to improve its performance. In both Britain and the USA they began to be targeted by instrument suppliers as potential purchasers. The availability of instruments (both actual and potential) made it possible to consider the adoption of technological processes which could be implemented only with their aid, either because important parameters could not be measured by the human senses, or because the reactions took place at a rate too fast for human control.

A close examination of the technology and performance of the industrial instruments of the period reveals that they could not, and did not, perform, as well as was claimed by

---

[43] Thwing Pyrometer Bulletin no. 10 (1938) (NMAH Archives, Smithsonian Institution, Collection 215, Instruments Society of America [ISA], Box 70).

[44] *Engineering Magazine*, **46** (1913), x.

their proponents. In the majority of applications automatic controllers were unable to offer the closeness of control achievable by operators. So although the introduction of instruments in the chemical industry was driven by a mixture of social and cultural factors, and by technological needs, I contend that until late in the period being considered the social and cultural factors were dominant. The social structure of the work place in America was based around the existence of a managerial bureaucracy. (It does not seem to matter whether this was 'scientific' or 'welfare' in ethos; in both cases it was accepted that managers controlled the work place). Instruments were able to provide the eyes and ears of managers in their absence and automatic control devices guaranteed that their instructions were carried out. In Britain, 'amateurism' continued as the prevailing characteristic, and "training for management was rare;" the work place was controlled by skilled workers who, even if they had wanted to, did not have the means or the power to introduce the use of instruments. And while this group might have welcomed measuring devices as a means of increasing their control over the product, they would have been much less likely to welcome automatic control devices that replaced workers. These differences in culture, I believe, partially explain the differences in the rate of adoption of instruments in the two countries.

# NORWEGIAN CAPITALISTS AND THE FERTILISER BUSINESS: THE CASE OF HAFSLUND AND THE ODDA PROCESS[1]

KNUT SOGNER
*Norwegian School of Management*
*PO Box 580*
*N-1301 Sandvika*
*Norway*

## Introduction

Norway is recognized as home to one of the most significant inventions in the history of agricultural chemistry, namely the electrochemical fixation of atmospheric nitrogen in 1903 that led to the foundation of Norsk Hydro. It is by and large unrecognized that another important invention in agricultural chemistry also took place in Norway, the Odda process for the production of phosphoric acid. The Odda process was meant for – and ideally suited to – the production of highly concentrated three-component fertiliser, containing nitrogen, phosphorus and potassium. It was invented in 1928 within the company Odda Smelteverk of Odda in western Norway. Odda is thus the name of a chemical process, of a company and of the town where the company is situated.

It is ironic that the Odda process is almost forgotten, even in Norway. This is especially so when we consider that although the electrochemical fixation of nitrogen was a scientific landmark that has been justly celebrated, laid the foundation for the largest Norwegian industrial enterprise for most of this century (Norsk Hydro), and made its

[1] This article is based on papers presented at conferences held at Oppdal, Norway ('Den norske historiske forenings årlige konferanse'), Rjukan, Norway (a joint Norwegian-German conference on innovation) and Strasbourg, France. A differently focused version is in preparation for the proceedings of the Rjukan conference, 'The rise and fall of Norwegian managerial capitalism. The case of Hafslund'. This article has benefited from comments from and conversations with several colleagues, notably Rolv Petter Amdam, Håkon With Andersen and Harald Rinde. I would especially like to thank Lars Thue for his incisive criticisms that encouraged me to investigate new sources and enhance my arguments, Yngve Nilsen for his thesis bringing the Kiær group to my attention, and Ketil Gjølme Andersen for several fruitful conversations about Norsk Hydro and the Odda process. Andersen and Gunnar Yttri are writing the history (in Norwegian) of Norsk Hydro's research and development.



*A.S. Travis et al. (eds.),*
***Determinants in the Evolution of the European Chemical Industry, 1900–1939,*** 239–256.

inventors Kristian Birkeland and Sam Eyde famous, it was soon completely eclipsed by the more economical Haber-Bosch ammonia process. Moreover, Norsk Hydro changed over to the Haber-Bosch process.[2] The Odda process, by contrast, invented in 1928 in the aftermath of I.G. Farben's introduction of processes for highly concentrated three-component fertiliser, Nitrophoska for short, turned out to be used on a much larger scale and for a much longer period of time. According to just one estimate, the Odda process was used in four countries, Norway, Germany, Holland, and France, for the production of 233,000 tons of three component fertiliser in the 1960-61 season, with a sales value of about $US 750 million in today's currency.[3]

Odda Smelteverk never used the process itself, however, but licenced it to I.G. Farben, Staatsmijnen of the Netherlands, as well as to Norsk Hydro. Odda stayed in its oldest businesses, that of producing calcium carbide and cyanamide. Neither did Odda's owner, the Hafslund group, use the Odda process. In 1937 it sold Odda Smelteverk along with the Odda process to the British Oxygen Company.

The aim of this paper is to understand the fate of the Odda process, and explain why it was invented at but never used by Odda Smelteverk and the Hafslund group. To achieve this, the paper is divided into three parts. The first part discusses the Hafslund group's strategic intent with respect to Odda Smelteverk and the Odda process. The second part looks into the Odda process itself, while the third part discusses why Hafslund did not exploit the Odda process.

**Part I**

Hafslund was a fairly new manufacturing company in the interwar period, starting out in 1919 by merging its electricity business with the remnant of a foreign owned carbide factory. Historically, however, Hafslund was already an important Norwegian industrial enterprise. It was formed in 1898 as the first main new Norwegian company to exploit hydroelectric power for industrial purposes.

Hafslund was very much the achievement of the Norwegian engineer Knud Bryn, who led the company from its inception until he retired in 1928. Bryn originally intended to start aluminium production in 1895, and had close contacts with the American aluminium company Pittsburgh Reduction. But in the end Pittsburgh and the Swiss aluminium company at Schaffhausen made a deal to keep Hafslund out of this business.[4] Bryn then used his German connection Schuckert – he had been their

[2] Kr. Anker Olsen, *Norsk Hydro gjennom 50 år. Et eventyr fra realitetenes verden* (Oslo: Norsk Hydro, 1955).

[3] VIM (Vestnorsk Industristadmuseum, Odda - The Museum of Industrial Places of Western Norway, Odda), OLA/154, Ea:2(4): 'Anslått produksjon av fullgjødsel i henhold til "Odda-prosessen" i sesongen 1960/61,' unsigned and undated. Estimated value in this source: 700 million Norwegian kroner (1961 level).

[4] Carl Just, *Aktieselskapet Hafslund 1898-1948*, (Oslo: Fabritius, 1948), pp. 38-40; HHO (Hafslund's historical archive), 1 (central archive), 'eldre arkivmasse, 0. Fellesklasse': 'Aktieselskapet Hafslunds

Norwegian agent, and Schuckert was involved in the building of the power station of Hafslund – to develop processes for the production of calcium carbide instead. Right from the start Hafslund was listed on the Oslo Stock exchange, but both the huge property crisis in Kristiania (now named Oslo, then and now the capital of Norway) after 1899 and uncertainties of carbide production discouraged Norwegian investors. Thus Schuckert became a bigger owner of Hafslund than it had bargained for, i.e. it came to own more than half the company, even though the carbide factory was owned as a separate company by a group with its formal headquarters in Switzerland.[5]

The optimistic plans for creating industry around and connected to Hafslund also had to be scrapped, for two reasons. First, there was a lack of capital. Second, and this was the factor that gave further direction to the development of Hafslund for some time, general use of electricity started to grow more rapidly than expected. This increased the demand for Hafslund's electricity, and also led to added interest in the electricity-producing potential of the Glomma river, where Hafslund was situated. New investors wanted to exploit two other major sites on the river for power generation, and Hafslund/Schuckert felt obliged to act and put these two power stations under the roof of Hafslund.[6] They thereby took control of the regional market for electricity. Hafslund thus became an energy company partly by accident, supplying electricity to several industrial companies and large areas of south-eastern Norway, the capital of Norway included. These activities kept Hafslund occupied until World War I.

During World War I, Hafslund was bought by a Norwegian group and gradually transformed into a manufacturing company. In a widely publicized speech in 1917, chief executive officer (CEO) Knud Bryn of Hafslund again signalled his strong interest in industrial matters, emphasizing both the need for national self-sufficiency in raw materials and the importance of industrial research and development.[7] From January 1919 Hafslund became a chemical manufacturing company when the Hafslund carbide-factory was bought from the Swiss group. Both this purchase and the purchase of Schuckert's shares in Hafslund were influenced by World War I, and especially the carbide factory's contribution to the German war effort, which was problematic in Norway.[8] The new owners, dominated by a group known as the "Kiær-Fearnley" trust

---

stiftelse' by Knud Bryn, 6.3.1928 (hereafter referred to as Bryn, op. cit.).

5 Bryn, op. cit. (4), p. 20.

6 Ibid., pp. 10-20.

7 John Peter Collett, "Videnskap og politikk. Samarbeide og konflikt om forskning for industriformål 1917-1930," unpublished hovedfag's thesis, University of Oslo, 1983, pp. 81-83.

8 See Bryn, op. cit., pp. 20-23.

at the time, also ran a carbide factory in mid Norway, Meraker Smelteverk. The executive officer of the latter, Iver Høy, was moved south to run the carbide factory at Hafslund, together with Meraker Smelteverk.[9]

From 1919 to 1928, Hafslund expanded as an industrial manufacturing company, not as an energy company. In 1924 the Hafslund-Meraker group founded a third company, Odda Smelteverk, to use the production facilities left there after the collapse of Swedish and British interests. Hafslund became the largest owner of Odda, Meraker's parent company took a big share, and the rest was divided among the owners of Hafslund and Meraker, the so-called Hafslund-Meraker group. The same owners took one third of the share capital in Tyssefallene, an energy company responsible for providing electricity to the Odda factories.[10] Hafslund took over Odda for two reasons. First, it controlled an important producer of carbide. Second, Odda had production facilities for the nitrogen fertiliser cyanamide, produced from carbide. Thus Odda specialized in fertilisers, while Hafslund and Meraker expanded their carbide sales by taking over Odda's quota in the international carbide syndicate.[11]

Hafslund thus acted according to a promising, yet troubling, situation regarding calcium carbide. On the one hand, due to the special situation during World War I and the general autarkic policies of the period, after 1918 several countries had built up their own capacities to produce calcium carbide, most notably Germany, an important outlet for Norwegian carbide before the war.[12] This led to a crisis of overproduction and lowered prices. On the other hand, partly in response to the lowered prices for calcium carbide, carbide became an attractive intermediate, and during the 1920s and 1930s several new and important processes and products using carbide were developed.[13] In Norway, as elsewhere,[14] there was a movement from World War I onwards to establish a national research policy for industries that were strategically important; an institute for carbide research was among the new institutions suggested. This never materialized, however, not least because of the crisis in the industry itself in the early 1920s. There

---

9 Just, op. cit. (4), p. 248.

10 HHO, 1, minutes of board meeting, meeting April 21, 1937, appendix: Heyerdahl's statement.

11 Bryn, op. cit., p. 24f. After a period of chaos among the carbide producers in Europe, 19 European carbide firms formed an international syndicate early in 1924. The Hafslund group, which was originally responsible for one third of the syndicate's total production, was actively involved in its formation, and Iver Høy became the syndicate's vice-president. This syndicate, modified several times, operated until World War II broke out.

12 Collett, op. cit. (7), pp. 218-9.

13 See contributions by Morris, and Travis, "Modernizing industrial organic chemistry," this volume.

14 See MacLeod's contribution to this volume.

were other factors: thus Hafslund supported the initiative, while another company, Elektrokemisk, a research-based company that founded several important new companies, among which Norsk Hydro is the most famous, was reluctant because of its own carbide research.[15] Little seems to have come out of Elektrokemisk's research, however. The Hafslund group followed two different paths: it used its new foothold in the fertiliser business to diversify into that business; and, as described below, it used its size to dominate international carbide sales.

The Norwegian carbide industry was resourceful and significant enough to influence international trade. The combination of Hafslund-Meraker-Odda created a formidable carbide producer. In the words of Knud Bryn in 1928: "Hafslund and Odda are completely superior to the other European producers of carbide, both because they have the cheapest energy and because they are located at sea-ports that are open the year round."[16] The Hafslund group dominated European carbide exports in the interwar period, forcing the international syndicate members to produce for their own home markets, while the Hafslund group, with its small home market, took about 50 per cent of the total export market.[17] Hafslund and its representative Iver Høy were skillful players in a tough game, and often the trump card in negotiations was the possibility of dumping the Hafslund group's large production on some producer's home market.[18]

For all practical purposes, Hafslund, Meraker and Odda were one and the same industrial company, even though they were run as different legal entities. Their common activities were also coordinated by legally binding commercial arrangements. In 1921 Hafslund and Meraker made a deal to avoid competition. Odda joined this arrangement when founded in 1924, and the three companies formed one entity inside the international carbide syndicate. They shared their common quota with 24/62 parts to Hafslund, 20/62 parts to Odda and 18/62 parts to Meraker.[19] Even after Meraker and Odda were sold by Hafslund, they continued to operate as one entity on the international market.

Hafslund as a company was an actor in a context characterized by Alfred Chandler as cooperative capitalism.[20] Cooperative capitalism is a kind of associative capitalism,

---

15 Collett, op. cit., pp. 216-220.

16 Bryn, op. cit., p. 33.

17 VIM, OLA/157, Ab, 1: Høy to Heyerdahl March 9, 1936.

18 Anders Rokne, *Odda Smelteverk A/S i femti år 1924-1974*, (Odda: Odda Smelteverk, 1974).

19 HHO, 1, minutes of board meeting, meeting April 21, 1937, appendix: Heyerdahl's statement.

20 Alfred D. Chandler, *Scale and Scope. The Dynamics of Industrial Capitalism*, (Cambridge, Mass.: Harvard/Belknap, 1990), and *idem*, *The Visible Hand. The Managerial Revolution in American Business*, (Cambridge, Mass.: Belknap/Harvard, 1977).

typified by companies connected through cartel agreements.[21] Hafslund was for a long time more or less a subsidiary of the German company Schuckert, later Siemens-Schuckert. Hafslund's technology – both for the production of electricity and the production of carbide – came from Schuckert.[22] Later, as an industrial company, Hafslund participated in two kinds of associations: as a company in a Norwegian group, and as a part of this group within international syndicates.[23] As a national energy company, Hafslund's activities were to a large extent regulated by different forms of contracts between Hafslund, other companies, communities, regions and the state.[24] Hafslund cannot be said to have sold its products on free markets.

Until it was reorganized in 1928, Hafslund functioned much like equivalent *German* companies. Hafslund collaborated by market-sharing activities, and made efforts to compete by developing new production processes. These competitive efforts were influenced by the managers' room for manoeuvre. Ever since its beginning, Hafslund had been concerned with huge tasks that demanded enormous managerial effort. During its early years as an electricity company, Hafslund built extensive distribution networks as well as large production facilities. This meant a lot of complicated deals, and CEO Knud Bryn led complex negotiations with local and regional authorities, as well as with the government. Bryn's most spectacular effort contributed to the government's fall. From 1903 to 1906 Bryn successfully led a group of land owners in an effort to regulate the water supply of Glomma, Norway's longest river, by regulating the level of water in Mjøsa, the biggest lake of Norway. A regulated Mjøsa more than doubled the amount of electricity Hafslund produced.[25] The government's controversial handling of the matter contributed to its downfall.[26]

It is possible to characterize Hafslund as one of those new enterprises of the second industrial revolution that demanded a totally new level of managerial effort. The term managerial revolution, associated with Alfred Chandler's analysis of the USA, possibly emphasizes the new role of management a little too much for the European situation.

---

21 J. Rogers Hollingsworth, Philippe C. Schmitter and Wolfgang Streeck, eds, *Governing Capitalist Economies*, (New York and Oxford: Oxford University Press, 1994).

22 Just, op. cit.

23 Just, op. cit., A. Høy, *A/S Meraker Smelteverk 1900-1950*, (Trondheim: Meraker Smelteverk, 1950); A. Andersen et al., *Smeltedigelen - en industrisaga. Odda Smelteverk gjennom 80 år*, (Odda: Nord 4 og Odda Smelteverk, 1980).

24 Just, op. cit.

25 Ibid., p. 98.

26 Lars Thue, *Statens kraft 1890-1947. Kraftutbygging og samfunnsutvikling* (Oslo: Cappelen, 1994), pp. 152-154.

But, as Chandler made clear in his book *Scale and Scope*, German and American industrial developments shared certain organizational similarities to the extent that both countries contained large hierarchically built companies. According to Chandler, British owners were much too concerned with maintaining organizational control – as opposed to control of markets – to fully exploit the new technological possibilities. It was a form of personal capitalism.[27] Hafslund took control of its regional market for electricity through intense managerial efforts, and can well be classified as a Norwegian example of the new managerial phenomenon.

Knud Bryn himself saw Hafslund's organizational structure as a German type. The company was led by the board of supervisors – or the 'Aufsichtsrat' as Bryn called it. He, as CEO, a supreme court lawyer, and the three executive officers of Hafslund, constituted a board of directors – Bryn's 'Vorstand.' This arrangement left room for managerial manoeuvre. The point of having so many executive officers was, as Bryn put it, "because the purpose is to give ... managers more independent responsibility."[28] The three executive officers each had the power of attorney. Within the scope given to them by the board of supervisors, the three could act on behalf of the company.[29] Two of the executive officers led the power business; while the third, Iver Høy, led the industrial manufacturing activity started in 1919, which from 1924 onwards meant all of the Hafslund group's three factories. Under Høy's leadership there developed a new trump card, namely industrial research and development.

**Part II**

The invention of the Odda process at Odda Smelteverk was thus no coincidence. It was the company's strategy to develop this large producer of the nitrogen fertiliser cyanamide also as a producer of *new* fertilisers. Odda produced 57,000 tons of cyanamide in 1928, mostly for export, and it had been experimenting with new processes for several years, and filing relevant patents, when it discovered an economical way to produce phosphoric acid.[30]

The phosphoric acid invention was also an outcome of the Hafslund Group's own strategy. In 1922 – before starting Odda Smelteverk – Hafslund considered the possibility that the Odda factory might produce new fertilisers that contained both nitrogen and phosphate, based on French processes.[31] Hafslund was thus part of a

27 See Chandler (1990), op. cit. (20).

28 Bryn, op. cit., p. 41.

29 HHO, 1: Board meeting June 7, 1928, HHO, 2 (Hafslund Manor), 'hylle nr. 46,' 'direktionsmøterne Hafslund 1914-19': 'Reglement for Adm.[...]komiteen 21.12.17'.

30 VIM, OLA/157.1, Za, 1: Report for the year 1928.

31 HHO, 2, 'hylle nr. 62,' Karbidfabrikken, diverse 1919-25: Høy to Hafslund's and Meraker's board of supervisors, September 11, 1922.

programme dedicated to manufacture of fertilisers containing highly concentrated levels of both nitrogen and phosphate. In March 1928 – seven to eight months prior to the invention of the Odda process – Knud Bryn, CEO of Hafslund, confirmed that the Hafslund group saw Odda's future as a producer of synthetic fertiliser: "The production will be even more profitable if the Odda factory is upgraded as a producer of synthetic fertiliser."[32]

Research starting from calcium carbide took two main paths. One went through acetylene, and another through (calcium) cyanamide. Both led to resins and plastics,[33] but the path through cyanamide also led to fertilisers, an active research field in itself. It is noteworthy that the formation of I.G. Farben in 1925, the combination of large German dye-manufacturers, immediately led to a new diversification strategy for the merged company. One important route that started from calcium carbide was the production of synthetic rubber. Another line started with synthetic ammonia (a very important product for I.G. Farben) and concerned new fertilisers. Under the research leadership of Alwin Mittasch, I.G. Farben tried to develop new fertilisers.[34]

Odda Smelteverk followed these technical developments concerning fertilisers closely. The inventor of the Odda process, chief chemist Erling Johnson, read about I.G. Farben's work on the three-component fertiliser, Nitrophoska: "I understood immediately the significance of Dr. Mittasch's work concerning the making of three-component fertiliser."[35] I.G. Farben had been able to dissolve the phosphate, so that nitrogen and potassium could easily be added. In Alwin Mittasch's words: "These highly concentrated mixed fertilisers are characterized by not being produced by a simple mix of its components, but by a chemical process and therefore including the components in their most intimate connection."[36] This was the first highly concentrated three-component fertiliser, which Johnson improved upon by inventing a simple and very efficient process. He used saltpetre (nitric) acid instead of hydrochloric acid to dissolve phosphate; he was then left with the desired mixture of phosphoric acid and ammonium nitrate. Excess calcium saltpetre (nitrate) was easily removed and, among several options, potassium could be added.

Johnson's own thorough account of his invention shows a man very well versed in his

---

[32] Bryn, op. cit., p. 46.

[33] Travis "Modernizing industrial organic chemistry," this volume.

[34] Morris and Reinhardt, this volume.

[35] VIM, OLA/156, 3: Johnson to Torvund, July 25, 1967.

[36] VIM, OLA/154, Fc:1(a): "Omkring den nyere utvikling av kunstgjødselindustrien. Nye fremgangsmåter til fremstilling av koncentrerte kunstgjødselblanding," lecture by E. Johnson, 1930, p. 178 (my translation from Norwegian).

scientific field.[37] He read international patents extensively, and knew exactly what he had achieved by inventing the Odda process: "in the patent literature one could not find any preferable way of doing it. I had to liberate myself from the current way of thinking."[38] Johnson knew the field of agricultural chemistry well because he had previously worked for the producer of fertiliser at Odda, The North Western Cyanamide Co., from 1916 to 1921. Of equal importance was his three years prior to this as assistant to professor Sebelien at the Norwegian Agricultural College, where his work concerned the effects of fertilisers.[39]

The simple answer to why Hafslund did not make use of the Odda process is that it was ready to be employed just when the world economy was at the bottom of its interwar crisis in 1930. Odda especially, and the Hafslund group generally, were hit by the crisis, so the companies themselves were not in good positions to make use of the process. However, before the crisis, serious plans were made in early 1929 to expand considerably Odda's production facilities. In February, Odda had prepared calculations that showed the process's excellent economics, and it made plans for building an ammonia plant, in cooperation with the American Cyanamid Company.[40] Odda had high hopes, but the plans were scrapped because of the crisis. Elias Kiær, chairman of the board of Odda, wrote to William Bell, head of American Cyanamid: We "note that owing to the general worldwide crisis you have – at least for the present – dropped the Odda Scheme. Personally I am rather disappointed as I had got into my head, that we might have built up something grand at Odda out of the combination of the Tysse and the Laatefoss powers."[41]

Kiær was clearly bitterly disappointed. He and his fellow owners did not possess the means themselves to exploit the Odda process commercially. The next step was to try to interest Norsk Hydro in the process. Norsk Hydro was an obvious choice, being a large producer of saltpetre, a main ingredient in the process. The relationship was clearly between one weak and one strong partner. As Kiær said to a fellow owner when the two companies reached an understanding that Norsk Hydro should test the Odda process out: "I believe the draft can not be accepted in the present form; we are selling ourselves during the period of trial to Hydro." Kiær believed Norsk Hydro should buy

---

37 Ibid.

38 VIM, OLA/156, 3: Johnson til Torvund, July 25, 1967.

39 Ibid.

40 Ibid.

41 SECK (Skogbruksmuseet, Elverum, Elias C. Kiærs kopibøker - the museum for forestry, Elverum, E.C. Kiær's copy books): Kiær to Bell, November 15, 1930.

Odda Smelteverk.[42]

It all benefitted Norsk Hydro, however.[43] While Odda was completely honest about its technical knowledge, Norsk Hydro tried to circumvent the Odda process patent. After negotiations fell through in 1933, three leaders of Odda summed up their bitter experiences: Norsk Hydro had not made any serious offer, but seemed to have wanted "to postpone the matter and to get hold of all possible information, information that is now used to circumvent the Odda process."[44] Even though Norsk Hydro clearly exploited the economic difficulties of Odda at the time, it is possible to understand why, since Odda's real plans had been to collaborate with American Cyanamid Company, in all probability as a competitor to Norsk Hydro. In any case, Norsk Hydro was in a very vulnerable position itself, also hit by the crisis; it was only in 1929 that it completed restructuring to produce nitrogen fertiliser by the Haber-Bosch process. In the end, Odda did very little. Odda Smelteverk was sold to the British Oxygen Company in late 1937, and the Odda process was licenced to I.G. Farben in 1938, to Dutch Staats Mijnen in 1939, and eventually to Norsk Hydro after the war.

The story could end here, with an admission that Odda was simply too small a company, and that it was the large companies such as American Cyanamid, Norsk Hydro and I.G. Farben that controlled the industry. There is a problem, however, in explaining why, from 1935 onwards, when the general crisis was over, and Hafslund and Odda made profits, no attempt was made to proceed with the Odda process. Inventing something like the Odda process had been a strategic goal of Hafslund for some time, but Hafslund did nothing to exploit a great in-house invention. Why?

**Part III**

The Hafslund group was not only hit by the general crisis; the group clearly was experiencing two profound changes: one of management and one of ownership. The Hafslund group of the 1920s was a mixture of two distinct and different trends of ownership and management. In 1928 Hafslund's owners finally gained control of the company they had bought ready-made in 1916. This took the control of company strategy away from its managers, and it happened at the time when the owners' outer strength had faltered somewhat.

The group that bought Hafslund in 1916 was dominated by "the Kiær-Fearnley trust,"

---

42 SECK (41): Kiær to J.B.Stang June 31, 1931.

43 Unpublished manuscript by K.G. Andersen (see note 1). Also: VIM, OLA/156: 'Produksjon av fullgjødsel i Norsk Hydro. En historisk oversikt' by Leiv Torvund, December 1983; Olsen, op. cit. (2), pp. 376f, 456 and 470f.

44 VIM, OLA/154, 3: 'Oversikt angående forhandlinger ført med Norsk Hydro [...] "Odda-prosessen",' October 1, 1934, by Høy, Bruce and Johnson.

known for their activities in the paper and pulp industry. Kiær and Fearnley were two families that, together with others, controlled many Norwegian industrial companies, and they were led by Elias Kiær as well as by two Thomas Fearnleys, senior and junior. At their height prior to the Russian revolution they had expanded greatly, owning a huge wood-industrial complex spanning many Norwegian, Swedish, Finnish and Russian companies and large areas of woodlands. They were, however, hit very hard in the first few years after the Russian revolution, and lost their largest enterprise, Dubrowka, in Russia, together with large parts of their invested capital.[45] At the end of the 1920s, the Kiær-family seemed to be harder hit than the Fearnley family, and their mutual operations do not seem to have been as well coordinated as previously.

Even though Thomas Fearnley jr. was an owner of both Norsk Hydro and Hafslund, he did not resolve the situation satisfactorily for both enterprises. Fearnley was a member of Norsk Hydro's board of supervisors, and had often been contacted by Hafslund to exploit this position.[46] What Fearnley's own interest in the situation concerning the Odda process was – as an owner of both Hafslund/Odda and Norsk Hydro – is difficult to say, but his real influence at Norsk Hydro must have been limited. We do not know to the full extent what went on between Norsk Hydro and Hafslund/Odda. However, Thomas Fearnley was involved in the negotiations up until July 1931, as was Elias Kiær and other Hafslund/Odda owners. Norsk Hydro seems to have been represented by its management, not its owners. And the management of Norsk Hydro was not particularly interested in buying the Odda factory, which was offered to them, although they were interested in the Odda process.[47]

Whether these negotiations failed because Thomas Fearnley owned only a small share of Norsk Hydro, or because the management of Norsk Hydro wanted to do nothing with Odda, both point to the failings of the kind of ownership coordination practiced at the Hafslund group. What was once a powerful group clearly no longer functioned the way it used to. The bitterness felt by the management of Odda toward Norsk Hydro reflected that. But Odda, through its association with Hafslund, was part of a modern industrial entity with managerial hierarchies that had accomplished a major breakthrough, the Odda process. The trouble was that it, too, changed, since Hafslund in 1928 was transformed into a different kind of company, one suited to the kind of personal and associative capitalism that the Kiær-group lived by.

Superficially, Hafslund became a dividend-yielding share; the company was reorganized in 1927 and 1928 so as to yield a constant yearly dividend of 5 per cent of

---

[45] Yngve Nilsen, Mange bekker små gjør en stor å. Treforedlingsbedriften A/S Dubrovkas etablering og vekst, unpublished hovedfag's thesis, University of Oslo, 1995, especially pp. 87-106.

[46] VIM, OLA/157, Ab, 1: Høy to Heyerdahl, March 9 1936.

[47] Oral information given by Ketil Gjølme Andersen (see note 1).

the value of the shares. Hafslund had always been concerned about the dividend, and a yield of 5 per cent or more was sought for. As early as 1900 the company's action reflected the stock market moves. When the stock market heard about the problems associated with getting the carbide production underway, Hafslund's share price fell dramatically, and Schuckert had to form a new ownership group to take over this business.[48] In 1919, Knud Bryn was instrumental in making the decision he so much regretted nine years later: Hafslund revalued the nominal share capital upwards, and was thus forced to pay a dividend of 5 to 7 per cent on this new and higher share capital. If this had not happened, Hafslund could have paid out "12 per cent dividend every year of the original share value, and even collected a total profit for all years ... instead one now has to trust cash-credit with our banks."[49] Thus the financial flexibility Bryn had as a manager for following his own strategies was diminished; he could no longer escape the watchful eyes of his owners.

Behind the changes at Hafslund were the company's economic problems and the ageing CEO, Knud Bryn, who was 73 years old in 1928. The reorganization seemed straightforward enough: there was centralization added to a general rationalization. In the words of company historian Carl Just: "the administration was greatly simplified, and the costs were cut down as much as was possible so the company could get out of its difficult situation."[50] In reality this meant the instigation of a cost control system by the new leader of the company, Hieronymus Heyerdahl, chairman of the board of supervisors. It seems this focus on the dividend dominated Heyerdahl's eighteen years in charge; the first thing Heyerdahl did when he resigned his position in 1947 was to criticize Hafslund for paying just 4 per cent dividend in 1946. Hafslund had – ever since Odda Smelteverk was sold in 1937 – paid out 5 per cent in dividend, and it was important to keep that figure constant.[51] Heyerdahl may not have won in 1947, but he made his mark on the company over the previous eighteen years. By selling Odda Smelteverk he enabled Hafslund to pay a 5 per cent dividend. In the process, Hafslund became, yet again, focused on its energy business, leaving chemical manufacture as a sideline.

The organizational changes in 1927-28 were profound. A new institution – the board of trustees, that is the owners' representatives – was initiated. They were to appoint members to the board of supervisors, the old Aufsichtsrat. Previously, the members of the old board of supervisors had been appointed directly by the owners. The biggest change to the daily operation of the company was that the board of directors was

48 Bryn, op. cit.

49 Ibid., p. 47.

50 Just, op. cit., p. 250.

51 HHO, 1, 'administrative organer': Heyerdahl to the chairman of Hafslund's board of trustees, May 1948.

abandoned, and the three executive directors lost their rights of attorney.[52] The functions of the board of directors were taken over by a new central office, placed directly under the board of supervisors and its head, Hieronymus Heyerdahl. Consequently Hafslund became directly under the owners' influence; Hafslund's managers were left little room for manoeuvre. For eight years the central office was placed directly under the board of supervisors, and even though the leader of the central office became executive officer in 1936, it seems that chairman of the board of supervisors Hieronymus Heyerdahl remained in charge until he retired in 1946. Not until then was the organization slightly expanded so that the executive officer could be relieved of detail work.[53]

The administrative changes in 1928 were the work of Hieronymus Heyerdahl,[54] and they were hotly disputed. The new organizational structure was sanctioned by the board of supervisors, with one negative voice. That lonely voice came, however, from the man who had built the company and its organization from scratch, Knud Bryn: "Bryn wanted added to the minutes that both the proposal for salary negotiations and the proposal for a new administrative order, according to his wishes, were prepared without his participation ... He could not, however, agree that the proposed administrative order was fortunate, but would not propose anything else or start a discussion because the matter has been sanctioned by the three executive officers."[55] Neither Bryn – nor anyone else – disputed Hafslund's need for cost restructuring. Hafslund had – like so many Norwegian companies in 1928 – financial difficulties, and a few years previously had established a savings committee.[56] But it is one thing to save. It is another thing to reorganize.

Unfortunately, we do not know exactly why Bryn was against the reorganization, but it was probably because of the centralization. The above analysis of how Bryn thought Hafslund's organizational structure functioned as a fairly decentralized whole while he was the leader is largely based on a 49-page history written by Bryn just after the new centralized organizational model had, against his wishes, been sanctioned. It is reasonable to see Bryn's history as a kind of interpretive testament, as an actor trying to stamp his view on the past. It should also be added that one of the executive officers, Iver Høy, who had accepted his loss of power of attorney, later asked for it to be

---

52 HHO, 1, board meeting, June 7, 1928; Bryn, op. cit., p. 33.

53 HHO, 1, board meeting, May 3 and 4, 1946.

54 HHO, 1, board meeting, September 9, 1927.

55 Ibid.

56 Bryn, op. cit., p. 32.

reinstated, without result.[57] Høy was also one of the three bitter Odda managers. One reason for these administrative changes was clearly Heyerdahl's insistance of 5 per cent dividend. Just after the board of supervisors had agreed upon the administrative changes, they decided in March 1928 – after strong argumentation from Heyerdahl – that they should pay 5 per cent dividend in the current situation – a situation that had motivated profound organizational changes and rationalization.[58]

The dividend matter – not really a crisis – was also behind the sale of Odda in 1937. Meraker Smelteverk had already been sold in 1928, but Meraker was the least important of the three companies of the Hafslund group. Odda was not sold without some discussion, and the chairman of the board of supervisors, Hieronymus Heyerdahl, was the active salesman. As one of the more reticent sellers, Elias Kiær, member of the board of supervisors at Hafslund and chairman of the board at Odda said: "The difference between the chairman of Hafslund's board and myself is essentially that Heyerdahl is a happy seller for the proposed price, while I am rather an unhappy seller."[59] The sale was not necessitated by either Hafslund or Odda's economic needs. As Heyerdahl himself said to Hafslund's board of trustees: "Even if Hafslund – as I have earlier made clear – will be able to pay for the great expenditure [of the modernization of the energy business] out of its own profits for the coming three years, that is without taking up loans, it is clear that the extraordinary cash [from the sale of Odda] will be very welcome."[60]

The financial situation at Odda was not any reason to sell, rather the contrary. The company was showing profits again, while both the production of carbide and cyanamide had been extensively modernized during the years of crisis.[61] Elias Kiær thought it sad to sell "just now as we can see the result of many years work. The factory is technically in tip-top shape and fully competitive. And the finances are so changed over the last few years that it is close to a fairy-tale."[62] Kiær believed the price "too low – we should [have had more] + the phosphate patent reserved for Odda Smelteverk's old owners,"[63] but he did not think that was possible, and referred to negotiations with

---

57 HHO, 1, board meeting, June 7, 1928.

58 HHO, 1, board meeting, March 15, 1928.

59 HHO, 1, board meeting, April 21, 1937, appendix: Elias Kiær's statement.

60 Ibid.

61 VIM, OLA/157.1, Za, 1: Yearly reports for the years 1932, 1933, 1934, 1935, 1936 and 1937.

62 HHO, 1, board meeting April 21, 1937, appendix: Elias Kiær's statement.

63 Ibid.

American Cyanamid Company and Norsk Hydro.[64] And so Odda and its important process was sold out of Hafslund. And even more importantly, the independent role of the large industrial organization of the Hafslund group disappeared.

The purpose here is not to depict Hieronymus Heyerdahl as a bad leader and Knud Bryn as a good leader, but to underline their different purposes. Heyerdahl was a business lawyer who reshaped Hafslund as a secure, dividend-yielding financial instrument. Knud Bryn was a representative of the new managerial class, looking to develop organizational competence and scope. Under him, and German and Norwegian owners, there developed a Norwegian entity that was very much like foreign big business; the Hafslund group arose as a diversified company with a footing in energy *and* industry. The Hafslund group was – alongside the cases of Norsk Hydro and Borregaard – yet one more example of big business and managerial capitalism in Norway. But, unlike Norsk Hydro, or the companies studied by Alfred Chandler, Hafslund disappeared as a big company.

These changes can only be understood as the work of Hafslund's owners more generally. The man with the real power was not Heyerdahl, but owner Elias Kiær. Kiær was not concerned about continuing the work done by Knud Bryn when Hafslund looked for a CEO to replace Bryn. Indeed, it seems Kiær was the one who suggested that Heyerdahl as chairman of the board should take over Bryn's position: "Should we not dare to wait hiring a new CEO; the work done by Bryn ... the last couple of years cannot be more than the chairman of the board can do when helped by a clever man to take care of the daily work; thereby we would achieve great savings and simplifications," he wrote to Heyerdahl in July 1927.[65] Heyerdahl thus clearly acted as representative of ownership, but also as an especially eager one. Though his cost-savings were generally appreciated, his intent of achieving a dividend of 5 per cent was not always approved, and Kiær in 1930 wrote to a fellow member of the board to say he would oppose Heyerdahl's proposal for a devaluation of the share value.[66]

The dividend matter, therefore, must be seen as the same kind of tool that large American corporations in this period used to control its divisions; it was a means to achieve control from above through financial measures. Indeed, after Elias Kiær had talked to one of Hafslund's executive officers in July 1927, he found the officer's thoughts about the need for further decentralisation, through decentralization of accounts, interesting.[67] The difference between the large American companies studied

---

[64] Ibid.

[65] SECK: Kiær to Heyerdahl 14.7.27.

[66] SECK: Kiær to J.B. Stang 8.3.30.

[67] SECK: Kiær to Heyerdahl 14.7.27.

by Alfred Chandler and the companies developed by Kiær and Heyerdahl was not decentralization through the creation of divisional autonomy, as in the USA, but the place where the level of corporate planning took place. Hafslund had been a kind of managerial-strategic centre within the Hafslund *group*. After 1928 the Hafslund group was reorganized so that Hafslund became, like Odda, an ordinary company. The two companies – Hafslund and Odda – may be seen as fairly equal to American divisions, but in the firms highlighted by Chandler it was a company's headquarters' (like Hafslund prior to 1928) that planned for the division; in the system developed by Kiær, it was he and his fellow owners, and what they could accomplish by themselves or through associations, that determined the future.

It was a deliberate policy to coordinate companies from the outside, a kind of personal capitalism that combined large operations as well as personal ownership associations. The alliances were made externally, through a personal owner network, not internally or vertically through managerial hierarchies. Thus it was Elias Kiær who took it upon himself, as owner, to organize new solutions; he was not responding to – or pushing – any plans involving managerial systems. To some extent this may be a typical Norwegian kind of capitalism.

Although this is just one example, it is an important one, because these owners operated within a system dominated by many families and spanning many sectors, with the paper and pulp sector being the most important. The rise of the Hafslund group in the 1920s showed how the industry could reorganize itself during the interwar crisis. The Norwegian historian Francis Sejersted has pointed out the difference between the Swedish and Norwegian banking system as one reason why Swedish companies have become larger than Norwegian ones. The Norwegian banking system, with its decentralized nature, was much harder hit by the interwar crisis than the Swedish banking system, and was not able to restructure Norwegian industry during the crisis. The Swedish banks used their large portfolios of troubled industrial companies to rationalize their structures; they created bigger companies,[68] thus implying an alliance between the Swedish financial sector and employed managers – squeezing the owners out. The case of Hafslund from 1919 to 1928 is an example of how the Norwegian industry itself – without help from banks – was able to coordinate itself, except that this reorganized group dissolved itself after the crisis was over for two reasons: the owners remained in control of the companies, and were themselves responsible for rationalization, shedding away managerial structures in the process, possibly being able to continue what may be a traditional Norwegian style of personal, associative capitalism.

The rise and fall of the Hafslund group seems to be the result of two processes. First, the Hafslund group was built by Norwegian engineers in a system of cooperative

68 Francis Sejersted, "Bank og samfunn," *Bergen Bank Kvartalsskrift*, **4** (1988), 95-103.

capitalism, a capitalism characterized by managerial autonomy, cartels and associative ownership. Second, during the years of crisis Hafslund was changed by Norwegian owners who focused on short term dividends and owner control. An important underlying factor was thus the combination of the Norwegian trust law with the owners wanting to control the company from above. If cartels and trusts of the Hafslund-Meraker-Odda type had been illegal in Norway the three companies would most probably have become one company in the 1920s. The three of them had to deal – or had great advantages from dealing – as one in the international economy, so there is every reason to believe another trust law could have brought them together. They functioned and acted as one company, and the only reason they were not brought together as one company must be because dealing with three companies gave the owners advantages of flexibility. They could – and they did – sell out individual companies. They were not forced to mould one big company, which probably was to their disadvantages in the period after the invention of the Odda process. Had the three companies kept together, they would have met the years after the crisis, and the post-war period, as a major corporation with the patent rights to a very important chemical process. Hafslund could have become what Norsk Hydro became, because it was only after World War II that the market for three-component fertiliser really took off.

**Conclusion**

In summary, during the interwar era of cartels and autarky, no single company could expect export success. And a small Norwegian producer of synthetic fertiliser would have had a tough job finding profits selling a new product worldwide. But from the mid-1930s, and of course after World War II, the situation began to change. Had the Hafslund group of 1928 – including both the Hafslund factory and the Odda factory – remained relatively intact, matters may have worked out differently.

The rise of the Hafslund group as an industrial company was based on two pillars. On the one hand, the company was owned by an ambitious group of proprietors that gradually expanded their industrial activities during the 1920s. On the other hand, Hafslund was based on a large managerial effort, and there was room for action from below. This room for action disappeared after 1928. By 1928 Hafslund was turned around. A company with considerable space for managerial action was transformed into a company focused on short term dividends and controlled by the owners. The room for initiative and technological development from below that had helped create innovations of the importance of the Odda process, disappeared. The room within the company did not disappear by itself, just by changing the organization, but the rationale behind the change in 1928 was the same as behind the sale of Odda Smelteverk in 1937: the dividend had to rise to 5 per cent of the share value. The Hafslund share had to become a predictable financial instrument inside a system governed from the outside of companies, by owners.

Does what happened at Hafslund after 1927 reflect larger cultural and institutional patterns of Norwegian industrial society? In the case of the functioning of the trust law, that allowed the cartel-like associations such as the Hafslund group, it does. In the more specific way of how Hafslund developed as a managerial company towards an owner-steered dividend conscious company, the answer is not so certain. What it does reflect, however, is the development of the once mighty Kiær-family. To achieve the kind of coordination from outside that Elias Kiær clearly was accustomed to, he had no qualms about eradicating the managerial structures of Hafslund when economic pressures necessitated change. But this process – added to the vain striving for an associative deal with American Cyanamid Company and Norsk Hydro over the Odda process – left Hafslund a rather weak company without the kind of managerial initiative that might have led to the commercialization of the Odda process.

# THE SWISS PHARMACEUTICAL INDUSTRY: THE IMPACT OF INDUSTRIAL PROPERTY RIGHTS AND TRUST IN THE LABORATORY, 1907-1939

JAKOB TANNER
*Forschungsstelle für Sozial- und Wirtschaftsgeschichte*
*University of Zurich*
*Rämistrasse 64*
*CH-8001 Zurich*
*Switzerland*

## 1. The development of the pharmaceutical industry

The rise of the health products and modern pharmaceutical industry based originally on coal-tar derivatives may be dated to the end of the 19th century. At that time, new medicaments were created out of a symbiosis of science and, the mainly organic, chemical industry which – due to the mechanisms of competition – were subject to ongoing changes and needs for improvement. The industrial historian Wolfgang Wimmer summarized his study on the "Gesundheitswesen und die Innovationen der Pharma-Industrie in Deutschland" [The Health Service and Innovations of the Pharmaceutical Industry in Germany] between 1880 and 1935 with a statement that harks back to the advertising slogans of the very companies he was writing about: "We almost always have something new available."[1] This not only refers to the rather hectic dynamics of innovation that are observed within such companies but also to an attitude that demonstrates the awareness of and tendency towards progress for the entire era. Representatives of the pharmaceutical industry were soon able to point with some pride to the fact that they were now fortunate enough to speak of "some diseases which a few decades ago were dangerous epidemics ... in the past tense."[2]

The era of medical products based on laboratory research began in the 1880s. At that

1 Wolfgang Wimmer, *"Wir haben fast immer was Neues": Gesundheitswesen und Innovationen der Pharma-Industrie in Deutschland, 1880-1935* (Berlin, 1994).

2 Hans Fehr, *Die chemische Industrie in Basel, Separatdruck aus den Basler Nachrichten* (Basle, 1954), p. 4.



*A.S. Travis et al. (eds.),*
***Determinants in the Evolution of the European Chemical Industry, 1900–1939,*** 257–271.

time, a new technological trajectory was embarked upon; that is, a new path-dependent corporate learning process leading to new production processes, product lines, enterprises, and markets, which became technologically institutionalized by a broad range of professional groups, allied with training and research programmes.[3]

Douglass North's concept of the "Second Economic Transformation" helps to explain why this new approach became generally accepted.[4] At the end of the 19th century, a phase of industrialization began which for the first time ever helped access the potential of science and research to create an artificial world that would change society and promote further growth. The fact that the Swiss pharmaceutical enterprises decided to follow this path as early as the 1880s has to be interpreted in the light of the crisis and structural change of the "Great Depression" and the by then dire competition from German enterprises. Swiss firms met these challenges mainly by specializing in high-quality niche-products with high profitability, which in turn was only possible thanks to a much more theory-based and, in the long term, increasingly systematic research.[5] In this manner, a continual long-term flow of innovations was set in motion from the 1880s onwards.[6] This process was supported by concomitant demands: for the emergence of large, modern, export-oriented enterprises such as those established at the end of the 19th century as joint-stock companies in the chemical and pharmaceutical sector; and social consolidation of the medical profession into an organized body that supported the new medications in the health product market.[7]

At the beginning of the 20th century, there were already a number of internationally renowned enterprises in Switzerland that had embarked on programmes of increased diversification and that began to produce drugs. First of all, there were the coal-tar dyestuff companies. They had enjoyed unparalleled growth since the 1860s and were now able to profit from the industrial manufacture of pharmaceutical preparations based

3 Giovanni Dosi and Richard R. Nelson, "Theorien der Evolution in den Wirtschaftswissenschaften," in Valentin Braitenberg and Inga Hosp, eds, *Evolution: Entwicklung und Organisation in der Natur* (Reinbek b. Hamburg, 1994), p. 219.

4 Douglass C. North, *Structure and Change in Economic History* (New York/London, 1981).

5 Tobias Straumann, *Die Schöpfung im Reagenzglas: Eine Geschichte der Basler Chemie (1850 - 1920)* (Basle, 1995), p. 119 ff.

6 The significance of entrepreneurial investment in innovation-oriented activities is analyzed by Gerry P. Sweeney, "Innovation is entrepreneur-led," in Gerry P. Sweeney, ed., *Innovation Policies: An International Perspective* (London, 1985), pp. 80 - 113.

7 Geoffrey Jones and Harm Schröter, eds, *The Rise of Multinationals in Continental Europe* (Aldershot, 1993); Claudia Huerkamp, "Ärzte und Professionalisierung in Deutschland: Überlegungen zum Wandel des Arztberufs im 19. Jahrhundert," *Geschichte und Gesellschaft*, **6** (1980), 349-382; Claudia Huerkamp, *Der Aufstieg der Ärzte im 19. Jahrhundert: Vom gelehrten Stand zum professionellen Experten: Das Beispiel Preussens* (Göttingen, 1985).

on the intermediates used in the manufacture of dyes. This was a new, promising, and clearly profitable, sales sector. Well established companies became the leaders because the development of drugs in an industrial laboratory was an uncertain and highly risky matter that presupposed a strong entrepreneurial streak and adequate financial backing.[8] Among the leaders were Durand & Huguenin and Ciba (or "Chemische Industrie Basel"), which had presented its first range of drugs at the Paris World Exhibition in 1889.[9] Sandoz's entry was more hesitant; although the company had experience with the manufacture and sale of pharmaceutical preparations around 1900, it finally joined the new market in 1917.[10] For this enterprise, we do have figures which show the share of pharmaceutical products in the total turnover.[11] From 1923 onwards, a strong growth of turnover in the drug sector is apparent. It rose from 1 per cent in 1923 to 15.2 per cent in 1930. The relative increase in significance of pharmaceuticals continued during the 1930s crisis and peaked, for the period under review, in 1938 at 28.6 per cent. In the first two decades after World War I, the growth of turnover in the pharmaceutical division was, therefore, still substantially above that of growth in other sectors of the group. A rapidly rising star among the newcomers was Hoffmann-La Roche, founded in 1896, which quickly gained acceptance with a range of products composed of renowned proprietary drugs.[12]

In addition to these already quite large companies, a considerable number of smaller and medium-sized enterprises were established that comprised the so-called "Schweizerische Spezialitätenindustrie" [the Swiss industry of pharmaceutical preparations]. These were small workshops and factories specializing in the manufacture of so-called 'Geheimmittel' or 'Patentmittel' (patent or proprietary

---

[8] Harm G. Schröter, "Unternehmensleitung und Auslandsproduktion: Entscheidungsprozesse, Probleme und Konsequenzen in der schweizerischen Chemieindustrie vor 1914," *Schweizerische Zeitschrift für Geschichte*, **44** (1994), 14-53.

[9] Nicolas Jaquet, *Die Entwicklung und volkswirtschaftliche Bedeutung der schweizerischen Teerfarbenindustrie* (Basle, 1923); *Ciba Builds to Serve*, (Basle [1959]), p. 49.

[10] Hans Fritz, *Industrielle Arzneimittelherstellung: Die pharmazeutische Industrie in Basel am Beispiel der Sandoz AG* (Stuttgart, 1992); Arthur Stoll, *25 Jahre Dienst an der Pharmazeutischen Forschung* (Basle, 1942) (Off-print of the Basler Nachrichten no. 135, 19 May 1942). Geigy evaluated the possibility of diversification into the pharmaceutical field at the turn of the century but decided to persevere in a wait-and-see attitude up to the end of the 1930s. See Alfred Bürgin, *Geschichte des Geigy-Unternehmens von 1758 bis 1939: Ein Beitrag zur Basler Unternehmer- und Wirtschaftsgeschichte* (Basle, 1958); Carl Faller, *Pharmaka* (Basle, 1953); for an overview see Schröter, op. cit. (8).

[11] Fritz, op. cit. (10), p. 227 f. Unfortunately, the figures for the 1902-1922 period are not reliable.

[12] Hans Conrad Peyer, *Roche: Geschichte eines Unternehmens, 1896 - 1996* (Basle, 1996); for 1905, the number of employees in chemical enterprises were as follows: 550 (Ciba), 300 (Sandoz) and 45 (Hoffmann-La Roche). See N. Wassilieff, *Die Arbeitsverhältnisse in den Basler chemischen Fabriken* (Basle, 1905), p. 3; and US Department of Commerce, *Chemical Industry and Trade in Switzerland* (Washington, 1929).

medicines).[13] They also increasingly benefited from the structural shift within the drug markets, or sold so-called 'contrefaçons,' that is, imitations or variations of scientifically developed drugs. A large number of those enterprises were able to establish individual proprietary brands, though they never achieved the same turnovers as the few really large companies. In Switzerland, the predominance of modern industrial enterprises in the pharmaceutical industry was quite significant. The large corporations domiciled in Basle represented about 90 per cent of Swiss pharmaceutical production in the 1930s. As more than 90 per cent of this overall production was exported, patent applications in important foreign markets provide an additional indicator of the relative significance of these large corporations.[14] A study of the United States market (the second most important export-market for the Swiss pharmaceutical industry),[15] shows that the largest Swiss enterprises accounted for a very high share of patents and, therefore, innovative activity, in the interwar years, and especially in the 1930s. "Switzerland followed the German pattern, with large firms dominating local technological activity in both the chemical and electrical fields (Ciba and Brown Boveri)."[16] For large firms outside the United States filing patents in North America the absolute highest value of 75.4 per cent came from the Swiss chemical industry during 1920-1939. Compare this with the electrical goods industry that represented the next largest group, with 48.6 per cent. The percentages of German patents were 66.4 for chemical and 57.1 for electrical goods.[17]

Throughout the war, upheavals, and crises, the large Swiss pharmaceutical companies increased the production of pharmaceutical preparations marketed under brand-names, invested heavily in rationalization of production facilities, and established an effective international marketing and propaganda apparatus. Marketing concentrated on so-called "ethical products:" drugs supplied directly to physicians through target-oriented product placement and global distribution systems.[18] By contrast, there was the so-called 'Laienwerbung,' or sales by advertisement, which was increasingly associated with

---

13 The big chemical corporations insisted upon a distinction between patented pharmaceuticals and so-called proprietaries.

14 Gianfranco Varini, *Die schweizerische chemische Exportindustrie im internationalen Wettbewerb* (Basle, 1958), p. 119.

15 Ibid., p. 121.

16 John Cantwell and Pilar Barrera, *The Rise of Corporate R & D and the Technological Performance of the Largest European Firms from the Interwar Years Onwards* (Reading: Discussion Papers in Economics no. 271, University of Reading, Department of Economics, 1993), p. 13.

17 Ibid., p. 31.

18 The British Productivity Council, *A Review of Productivity in the Pharmaceutical Industry* (London, 1955), p. 1 f; Ciba, *Geschichte und Entwicklung der Gesellschaft für Chemische Industrie in Basel, 1864-1926* (Basle, 1926), p. 67; Fehr, op. cit. (2), p. 6; *Ciba Builds to Serve*, op. cit. (9), p. 49.

heavy promotional advertising. Many renowned companies emphasized that they did not resort to such methods.[19] The success of this expansion strategy resulted in the following: the share of chemical exports within Swiss total exports increased from 4.6 per cent in 1911 to 8.8 per cent in 1930, and doubled in 1939 to 19.7 per cent.[20] In the same period (1906-1938), the share of pharmaceutical preparations in the overall export of the chemical industry almost doubled, from 12 per cent to 23 per cent.[21]

This rise of the Swiss pharmaceutical industry was linked to a substantial structural shift within drug markets. Traditional remedies and in particular the so-called 'Geheimmittel' (patent medicines) came under pressure from authorities. In various cantons, advertising for these products was prohibited or severely restricted, which diminished sales opportunities for such products. Parallel to this, opiates – till then the most effective analgesics and indispensable all-purpose medicines for most of the population – began to meet with a lot of scepticism and even outright rejection. Although they had already been subjected to certain controls since the end of the 19th century, they were now declared illegal within the scope of the Den Haag International Opium Convention of 1912 by an increasing number of countries. This repressive drug legislation mirrored the dramatic loss of prestige of these substances. In particular, in Switzerland the pharmaceutical industry had no real understanding of this negative evaluation and its representatives remained firmly engaged in the production of alkaloids, resisting the policy of prohibition. In spite of their vehement opposition, the Swiss finally introduced a law prohibiting such drugs in 1924-25. This was at a time when the pharmaceutical industry had began to introduce an increasing number of synthetic substitutes and novel substances into the market. In 1924, the annual report of the Schweizerische Handels- und Industrieverein (SHIV) stated "with satisfaction that ... the development in the field of proprietary pharmaceutical preparations continued to show good progress. Thanks to their excellent properties, they increasingly enjoy a good reputation all over the world, which was moreover increased by the introduction of new preparations in the report year."[22] In addition, the report explained; "the fact that the Swiss chemical-pharmaceutical industry was not only able to maintain its position in the main sales markets in spite of all these obstacles but also managed to expand in a certain direction mainly due to the quality of the products and in particular, considering pharmaceutical preparations, to the fact that they are based on a strictly scientific

---

19 Hartmann Rordorf, *Über die Entwicklung der chemisch-pharmazeutischen Spezialitätenindustrie in der Schweiz, 1875-1925* (Wohlen, 1927); Fritz, op. cit. (10); Peyer, op. cit. (12).

20 Varini, op. cit. (14), p. 249.

21 Ibid., p. 25.

22 Annual report of the *Schweizerische Handels- und Industrieverein (A.r. SHIV)* (1924), p. 187; also see *Handbuch der schweizerischen Volkswirtschaft*, **1** (Berne, 1939), 329.

basis."[23] This development continued throughout the entire interwar period.

**2. Patent legislation and patenting in Switzerland**

The more important that scientific preparations for products and process innovations, market access and, in the end, entrepreneurial viability became, the more significant became legal regulation of so-called immaterial or intellectual property rights. The initiation of the "inventors' industry"[24] provided the companion piece to the establishment of proprietary products. Such procedures were treated by the so-called 'Immaterialgütertheorie' (immaterial goods theory) which was established in 1910 by German patent lawyer J. Kohler and later developed into present intellectual property rulings.[25] This theory and the respective regulations enabled an integration of the scientific and aesthetical dimension of the production of goods into the overall economy. Thus, this approach strove to integrate specific know-how and a formulaic language into a market-price mechanism. Here, a self-referential and autopoïetical link between a "production of laws" and the creation of aesthetic and visual forms of communication in a large modern corporation can be perceived.[26]

At first, Switzerland was considered an exception.[27] For, since the initial days of the dyestuffs industry around 1860, the chemical companies domiciled in Switzerland profited in a rather informal way from the know-how of foreign and, in particular, German and French companies.[28] This was attractive because Swiss industrial

23 *A.r. SHIV* (1924), p. 182.

24 North, op. cit. (4), p. 171 f.; R. Nelson, M. Peck and E. Kolachek, *Technology, Economic Growth and Public Policy* (Washington, D.C., 1967).

25 Alphonse Gemuseus, "Das schweizerische Patentgesetz in seiner Anwendung auf chemische Erfindungen," Doctoral Thesis, University of Berne, 1930, p. 7.

26 Michael Hutter, *Die Produktion von Recht: Eine selbstreferentielle Theorie der Wirtschaft, angewandt auf den Fall des Arzneimittelpatentrechts* (Tübingen, 1989).

27 Jakob Tanner, "Property rights, Innovationsdynamik und Marktmacht: Zur Bedeutung des schweizerischen Patent- und Markenschutzes für die Entwicklung der chemisch-pharmazeutischen Industrie (1907-1928)," in Andreas Ernst and Erich Wigger, eds, *Die neue Schweiz? Eine Gesellschaft zwischen Integration und Polarisierung (1910-1930)* (Zürich, 1996), pp. 273 - 303; Straumann, op. cit. (5), p. 132 f; p. 153.

28 Tanner, op. cit. (27); different aspects of the impact of patent legislation on the pharmaceutical industry are analysed in *Handbuch der schweizerischen Volkswirtschaft*, **1** (Berne, 1955), p. 580; Jacob Flubacher, "Die chemische Erfindung unter besonderer Berücksichtigung der Schweiz," Doctoral Thesis, University of Basle, 1933; Carlo Steffen, "Erfindungen von chemischen Verfahren und Arzneimitteln nach schweizerischem Patentrecht," Doctoral Thesis, University of Bellinzona, 1945; and Paul Graf, *Die Normen des schweizerischen Patentrechts für chemische Erfindungen, Arzneimittel-, Lebensmittel- und Textilerfindungen* (Zürich, 1935). For the textile industry, a so-called "textile paragraph" was created in 1907, which excluded certain inventions touching upon textile finishing from patenting rights; it remained valid until 1954. See Institut für gewerblichen Rechtsschutz, ed., *Kernprobleme des Patentrechts: Festschrift zum 100-jährigen Bestehen eines eidgenössischen Patentgesetzes* (Zurich, 1988).

enterprises (to quote Alfred D. Chandler) were not "first movers" but subsequent "challenger companies" which tended to copy the German approach from the 1860s.[29] They did not specialize in innovation but imitation, and copied what others had developed, quickly and efficiently, making great profits in the process.[30] Legally this was possible because Switzerland and the Netherlands were the only European countries which refrained from introducing national patent systems for inventions to assure the right to carry on a trade or business.[31] The protocol of a conference held within the scope of the Zurich National Exhibition in 1883 provides a good synopsis of the discussions held in Switzerland on the question of patent laws and the protection of industrial property. At that time, 170 representatives of national industry and trade convened in the Zurich town-hall. Prominent representatives of the chemical industry[32] defended themselves against the allegation (supported by the textile industry) that Switzerland was a "nation of industrial robber barons."[33] The representatives of the chemical industry were successful, and the new law of 1888 continued to exclude chemical inventions.

The continued appropriation of inventions protected through foreign patents by the Swiss chemical and pharmaceutical industry was above all criticized by the Germans as outright entrepreneurial piracy and engendered vehement protests. As in other instances, Switzerland was forced from outside its borders to change its laws, and was threatened by retaliations if it failed to do so. When a new trade agreement was signed between Germany and Switzerland in 1904, it contained an additional clause specifying that Switzerland would have to create a legal basis for patent protection that included the chemical and pharmaceutical industry before the end of 1907. What happened then is exemplified by the discussion held in the *Zeitschrift für angewandte Chemie* (Journal of Applied Chemistry). During the preparatory phase of the 1906 revision of the Federal Patent Act, Arthur Eichengrün, a renowned chemist working for Bayer in Elberfeld/Wuppertal, spoke of a "special industry pillaging the property of others" and

---

29 Straumann, op. cit. (5), p. 111.

30 Eric Schiff, *Industrialization Without National Patents: The Netherlands 1869-1912; Switzerland 1850-1907* (Princeton, 1971); Edith Tilton Penrose, *The Economics of the International Patent System* (Baltimore, 1951); see also Peter Zweifel and Gabriella Pedroni, *Innovation und Imitation: Eine wirtschaftspolitische Gratwanderung* (Basle, 1985).

31 In the Netherlands, there was no patent legislation between 1869 and 1912.

32 On behalf of the chemical industry of Basle, the conference was attended by Dr. Annaheim (Gewerbeverein Basel); R. Bindschedler (head of Bindschedler, Busch & Cie); a Sandoz representative; M.P. J. R. Geigy-Merian (representative of the Basle Chamber of Commerce); Robert Gnehm (Ciba representative); Dr. Juwalta and Sarauw (Durand & Hugenin), and J. Weinmann, chemist. See *Protocoll des Schweizerischen Congresses betreffend die Frage der Einführung des Erfindungs-Schutzes*, Zürich, 24/25 September 1883 (Zurich, 1883).

33 Ibid., Votum Steiger-Meyer aus Herisau, p. 86.

declared his "vivid displeasure" with the Swiss government "which even supported these doings."[34] He attacked the Swiss industry which "did not find its equivalent anywhere else in the world" and spoke of an "industry which regards any invention or discovery touching upon the field of medicines, no matter where or when they were or shall be made, without so much as a by your leave as its property ... Without respect of ownership or proprietary rights, patents or laws, it effortlessly plucks the fruits of the tree of pharmaceutical science which others planted and cared for, though only the ripe and easily accessible, of course."[35] Besides, the German scientist had become aware of the fact that "the Basle chemical factories had not only spoken up for patent protection of pharmaceutical products but had also demanded in a carefully reasoned motion submitted to the Swiss Federal Council" a quick realization of such measures.[36] In fact, the most innovative Swiss companies had changed their policy as early as 1894 and decided that patent protection of chemical procedures was to their mutual advantages. This position was now consolidated, and in 1907, shortly before the ultimatum expired, Switzerland finally introduced a patent act.[37]

This law supported the product and process innovations of the chemical-pharmaceutical enterprises whose sales of pharmaceuticals now increased decade after decade with an amazing (absolute as well as relative) regularity. Already in 1911, the SHIV Annual Report declared: "It may be due to the continually increasing competition with commercial preparations as well as the new approaches of modern therapy that most factories are reluctant to accept additional preparations and increasingly turn towards the manufacture of prepared medications."[38] This shift helped the Swiss chemical industry to rid itself of its rather dubious reputation as a "'pirates' and foreigners' industry." However, the same SHIV report also mentioned 'outsiders' which unsettled "the market by undercutting offers ... which were, however, often only possible to the detriment of quality ... Even favourite proprietary articles had to fight against a competition which was not above reproach," and it had become clear "that the undeserved reputation of the Swiss chemical-pharmaceutical industry to tend towards 'contrefaçon' could to date not be entirely refuted."[39] World War I changed many

---

34 Arthur Eichengrün, "Das Schweizer Patentgesetz und die Schweizer Contrefaçons," *Zeitschrift für angewandte Chemie*, **19** (7 December 1906), 2022 f.

35 *Zeitschrift für angewandte Chemie*, **19** (1906), 1925.

36 Ibid., p. 1925.

37 *Handbuch der schweizerischen Volkswirtschaft*, **1** (Berne, 1955), 580. Even after 1907, the commodity itself (pharmaceuticals, food, beverages, foodstuffs) remained outside the domain of patent regulation.

38 *A.r. SHIV* (1911), p. 129.

39 Ibid.

imitations were now brought against their foreign competitors. On the basis of a strong competitive position in international markets, based on scientific research and development, leading Swiss enterprises were able to increasingly profit and benefit from the patent and proprietary-rights protection for their own entrepreneurial policies and objectives.[40]

If we relate these statements on the development of the industry with patents granted in Switzerland, we can readily understand why the number of patents doubled between 1907 and 1913. A substantial part of this development was due to the chemical industry. In 1914, however, this development suffered a strong set-back, and the number of patents applied for and granted in 1913 was not to be surpassed until 1928.[41] As for the Basle chemical companies, the four most important dye factories (Ciba, Geigy, Sandoz, and Durand & Huguenin) were granted a grand total of 270 patents between 1908 and 1917, while Hoffmann-La Roche acquired 73. Between 1920 and 1930, Ciba increased its number of patents by an impressive 580, while Geigy, Sandoz, and Durand & Huguenin filed 181 patents in the same period, and Hoffmann-La Roche filed 88. Some 30 per cent of the patents that the three drug-manufacturing companies (Hoffmann-La Roche, Ciba, and Sandoz) acquired were explicitly declared to be pharmaceutical innovations: at Hoffmann-La Roche, this share amounted to 97 per cent, and at Sandoz to 71 per cent.[42] The patent protection of production in the chemical field reflected a tendency towards a strategic specialization of product development. After 1918, the newly established community of interests created by Ciba, Geigy and Sandoz, the I.G. Chemie, consolidated a cartel-type division of markets; at the same time, there was a definite prolongation of the patent-protection period.[43] All in all, the patenting of Swiss inventions was much more regular in the 1910s and 1920s than that of foreign companies, which suffered a substantial decline during and immediately after the war years. A comparison of Swiss and German patents is particularly interesting, for there was an obvious reverse development – the patent offensives of the Swiss "inventors' industry" reflected declines of the respective German companies, and vice versa.[44] The

[40] Hansjörg Siegenthaler, "Grosse Depression und erfinderische Aktivität in der Schweiz," in Harald Winkel, ed., *Finanz - und wirtschaftspolitische Fragen der Zwischenkriegszeit* (Berlin, 1973), pp. 175-208.

[41] For the development after 1928 see Ibid., table, p. 189.

[42] *Patente. Jahreskatalog*, published by the Eidgenössisches Amt für geistiges Eigentum, Statistik der Patente, 1908-1917; 1920-1929.

[43] Paul Graf, *Der Erfindungspatentschutz der Schweizer Industrie*, (off-print of: Industrielle Organisation no. 4, 1936), p. 7.

[44] *Statistische Jahrbücher der Schweiz*; it would be informative to compare the patenting activities of Swiss companies in foreign countries.

following graphs show the development of Swiss and foreign patents between 1925 and 1934:[45]

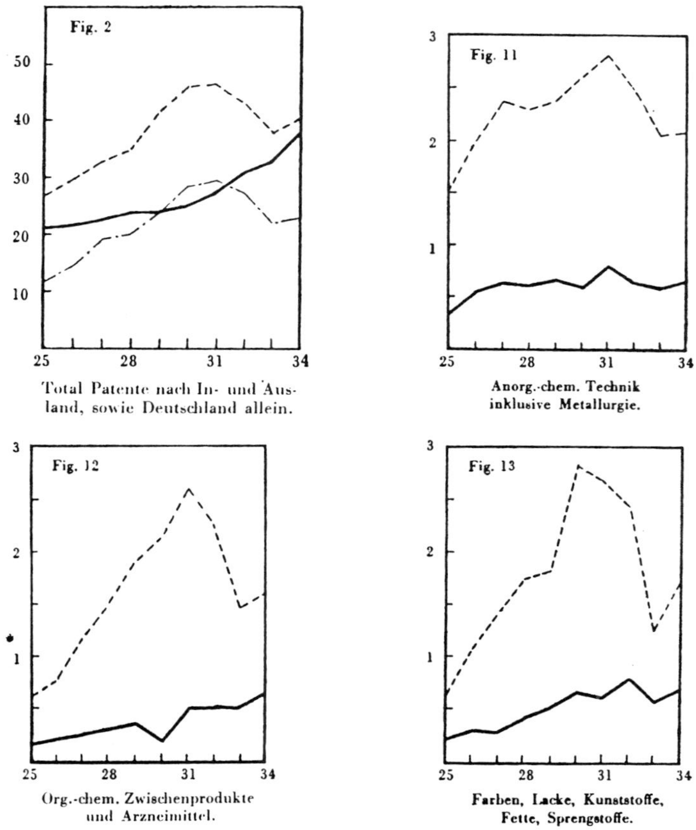


Patents, in hundreds, for the period 1925-1934. Solid lines (————) show total Swiss patents; dashed lines (---------) show total foreign patents. Top left, total patents, Switzerland, other countries, and Germany (-·-·-·-·-·-). Top right, inorganic chemical processes, including metallurgical. Bottom left organic intermediates and pharmaceuticals. Bottom right, dyestuffs, laquers, plastics, fats and explosives.

The curves show that the Swiss chemical industry was much more resistant to crises concerning its inventions than were its competitors abroad, where the largest part of the foreign patent rights of the Swiss industry were located.[46]

45 Graf, op. cit. (43), pp. 3 ff.

46 *Ibid.*, p. 9.

**3. The impact of industrial property rights on the entrepreneurial growth and structural change in the pharmaceutical industry**

In Switzerland, private companies in the pharmaceutical industry at first received substantial scientific support from state institutions (in particular the Swiss Federal Institute of Technology, or ETH, in Zurich, and the cantonal universities) which undertook large-scale research closely linked to industrial objectives. The scientific stimuli necessary for the establishment of a successful company came almost exclusively from the colleges and universities and in particular from the ETH, where a larger number of chairs in chemistry, pharmacology and physiology were endowed at the turn of the century.[47] After 1900, a chemical historian has noted, "this relatively young industry could – due to the enormous and quick progress in physiology and pharmacology – no longer rely on scientific results from outside."[48] The creation of in-company research capacities, and the installation of in-company laboratories and scientific departments, clearly shows that the competition became more fierce even among the Swiss. In a steadily more competitive private industry, the "public pool" of knowledge which state educational and research facilities had provided was replaced by a production of commercially utilizable knowledge directly controlled by corporate management. This process was part of a more general problem touching upon the standardization and classification of data as well as a control of information flow by the industrial companies.[49]

The more innovative a company became, the more it tended to subject that part of its resources which could be damaged or threatened by the competition to its own control structure. The principal resource was knowledge, which in the wrong hands could be copied. This leads us to the important question of the relationship between the tacit and the public elements of technology. For whereas the tacit element "is embodied in the organizational routines and collective expertise or skills of production teams," the public element "is an expression of generic knowledge, or of knowledge which is potentially generic ... and a latent public good."[50] Modern industrial enterprises try to protect this codifiable knowledge, which is susceptible to imitation, by means of

---

47 Alexander Tschirch, *Das pharmaceutische Universitätsinstitut und das akademische Studium der Pharmaceuten in der Schweiz, Deutschland und Österreich* (Berne, 1891).

48 Heinrich Lehmann, "Von der chemisch-pharmazeutischen Industrie Basels," reprinted from: Albin Bruckner, ed., *Wirtschaftsgeschichte Basels* (Zurich, 1947), p. 2.

49 For an international, comparative perspective see Goef Bowker, "Der Aufschwung der Industrieforschung," Michel Serres, ed., *Elemente einer Geschichte der Wissenschaften* (Frankfurt a.M., 1995), p. 864; for a general overview of the organization and structure of research and development see Markus Heiniger, *Vorüberlegungen zu einer Geschichte der Forschung und Entwicklung (F&E) in der Schweiz, 1930-1970* (Berne: Bundesamt für Statistik, 1980).

50 John Cantwell, *Multinational Corporations and Innovatory Activities: Towards a New, Evolutionary Approach* (Reading: Discussion Papers in Economics no. 172, University of Reading, Department of Economics, 1993), p. 6f.

patenting and secrecy. But the decisive immaterial or intellectual resource for the growth of firms are not patents, brand names, and industrial secrets, which provide temporary advantages, but the tacit knowledge which is strictly specific to every firm and cannot be fully copied by others.[51] This implicit knowledge is, however, the precondition "to recognize new technological opportunities through monitoring advances in generic knowledge made elsewhere, and increasing the ability of the firm to imitate the achievements of others more rapidly."[52]

How important this ability for rapid imitation and innovation was becomes quite obvious if we consider a hidden characteristic of the innovation dynamics of the pharmaceutical industry. This relates to the matter of equivalence of products. According to this, the increasing "flood of medicines," as it was soon to be criticized, was not simply the result of "pure science" and revolutionary insights. Rather, as a 1927 standard work on the principles of medicines stated, the "possibility to create equivalent rival preparations in a random number [had] created the appearance that modern synthetic chemistry ... created a truly enormous amount of new drugs." The author, Sigmund Fränkel, observed that "only a very few active basic substances had actually been found" and that the "flood of new substances" was the result of a competition which forced inventors and manufacturers to continually invent new variants and "provide them with the semblance of effectiveness in order to be competitive."[53] The practice of patenting, therefore, did not simply serve to protect novel product and process ideas but – along with the registration of new brands – proved suitable to provide variations of the already known with the semblance of something new. In many instances, research and development were thus linked to the development of sales strategies and demand by way of the patent and proprietary rights law. The necessity to utilize these products commercially, and the dynamics of market expansion, substantially affected the preconditions of laboratory research.[54]

Here, the diversification of the product range also resulted in a specialization of research departments; the outcome was a division of these activities according to scientific disciplines. At Ciba, for instance, this process set in after 1900, when a pharmaceutical production division began to emerge.[55] In 1908, the first medical staff were hired, and during World War I, a large biological department was created and

51 Ibid., p. 11.

52 Ibid., p. 9.

53 Sigmund Fränkel, *Die Arzneimittel-Synthese auf Grundlage der Beziehungen zwischen chemischem Aufbau und Wirkung: Für Ärzte, Chemiker und Pharmazeuten* (Berlin, 1927), pp. 5 ff.

54 For an introduction see: Ulrike Felt, Helga Nowotny and Klaus Taschwer, *Wissenschaftsforschung: Eine Einführung* (Frankfurt a.M./New York, 1995).

55 Ciba, ed., *Gesellschaft für Chemische Industrie in Basel 1884 - 1934* (Basle), p. 29.

divided into three areas: pharmacology, physiology, and bacteriology. During the 1920s, these research activities were expanded, and in 1936 moved into new premises that also boasted endocrinological and histopathological laboratories.[56]

These laboratories were established by all large enterprises in the pharmaceutical industry and were generally subjected to strong pressure to succeed. At the same time, research was increasingly subjected to standardized routines and thus somewhat trivialized.[57] Basically, researchers were able to achieve commercially viable results in two ways. Firstly, they could concentrate on the synthesis of naturally occurring substances; here, the 'artificial' component was restricted to process innovations, that is, to the artificial manufacture of natural substances which were to be copied and if necessary freed of negative side-effects and made as perfect as possible. This was a question of developing cost-effective technical and – if possible – industrial processes for replicating natural substances which were available only in low amounts due to their rarity. Wherever mass-production processes for these substances created a fall in prices, as was the case at Hoffmann-La Roche from the mid-1930s regarding vitamins, new possibilities for marketing and product innovation were created as well. The second way consisted in the creation of totally new substances or the study of effects of known substances. Once classes of substances or "chemical individuals" were found which had a physiological effect due to their molecular structure they had to be demonstrably useful and pass expensive screening procedures and clinical tests. Only a few actually managed to do so.[58]

**4. Trust in the laboratory**

All in all, laboratory work was characterized by a large amount of uncertainty.[59] Therefore, the communication and interaction of highly specialized scientists largely relied on trust.[60] Chemist Carl Djerassi, "father of the 'anti-baby' (contraceptive) pill," included in his novel *Cantor's Dilemma*, a 1983 quote from the editor of the *New England Journal of Medicine*, Arnold S. Relman: "It seems paradoxical that scientific research, in many ways one of the most questioning and sceptical of human activities,

---

[56] "50 Jahre Biologische Abteilung der Ciba," *Ciba-Blätter* 1966/204, pp. 16 ff.

[57] Bowker, op. cit. (49), p. 863.

[58] In the 1950s, Varini, op. cit. (14), p. 160, estimated the ratio between active substances and pharmaceuticals to be 1000: 1.8.

[59] Bruno Latour and Steven Woolgar, *Laboratory Life: The Social Construction of Scientific Facts* (Beverly Hills, [1986]); see also Bruno Latour, *Science in Action: How to Follow Scientists and Engineers Through Society* (Cambridge, Mass., 1987).

[60] Edmund D. Pellegrino, Robert M. Veatch and John P. Langan, eds, *Ethics, Trust, and the Professions. Philosophical and Cultural Aspects* (Washington, DC, 1991); Bernard Barber, *The Logic and Limits of Trust* (New Brunswick, New Jersey, 1983).

should be dependent on personal trust. But the fact is that without trust the research enterprise could not function."[61] This not only applies to the effective but also to the cognitive aspect: trust allows us to bridge the uncertainty engendered by the behaviour of other people, to reduce complex matters, to speed up the circulation of information, and promote the division of work into specialities.[62] Trust is, therefore, an important prerequisite for successful learning processes within any research and development department. In an industrial research laboratory, cooperation-inducing trust is, however, in an explicit if tense relation to the individualizing influence of competition. As regards the performance capacity of a group, the two behavioural modes are ambivalent as both may promote or hamper innovative research. In addition, there is a trade off between the two prerequisites; that is, the emphasis of a competitive aspect may create suspicion while the competitive climate may suffer due to too much trust (as a prerequisite for an operative competition of ideas). However, the respective conditions are everything but constant. Judging by personal experience, the tendency of individuals to present brilliant novelties and original solutions and approaches may increase while a creativity-hampering emotional catch-as-catch-can climate may well result at the same time.[63] Indeed, specialization on the basis of a division of labour and the simultaneous necessity of 'stabilizing' the object of scientific research require the most varied forms of cooperation; the success of the individual is thus bound to that of the group.

The question of the "subject of the research" is also important for a historical analysis due to another issue: the more complex research and development processes become, the more the 'inventor' as an individual becomes mere fiction. By contrast, there was a strong tendency to address successes which could not have come into being without an osmotic exchange and circulation among ideas as individual and to personalize the "glory of research." For even though patenting had increasingly become a matter handled by the corporate effort since the end of the 19th century, it was necessary to create a system of immaterial and material gratification based on individuals. Conveyed by a cooperative maximization of own benefits, this again stimulated informal cooperation and thus freed "positive external energies" on a corporate level.[64]

---

61 Carl Djerassi, *Cantor's Dilemma* (Harmondsworth, 1991).

62 Niklas Luhmann, *Vertrauen: Ein Mechanismus der Reduktion sozialer Komplexität* (Stuttgart, 1973).

63 Michel Callon, "Four models for the dynamics of science," in Sheila Jasanoff, Gerald E. Markle, James C. Petersen and Trevor Pinch, eds, *Handbook of Science and Technology Studies* (Thousand Oaks, 1994), p. 29-63.

64 The success of Basle was due to the comparative advantage realized by efficient networks and synergetic effects. The transfer of scientific knowledge and technical know-how favoured the emergence of a regional, transinstitutional research community that was able to counterbalance the pressure of competition through cognitive and political cooperation.

Here, the laboratory is once again the source of justification. The new remedies sold best if the public's belief in the effectiveness and supremacy of scientifically based biomedicine was intact. Therefore, new non-verbal, visual promotional procedures were developed, along with product innovation and diversification based on patent rights, which provided the product with its very own pictographs and facilitated its identification. Once registered, the proprietary article represented the scientific renown of the product and the company. A Ciba advertisement of 1939 said as much: "The Ciba brand states that the products which carry it are manufactured according to scientific principles and meet all respective technical or hygienic requirements."[65]

This was one of the principles upon which a positive interaction between the pharmaceutical enterprises and the health sector was built since the turn of the century: for without the (continuing) process and product innovation in the pharmaceutical field, medical science would not have been able to satisfy society the way it actually did in the 20th century – and without the concomitant diversification and intensification of prescriptions the sales markets of the industrial enterprises could not have expanded throughout the entire century with their quite impressive growth rates. The position of Switzerland in this process, which in the interwar period began to spread to all industrialized nations, was to a large extent due to the establishment of large-scale enterprises which maintained a competitive edge through specialization. These companies assured their markets on a global scale due to their patent and brand measures and thus consolidated their abilities to initiate successful, profit-oriented learning processes. In this manner, they managed to translate immaterial property rights into entrepreneurial property power, and to hone it into a weapon of competitive power within a national as well as international context. If, as Gianfranco Varini in his 1958 dissertation asserted, "the market position of the patent holder ... not only depends on a prohibition of imitation but also on a negotiatory strength based on the number of his patents," the patent law was a factor which furthered the entrepreneurial process of concentration, for it allowed strong patent holders to force "agreements muzzling the competition, block licensing, the obligation to cross license any future patents of the licence holder, etc." upon their competitors. "The Swiss Patent Act also works as a monopoly by furthering concentration tendencies in various respects."[66] The patent and brand law thus turns out to be a medium of amplification and consolidation furthering those tendencies which were essential constitutive elements of the interwar corporate learning process.

---

[65] Ciba, *Aus der Entwicklungsgeschichte der Gesellschaft für Chemische Industrie Basel* (Basle, 1939), cover page.

[66] Varini, op. cit. (14), p. 188. See also M. Börlin, "Die volkswirtschaftliche Problematik der Patentgesetzgebung," Doctoral Thesis, University of Zurich/St. Gall, 1954.

# Part 5

## State Intervention and Industrial Autarky

# TECHNICAL CHANGE IN THE ITALIAN CHEMICAL INDUSTRY: MARKETS, FIRMS AND STATE INTERVENTION

ROLF PETRI
*Martin-Luther-Universität*
*Institut für Geschichte*
*Kröllwitzer Straße 44*
*D - 06120 Halle (Saale), Germany*

The initial technological backwardness of the Italian chemical industry is here explained by the country's economic development. Prior to 1930, chemical production was based on the needs of agriculture and traditional manufacturing. Lack of energy and raw materials prevented the development of a modern organic chemical industry. However, in the 1930s protectionism and State intervention accelerated R&D, and the transfer of technology from other countries led to a number of advances in modern processes. Though many of these new processes were not at first viable they did provide the basis of scientific knowledge, technical skills and practical experience that facilitated the extraordinary growth of chemical production in the post-war period.[1]

## 1. A profile of the sector

During the first decade of the 20th century Italy concluded her primary phase of industrialization and modern economic growth. Subsequently, structural changes and industrial advances, though proceeding, could not quickly overcome the backwardness of various economic sectors and regional areas. Indeed, it was only around 1960 that Italy achieved the full economic and social patterns characteristic of industrialized nations. Certainly it would not have been possible to attain this result without the advancement of the so-called modern branches, such as chemicals. Chemical industry became a prominent component of a more general "economic miracle," that outside Italy has rarely been acknowledged.[2]

---

[1] Abbreviations for archival sources used in this paper: AConf = Archivio della Confindustria, Roma; ACS-IRI = Archivio centrale dello Stato, Fondo Iri, Roma; AST = Archivio di Stato Terni; ASV = Archivio di Stato Venezia; ATdR = Archivio Thaon de Revel, Torino; BA = Bayer Archiv, Leverkusen; BASF = BASF-Archiv, Ludwigshafen; BuA = Bundesarchiv, Koblenz; HA = Hoechst-Archiv, Frankfurt.

[2] See, however, in the English language: Alexander Gerschenkron, "The industrial development of Italy: a debate with Rosario Romeo," in *idem*, *Continuity in History and other Essays* (Cambridge, Mass.: Harvard UP, 1968), pp. 72-89; Gianni Toniolo, *An Economic History of Liberal Italy 1850-1918* (London New York: Routledge, 1990); Vera Zamagni, *The Economic History of Italy 1860-1990* (Oxford: Clarendon Press, 1993); *idem*, "The Italian 'Economic Miracle' revisited," in Ennio Di Nolfo, ed., *Power in Europe? Great*



*A.S. Travis et al. (eds.),*
***Determinants in the Evolution of the European Chemical Industry, 1900–1939,*** 275–300.

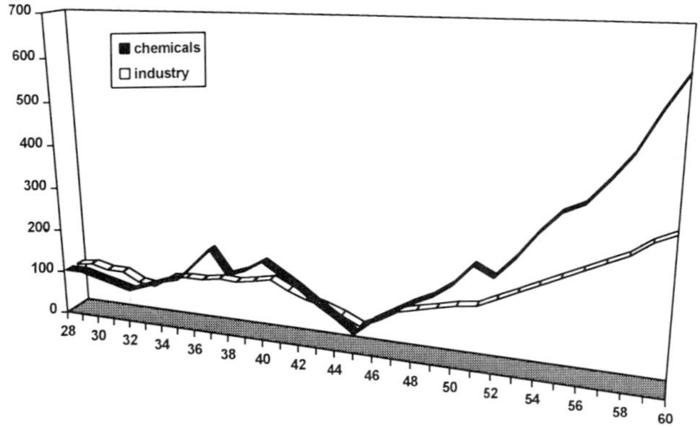


Figure 1 Italy 1928-60: production indexes of chemical industry and industry as a whole (1928 = 100 per cent; source: G.B. Aldo Trespidi, *Realtà e prospettive dell'industria chimica in Italia* [Roma: Editrice sindacale italiana, 1967].)

In the period 1896-1925 the average annual growth of chemical production was 10.5 per cent in constant prices: 16.1 per cent during the first wave of accelerated industrialization before 1908, 5.0 per cent in the pre-war period 1908-13, and 6.6 per cent between 1913 and 1925. According to the Organization for European Economic Cooperation statistics, the proportion of chemicals in overall production of manufactured goods rose from 4 per cent in 1901, to 7 per cent in 1913 and 13 per cent in 1929; in 1937, finally, it represented an 18 per cent share of industrial production.[3] During the 1929-32 crisis period, the industrial sector as a whole performed somewhat better than chemicals, but after 1933 the latter consistently outperformed the average growth rate of industrial production, with the exception of 1945. Figure 1 shows that the performance of chemicals was very dynamic in the pre-World War II and early war period (1934-41), as well as during reconstruction (1946-50) and, in a particularly striking way, in the era of the so-called Italian economic miracle (1951-63). During those years the average increase in manufacturing industry was 8 per cent, whereas the

---

*Britain, France, Germany and Italy and the Origins of the EEC 1952-1957* (Berlin/New York: de Gruyter, 1992), pp. 197-226; Nicola Rossi and Gianni Toniolo, "Catching up or falling behind? Italy's economic growth 1895-1947," *Economic History Review*, **45**, 3 (1992); Giovanni Federico, "Italy, 1860-1940: A little-known success story," *Economic History Review*, **49**, 4 (1996), 764-786; *idem*, ed., *The Economic Development of Italy Since 1870* (Aldershot: Elgar, 1994).

3 Toniolo, op. cit. (2), p.8; "Statistiques industrielles 1900-1955," *Bulletins statistiques de l'OECE* (Paris: OECE, 1956); V. Paretti and G. Bloch, "La production industrielle en Europe Occidentale et aux Etats Units dal 1901 à 1955," *Moneta e credito*, **4** (1956), 286.

growth rate of chemicals was almost 17 per cent. It seems quite reasonable to assume that such an extraordinary display of vitality could not have been achieved without a prior accumulation of basic expertise.

### *1.1 Production, domestic demand and exports*

The production figures in table 1 illustrate how the structure of the chemical industry was determined by the ongoing macroeconomic influence of farming and traditional industries. Up to World War I Italy's industrialization process had been driven mainly by textiles. For agriculture, even in the interwar period, there was an enormous need for modernization. The endemic underemployment of labour in agriculture depressed wages and restricted the domestic market for capital goods as well as the demand for manufactured consumer goods. During the emergence of the chemical industry these external obstacles determined the nature of its structure. The most typical example was the leading chemical firm Montecatini. Established in 1888 as a mining company, it diversified towards chemicals only after 1910, mainly by expanding the mining of pyrites and sulphur during World War I and by absorbing in 1920 the leading superphosphate fertiliser producers Unione concimi e prodotti chimici and Società prodotti chimici colla e concimi. After having increased the labour force from 2,500 in 1910 to 18,000 in 1929, the firm produced 55 per cent of national sulphuric acid output, 62 per cent of superphosphates, 65 per cent of copper sulphate (the principal fungicide of Italian agriculture, essential for protecting vineyards), 66 per cent of nitrogenous fertilisers and almost 80 per cent of nitric acid. Later it also accounted for a substantial share of calcium cyanamide.[4]

The company gained from the efforts undertaken by Mussolini's government to raise the productivity of agriculture by means of organization, mechanization and greater use of manufactured fertilisers. Since phosphates and potash had to be imported and the price dictated by the potash syndicate was considered to be exorbitant, the government fostered the development of calcium cyanamide and nitrogen fertilisers. It is symptomatic of adjustment of technical change to the requirements of agriculture that the only major invention of the 1920s that in some way preceded diversification towards organic chemistry was the high pressure synthetic ammonia process. The ammonia process – a modified Haber-Bosch process – had been developed by the company's later chief engineer Giacomo Fauser even before he was hired in 1921. Although after 1930 a substantial diversification into other chemical branches took place, the production of fertilisers and fungicides remained the most important feature of growth before 1936. During the first half of the 1930s the company's work force increased to more than 40,000 workers.

The historical development of Italy's leading chemical company, as well as that of this sector and the economy as a whole, perpetuated the specialization in inorganic chemistry

---

4 Franco Amatori, "Montecatini: un profilo storico," in Franco Amatori and Bruno Bezza, eds, *Montecatini 1888-1966. Capitoli di storia di una grande azienda* (Bologna: Il Mulino, 1990), p.45.

Table 1. Italy 1921-41: selected products of the chemical industry (1,000 t, unless indicated; *= 1929; **= 1942; ***= 1944; **** = 10,000 hl; - = unknown)

| | 1921 | 1926 | 1931 | 1936 | 1941 |
|---|---|---|---|---|---|
| Sulphuric acid | 672 | 1367 | 1012 | 1532 | 1818 |
| Hydrochloric acid | 14 | 46 | 35 | 45 | 65 |
| Nitric acid | 6 | 11 | 60 | 194 | 467 |
| Caustic soda | 26 | 103 | 117 | 140 | 198 |
| Sodium carbonate | 27 | 147 | 181 | 295 | 390 |
| Sodium sulphate | 8 | 17 | 18 | - | 26 |
| Copper sulphate | 95 | 100 | 79 | 124 | 61 |
| Carbon disulphide | 3 | 11 | 12 | 28 | 62 |
| Aluminium sulphate | 2 | 4 | 3 | 8 | 20 |
| Oxygen (10,000m$^3$) | 259 | 781 | 738 | 1577 | 2752 |
| Hydrogen peroxide | 0.04 | 0.1 | 0.4 | - | 2 |
| Citric acid | 0.8 | 2 | 2 | 1.8 | 2.2 |
| Tannin extract | 17 | 50 | 49 | 66 | 84 |
| Lead oxides (red lead, etc.) | 2.6 | 6.5 | 4.4 | 7.6 | 4.4 |
| Zinc oxide | 1.8 | 4.9 | 3.4 | 10.2 | 7.9 |
| Superphosphate | 859 | 1475 | 792 | 1366 | 1136 |
| Ammonium sulphate | 6 | 61 | 138 | 148 | 316 |
| Calcium cyanamide | 21 | 45 | 85 | 197 | 186 |
| Calcium carbide | 32 | 62 | 76 | 156 | 215 |
| Ammonia | 0.03 | 16 | 50 | 77 | 167 |
| Acetic acid | 0.2 | 1 | 3 | 9 | 34 |
| Butyl/methyl alcohol**** | - | - | 50* | 110 | 111 |
| Glycerol | - | - | - | 3.5 | 4.2 |
| Acetone | - | - | - | 2.4 | 3.5 |
| Tar | 27 | 65 | 93 | 90 | 107 |
| Benzene | - | - | 3.2* | 6 | 17 |
| Toluene | - | - | 0.3* | 0.8 | 2.3 |
| Synthetic organic dyes | 3 | 7 | 5 | 9 | 16 |
| Phenol for 'Bakelite' | - | - | - | 1 | 4.5 |
| Urea for Caurit | - | - | - | 0.6 | 2.8 |
| Amino and phenol plastics | - | - | - | ≈3 | ≈15 |
| Plexiglas (1,000m$^2$) | | | | | 15 |
| Synthetic rubber | | | | | 3.3** |
| Synthetic fibres | | | | | 0.07*** |
| Paper cellulose | 0.06 | 2.9 | 6 | 24 | 63 |
| Artificial silk | 1.5 | 17 | 34 | 39 | 57 |
| Other artificial fibres | | | 2 | 53 | 144 |
| Aluminium | 0.7 | 1.9 | 11 | 14 | 48 |
| Zinc | 0.5 | 7.6 | 17 | 29 | 40 |
| Magnesium | | | | | 2 |

Sources: ISTAT, Sommario di statistiche storiche 1861-1965; Annuario Statistico 1944-48; ACS-IRI, numerazione nera, b 83, Relazione riassuntiva piani autarchici; Guida dell'autarchia 1942-43 (Milano, 1942); Statistisches Reichsamt, Das italienische Wirtschaftspotential im deutschen Machtbereich (1944); AConf, fondo Balella, b 76; Bundesarchiv, R3, 211 Amtsgruppe Textilrohstoffe; ASV Camera di commercio, 1933-III-1, f 28.

and the extraction of chemical compounds, tannin, and fats from vegetable and animal materials. Besides agriculture, traditional industrial branches such as leather and textiles, oil mills, foodstuffs, glass and paper factories provided the major part of the demand. Carbon disulphide was originally used as a solvent in agroindustrial processes, prior to being employed in the production of war gases and the manufacture of rubber and artificial silk. Aluminium sulphate was utilized in dyeing; caustic soda was required by producers of soap, paper, artificial textiles and pharmaceuticals; sodium sulphate was important in the manufacture of artificial textile fibres, glass, paper and leather products; sodium carbonate was essential to a number of processes used by the glass, textile, paper and dyestuff industries. While citric acid extraction stagnated due to the widespread use of substitutes, tannin exports continued to perform well.

One may argue that the importance of Italian textiles should have stimulated the development of synthetic dyes. That had happened in other European countries, where synthetic dyes were the forerunners of a wide range of organic products. In Italy, the chemical industry started late when compared with foreign competitors, and had to face high costs and lack of skills. Dyestuff imports therefore continued to be substantial. The state, well accustomed to protecting and subsidizing farming, steelworks, shipbuilding and many other sectors, did not make any attempt to protect the chemical industry. Chemical production was considered to be an auxiliary to agriculture and traditional industries.

Only during World War I did the forced interruption of external trade and the war effort induce the government to encourage inventors and entrepreneurs to enter the domain of electrochemical applications and certain other chemical processes. New firms were founded, factories erected and processes introduced in order to obtain substitutes, explosives and toxic gases. Chemists gained new recognition and national prestige as inventors, engineers, entrepreneurs, managers and bureaucrats. The end of the war and the reinstatement of free-trade principles in chemicals ended this temporary boom. However, the gradual return of protectionism after 1925 and the implementation of an autarky policy in 1935 offered another opportunity for the chemical industry to overcome its backwardness. The 1930s saw the introduction of a broad range of new processes in the field of organic chemistry (see table 1). Many processes were investigated, and brought on stream before or during World War II. Also, the production of light metals to replace copper was encouraged by autarky, as well as by the preparation for war. Thus, it is necessary to focus on the period 1930-43 when dealing with technical change.

Another important branch of the Italian chemical industry was that of fibres, such as artificial silk derived from cellulose and a casein-based fibre called Lanital. Their dynamic development partially replaced the traditional textiles as a leading export industry, while artificial fibres also replaced significantly cotton, wool and jute imports. More than 80 per cent of the overall output was produced by SNIA Viscosa and CISA Viscosa, two associated firms dominated by the same network of interlinked international interests (Courtaulds, AKU, Vereinigte Glanzstoff, etc.). Foreign companies held important shares in other firms too, such as Châtillon, Bemberg Italia and the Rhodiaceta Italiana (founded by Montecatini). Moreover, artificial fibres were also produced by Gerli, Orsi Mangelli, Manifattura Casale and Pirelli.[5] The predominant foreign influence in this sector might have provoked political criticism, if it had not been mitigated by the convincing patriotic image of the SNIA's chairman Franco Marinotti, and the fact that foreign ownership did not actually interfere with nationalist purposes. But the involvement of foreign capital must principally be explained by the macroeconomic importance that these goods gained in the course of the 1920s and 1930s. Artificial textile fibres, auxiliaries and manufactures alone represented 10.5 per cent of Italian exports for 1939. On the whole, the favourable export-import balance of chemical products (exports 1,778 million lire, and imports 711 million lire), was mainly due to artificial textile fibres. However, even when the export of fibres and the import of cellulose (for fibre production) are left out, the balance was still biased towards chemical exports, 639.7 million lire (equal to 5.9 per cent of total exports), while imports amounted to 462.5 million lire (4.1 per cent of total imports; current prices).[6] Was, therefore, the chemical business a major contributor to the Italian economy as a whole? It is probably impossible to give a definite answer, because the dynamic advantages and disadvantages deriving to the other branches and sectors from the domestic production of chemicals, and the additional need of some raw materials, should be taken into consideration as well. One has to bear in mind that coal, coke, mineral oil and derivatives represented 20.6 per cent of total imports, and that the energy and carbon-based raw material demand of the chemical industry was by no means insignificant. The lack of energy remained the ultimate constraint of the Italian industrial economy up to the 1950s, when the relative cost of petroleum began to decrease. It may therefore be assumed that the basic technological capabilities to launch a modern chemical industry had already been accumulated by the late 1930s, and that in the first place constraints of war economy and difficulties of reconstruction were responsible for the postponement of its very dynamic growth.

### *1.2 Firms, chemists, and marketing organization*

The 41 enterprises listed in table 2 represent only 0.7 per cent of the 5,793 firms that in 1941 belonged to the statistical categories of chemicals, artificial textiles, and rubber and plastics.[7] These firms employed 281,828 workers, i.e. on average 48.6 workers per

[5] BuA, R 3, 211; ACS-IRI, Rossa, 15

[6] Statistisches Reichsamt, *Der Außenhandel Italiens 1939-1943* (1944); Nicola Rossi, Andrea Sorgato, and Gianni Toniolo, "I conti economici italiani: una ricostruzione statistica 1890-1990," *Rivista di storia economica*, **1** (1993).

[7] A comparison between table 2 and the overall figures for the sector can only provide approximate evidence. Since the entries in the table do not entirely match the statistical classifications adopted for the sector, the number of firms is not exactly comparable.

firm.[8] If one bears in mind that Montecatini alone accounted approximately for a quarter, and the larger enterprises certainly more than half, of the work-force, the division of the sector can be clearly perceived: on the one side large and medium-sized companies; on the other side myriads of small and very small firms that extracted perfumes, colours, pigments, tinctures, sweeteners, tannin and animal or vegetable fats from biological or mineral matter, manufactured wax candles or soaps, and processed rubber, camphor or other imported materials for the needs of local markets. For the period under investigation, it is not arduous to presume that the former were the protagonists of technical change, while the latter represented the most traditional segment. The small firms displayed a very low level of labour productivity and suffered from an almost general lack of scientific support. The only noticeable change occurred when the consortia and the fascist *corporazioni* imposed the unification of technical standards in order to simplify the organization of distribution. These interventions attenuated the technical and regional fragmentation both of production and of marketing.

Only Montecatini approached the typical pattern of the big European chemical enterprises which - like ICI, I.G. Farben and Rhône-Poulenc - had diversified into almost all principal branches of modern industrial chemistry. Also, the state-owned industries operated in numerous branches, but they were not incorporated under a unified command. The organizations AGIP and IRI were completely separate, and the affiliate firms of IRI, although subjected to political supervision, were nevertheless too independent to adapt to the uniform organizational structure of an integrated corporation. Montecatini's range of diversification, on the other hand, suffered from serious shortfalls in organic chemistry. However, the company was powerful enough to prevent attempts by smaller firms to diversify towards the products that it was not able or willing to produce. In order to close the technological gaps, on several occasions the state-owned firms had to organize resistance against Montecatini's interference.

Some of the specialized firms, such as SNIA and CISA Viscosa, SAVA and Solvay, were controlled by foreign capital. Other medium-sized enterprises operated in specific areas such as pharmaceuticals, detergents, paper and vegetable fats. They were run by majority shareholders, individual owners or family representatives (Burgo, Gaslini, Montesi, Parodi Delfino, Saronio, etc.), who were not only the owners and heads of management, but often also graduate chemists or engineers. The chemical productions of big industrial enterprises, such as FIAT and Pirelli, were geared to goods not yet offered on the domestic market, rather than towards generic diversification strategies. Italgas, SAFFA and Stabilimenti di Rumianca were specialized enterprises (the former two remained marginal to the core of the sector), which, thanks to the uncommon lack of a clearly discernable majority shareholder, are classified, somewhat generously, as

8 ACS-IRI, numerazione nera, b 81, f Appunti e relazioni varie 1943-45.

Table 2: Chemical companies operating in Italy (with a capital stock ≥30 million lire in 1942;*= 1939 data.)

| *Company* | *(branch) [main shareholders]* | *capital: million lire* |
|---|---|---|
| Montecatini | (inorganic and organic chemicals, non-ferrous metals) [see text] | 1600 |
| Terni | (electricity, steel, explosives, fertilisers, heavy chemicals) [IRI] | 1200 |
| SNIA Viscosa | (artificial fibres, heavy chemicals) [Courtaulds etc.] | 1050 |
| ANIC | (oil refining, synthetic fuels, chemicals) [AGIP, Montecatini] | 750 |
| Italgas | (gas, organic intermediate products) [public company] | 541 |
| AGIP | (oil refineries, fuel, methane gas, heavy chemicals) [State] | 500 |
| Pirelli | (rubber and rubber products) [Pirelli family] | 500 |
| SAIGS | (synthetic rubber, organic chemicals) [IRI, Pirelli] | 500 |
| Ammonia deriv. | (ammonia fertilisers, heavy chemicals) [Montecatini] | 300 |
| Solvay & Co. | (alkali products: caustic soda, etc.) [Belgian Solvay] | 300* |
| CISA Viscosa | (artificial fibres, heavy chemicals) [SNIA, Italrayon] | 257 |
| INA | (alumina, aluminium) [Montecatini] | 240 |
| Burgo | (paper, chlorine, cellulose) [Luigi Burgo] | 215 |
| SA Cokitalia | (tar distillates, aromatics) [Montecatini, Italgas] | 200 |
| Châtillon | (artificial textile fibres) [syndicate of the wool industry] | 175 |
| ACNA | (dyes, pharmaceuticals, explosives) [Montecatini, I.G. Farben] | 150 |
| Indus. zuccheri | (methylated spirit, etc.) [Piaggio family] | 150 |
| Monteponi | (zinc, lead, etc.) [Montecatini] | 150 |
| Montevecchio | (zinc, lead, sulphuric acid, etc.) [Montecatini] | 150 |
| Rumianca | (alkali products, inorganic chemicals) [public company] | 150 |
| CCS | (soda, chlorine, cellulose, toxic gases, etc.) [IRI] | 141 |
| Dinamite Nobel | (explosives, heavy chemicals) [Montecatini] | 100 |
| SA Gaslini | (vegetable fats, essences, etc.) [Gerolamo Gaslini] | 100 |
| SAFFA | (matches, sulphur products) [public company] | 100 |
| SAVA | (alumina, aluminium) [AIAG-Alusuisse] | 100 |
| SMI | (non-ferrous alloys) [Orlando, Pirelli, Zanoletti, etc.] | 90 |
| Distillerie ital. | (methylated spirit, etc.) [Eridania sugar industries] | 78 |
| Farmitalia | (pharmaceuticals) [Montecatini, Rhône-Poulenc] | 65 |
| Mira Lanza | (soaps, detergents) [Magni, etc.] | 60 |
| BPD | (explosives, byproducts) [Leopoldo Parodi Delfino] | 50 |
| Carlo Erba | (pharmaceuticals, dietetics, etc.) [Erba family, Saronio] | 50 |
| ROMSA | (oil refineries, fuels, heavy chemicals) [AGIP] | 50 |
| Vetrocoke | (tar, aromatics, ammonia, fertilisers, Plexiglas) [FIAT] | 50* |
| Ledoga | (pharmaceuticals) [Lepetit, Dollfus, Gansser] | 48 |
| FRGomma | (rubber and rubber products) [Pirelli] | 36 |
| Caffaro | (heavy chemicals, pigments) [linkage with Rumianca] | 34 |
| Monte Amiata | (mercury) [IRI] | 32 |
| L'Appula | (essences) [Montecatini] | 30 |
| Aniene | (alkali products) [Solvay] | 30* |
| Dist.Cavarzere | (methylated spirit, etc.) [Ilario Montesi] | 30 |
| SCIA | (soda, chlorine, cellulose) [IRI] | 30 |

Sources: *La chimica e l'industria* (1942); Archivio centrale dello Stato, fondo IRI; Archivio di Stato Venezia, Camera di commercio, 1933-III-1, 117; B.Flury-Nencini, "La società Solvay," *Liburni civitas* , **5-6** (1939).

public companies. The truth is that both the dimension and the functioning of the Italian capital market and stock exchange did not favour the formation of public companies in the present sense of the term. In the case of Italgas, for example, a small group of banks, insurance companies and industrialists established a rigid supervision based upon informal agreements, while the mass of 18,000 individual shareholders had virtually no influence.

The foregoing, if considered against the background of rather slowly growing domestic markets and declining international trade, may help to understand why Italy was not the most appropriate environment in which to apply "Chandlerian" principles of organization. The only firm that seemed to possess the necessary prerequisites was, once again, Montecatini. The capital of the company was not controlled by a single majority stockholder, but by a sort of financial and industrial oligarchy (the holding Bastogi, electric companies, industrialists, banks, insurances, etc.), which nevertheless delegated almost all the decision and management power to the president, Guido Donegani. Notwithstanding the evident separation of ownership and management, the considerable scale of production and the wide-ranging diversification, Franco Amatori has identified "a rigid concentration of decision-making." Diversification was carried out by a number of affiliate companies, the directors of which were recruited from the inner circle of the chairman's staff. In spite of their formal autonomy, they behaved as loyal executors of Donegani's orders rather than managers of independent firms or directors of divisions. Certainly, diversification was not followed by the creation of a multidivisional structure.[9]

Not only Montecatini, but almost all Italian chemical companies were run according to a pyramid-shaped command model in the technical, administrative and accounting fields. Sometimes the term management was synonymous with proprietors, but more often it was handled directly by the engineering personnel, in order to put into place a strategy for rationalization designed almost exclusively to improve technical performance.[10] This schema was, after all, not as irrational as it might seem. In the Italian market the following factors acquired increasing importance: state-control over foreign trade; corporatism and cartels; market shares worked out by agreement between firms; raw materials and machinery allocated by state and *corporazioni* (i.e. organizations that included the firms, engineers, and workers representatives and were subjected to political supervision); price fixing for commodities, capital inputs and salaries; exemption from taxes and customs duty subsidies, and, lastly, state contracts. By far the strongest demand came from large scale industry and the state; that is, it was partly oligopolistic and partly public. The distribution of fertilisers and fungicides was

---

9 Amatori, op. cit. (4), pp. 55-56; and Richard A. Smith, "Montecatini: Italy's chosen instrument," *Fortune*, **4** (1957), 272.

10 Giulio Sapelli, "Gli organizzatori della produzione tra struttura d'impresa e modelli culturali," in *Storia d'Italia. Annali*, (Torino: Einaudi, 1981), vol. 4, pp. 637-638.

managed by the *corporazioni* of the farmers; in addition the customers in the small and medium sized industries were organized in consortia. The domestic market for chemical consumer goods, on the contrary, was hampered by agricultural underemployment and low-wage policies. In such a highly regulated market it was fairly normal to think that the best strategy for marketing was to combine political and corporative lobbying with technical performance. Once the strategy was combined in an efficient way, the firm did not have to worry too much about the costs of the technical procedures, because the subsidies and the rules of a cartelized oligopolistic market allowed profitable pricing.

Within the firm's administration and organisation chemists and engineers had, therefore, a pre-eminent rôle. It may be worth examining briefly the mentality of the leading technical personnel at that time. It is true that only in the 1930s did government and industry begin to promote systematic R&D programmes in some way comparable to those in the leading industrial countries. But, despite this, there was a generation of engineers and chemists holding management positions and with a good scientific background in experimental chemistry. The university and polytechnic departments of industrial chemistry, where these men had received their training, and to which World War I had given a decisive boost, not only gave them a solid scientific education but also transmitted the industrial patriotism of those years.[11]

One can therefore well understand how it was that in the 1930s many engineers saw business ethics and professional commitment perfectly well represented by an industrial policy that was not concerned so much with costs but rather with the 'moral' justification of every technical achievement deemed to be valid for the future of the nation. Many of Italy's best chemists and engineers were engaged in political and administrative structures. And what was true for the chemists and engineers of the private sector held even more for the technical and managerial staff of the state-owned industries. The chemist Francesco Giordani, vice-president (1937-39) and president (1939-43) of the State-holding IRI, contributed decisively to the design of industrial policy, promoted research and development in the alkali and paper industries, brought on stream, together with Pirelli, one of the first advanced processes for industrial polymers, counter-balanced the monopolistic ambitions of Montecatini, and participated more generally in shaping the structure of industrial change. In the period 1947-50 he was an alternate executive director of the International Bank of Reconstruction and Development. In 1956 he returned to the chair of the National Research Council (which he held in 1943-44), promoting regional development policies and nuclear energy. Giordani was a prominent representative of the nationalist and industrialist culture of Italy's leading chemists in the first half of the

[11] Roberto Maiocchi, "Il ruolo delle scienze nello sviluppo industriale italiano," in *Storia d'Italia. Annali*, vol. 3 (Torino: Einaudi, 1980); Valerio Castronovo, ed., *Il Politecnico di Milano nella storia italiana (1914-1963)*, (Bari-Roma: Laterza, 1989); Luigi Cerruti, *La cultura dei chimici*, in Corrado Stajano, ed., *La cultura italiana del Novecento* (Roma-Bari: Laterza, 1996), pp. 137-152.

century.[12]

***1.3 Guide-lines for state intervention***

The state intervention after 1934 is understood by the high degree of convergence between the strategic purposes of the chemical industry, the ideas of the technocratic managers of state-owned industries, and the political guidelines of the fascist government. More than a dozen representatives of the chemical industry – chemists and top managers – were members of the *Camera dei fasci e delle corporazioni* or of the *Senato*. The president of Montecatini, Guido Donegani, together with the other *senatori* and *consiglieri nazionali*, Pirelli, Marinotti and Parodi Delfino, also belonged to the national council of the industrialists mighty confederation, the *Confindustria*. This was only the institutional surface of a more substantial grouping and general cooperation between this particular branch of industry (the same holds for other industries) and political power. The regulation and allocation of resources generated conflicts between competitors, compelling sometimes the *corporazioni* and at other times the authority to mediate and to arbitrate. This was by no means a simple task. Seen from the standpoint of the firms, the so-called fascist economy was not a rough command system, but rather a vertically and horizontally interconnected network of formal and informal linkages which was impossible for one single actor to dominate. It is nevertheless important to emphasize that apart from all conflicts the strategic goals of chemical industry and state were in good agreement. State intervention was based on the agendas of chemists, industrialists and technocrates as well as on the decisions of Mussolini and his ministers.

The ultimate justification for the industrialists consent to the "autarky plans" (which had to a large extent been drawn up by the leading men of industry) was that the autarky strategy of the 1930s represented the current variant of a deeply rooted neo-mercantilist approach to industrialization, rather than a contingent byproduct of fascist foreign policy.[13] The rationale was substantially that of Friedrich List's "theory of productive forces:" in order to improve long-term economic prospects and the proper status in international competition, a "late coming" nation must not blindly obey the rules of comparative costs, which reveal only short-term advantages. It must invest in industries, technologies, and human capital leading to improvements in overall productivity, even if the initial costs may be high. In the interwar period, this argument was supported by a Schumpeterian-like theoretical framework, according to which the state, the *corporazioni*, and the economy as a whole, in the face of the inherent uncertainty of economic development, had to behave like an individual entrepreneur

[12] Laura Scalpelli: "Francesco Giordani," in Alberto Mortara, ed., *I protagonisti dell'intervento pubblico in Italia* (Milano: Angeli, 1984); Barbara Curli, "Francesco Giordani," *La città nuova*, **4-5** (1994), pp. III-XVI.

[13] Theodor Pütz, "Autarkie," in *Handwörterbuch der Sozialwissenschaften*, vol. 1 (Stuttgart/Göttingen Tübingen: Fischer-Siebeck-Vandenhoek, 1956), pp. 520-524.

gambling on the long-term benefits of (technological) innovation. While future benefits could not be predicted, in the short period costs of innovation might have caused inefficient allocation of 'given' resources. As professor Di Fenizio taught his students, the principles of indeterminism were particularly cogent to the economics of (organic) chemicals, because the complexity of the interdependent production lines, and the related calculation of scales of operation, generated unstable hierarchies between "main products" and 'byproducts,' leaving ample space for the creation of new opportunities by technical innovation.[14]

The preferences of industrial policy have to be seen from this angle, and not only from the more superficial perspective of war preparations. In practice, however, for the bulk of chemical branches the two motivations made little difference. The replacement of imports was of course welcomed, because it obliged the state to foster the production of new synthetic goods. Insistently claimed, rather than passively welcomed, this policy led to the escape from the 'dictates' of international cartels and the decision to confine foreign capital to spheres where it could not hamper the acquisition of technical capabilities.

State intervention was not limited to the defence of the chemical industry's common interests, but also participation in its internal disputes. One of its most important functions was the counterbalance of Montecatini's power. Mussolini was aware of the monopolistic position that the leading chemical company had achieved during the 1920s in the field of nitrogen fertilisers, the increasing use of which was an essential feature of his agricultural programme.[15] Thus the state favoured the entry of other companies, such as FIAT's affiliate Vetrocoke, which integrated towards synthetic polymers, competing with Montecatini for this market too. It is difficult to say if this line of intervention was harmful or beneficial to the sector's development. On the one hand, an additional strengthening of Montecatini's position might have led to large-scale production, especially in branches where subsidies encouraged the survival of inefficient factories. On the other hand, an unopposed predominance of the company might have prevented others from undertaking the innovations that the leading producer was not capable, or willing, to undertake. Presumably, without state intervention progress in the strategic field of industrial polymers would have been injuriously slow.

## 2. Technical change

The political emphasis on invention and technology was underscored by the fact that the first chairman of the Government Committee for Scarce Raw Materials and Substitutes was the pioneer of radiotelegraphy, Guglielmo Marconi. Besides this

---

14 Ferdinando Di Fenizio, *Lezioni di economia industriale* (Milano: Servizi editoriali del Guf, n.d., [1941-43]), pp. 77-106; on the same argument, see Theodore J. Kreps, "Joint costs in the chemical industry," *Quarterly Journal of Economics*, **44** (1930), 416 and 443.

15 BA, Akten der Direktionsabteilung, 424, SA Magnesio italiano Sulcis.

committee, which obviously had to deal with chemistry too, various other ministerial boards and subordinated administrations carried out or stimulated R&D on fuels, explosives, toxic gases, gas masks and similar war related topics, and collaborated with the Ente nazionale italiano di unificazione (UNI) in order to determine quality standards for the suppliers of the armed forces. In reality, these initiatives did not bring about a coherent programme of public research. Relevant work was, however, done by the Centro chimico militare in the field of toxic and smoke-forming gases and their antidotes, and by the Ministry of Corporations (i.e. Economy), which created a division for the study of petroleum and liquid fuels. One of its departments, the Ufficio combustibili, coordinated private research activities with those of the Reale stazione sperimentale per i combustibili.[16] The *corporazioni* instituted research committees on specific issues too. The state-holding IRI stimulated the technical change of single branches, promoting the Società per lo sviluppo della cellulosa (jointly with Cartiere Burgo) and the Istituto per lo studio della gomma sintetica (with Pirelli).

In Italy the chemical industry remained distant from academic cooperation, as already achieved in Germany, America, Japan, France, etc. Nevertheless, some substantial improvements did occur. Situated at the very core of cooperation between science and practice, the industrial chemical department of the Politecnico di Milano was the best funded research institute.[17] Many other universities and polytechnical high schools established more durable relations with industry. Previously the scarce bilateral coordination had been related to the deficiency of private industrial research. Now the requirements of autarky stimulated a major spread of private R&D offices and laboratories, such as: the Istituto scientifico per le ricerche e sperimentazioni chimiche of the Montecatini; the Laboratorio ricerche e Tintoria sperimentale of the ACNA, and the similar arrangements of Bianchi & Co. and the Saronio company; the Centro sperimentale e di ricerche of SNIA Viscosa; the Laboratorio centrale di studi of the Bombrini Parodi Delfino BPD; the Laboratorio ricerche e fabbrica catalizzatori of the ANIC; the Bicocca-laboratories of Pirelli; and the Società italiana ricerche industriali SIRI, which had been founded by Luigi Casale, becoming the earliest example of specialization in chemical R&D and technical consulting in Italy. With the exception of the latter two, which many years before had begun to carry out systematic R&D, most of the other laboratories either had not existed at all, or had limited their occasional activity to contingent issues. Only in the 1930s did most firms begin to invest seriously in research.

Did the new activities instantly alter the quantitative indicators of innovation? Trends in

---

16 Camera dei fasci e delle corporazioni, Annuario 1939-40 (Roma: Tipografia della Camera, 1939), pp. 338-339; *La chimica e l'industria*, **8** (1941), 343; Felice Guarneri, *Battaglie economiche fra le due guerre* (Bologna: Il Mulino, 1988), pp. 557-560.

17 Roberto Maiocchi, "L'attività di ricerca nel Politecnico di Milano tra le due guerre," *Storia in Lombardia*, **3** (1989), 49-52.

patent registration suggests not. Compared to the 1937 data, the number of chemical patents registered by Italian individuals and corporate bodies increased by 23.3 per cent in 1938-40 and by 49.1 per cent in 1941-43; the 'national' share of chemical patents registration rose from 17.5 per cent in 1937-40 to 26.7 per cent in 1941-43 and to 31.1 per cent in 1944-47.[18] But the absolute variation was limited to some dozen or so, and this is certainly not enough to be interpreted as a radical shift towards innovation. Research was not yet a matter of widespread entrepreneurial behaviour, but was conducted by a small group of firms in a limited number of branches, and often adjusted to adaptation and imitation rather than invention. The structural constraints deriving from Italy's delay and the weakness of the consumer-oriented manufactures were reinforced by a war economy that concentrated "innovation on an extremely narrow range of priorities."[19] The really interesting question is whether the few priorities could permanently affect the sector's evolution, or not.

It has been argued that the autarky policy meant voluntary isolation from international technological progress.[20] At least for applied industrial chemicals this is by no means true. The political purpose was the autarkic *product*, not autarkic *technology*. Indeed, wherever possible attempts were made to innovate through imitation or through technology transfer. The latter was carried out by acquiring patents and licences, or working joint ventures with foreign companies. Before and even during World War II technical journals such as *La Chimica e l'Industria, La Metallurgia Italiana*, *Alluminio* and *Gomma* disseminated foreign industrial applications as well as they could. Technological transfer from Italy to the rest of the world was essentially limited to the Fauser patents for electrolysis of water, water gas or coke oven gas, the production of synthetic ammonia and nitrogen fertilisers, and to the foreign licences that the SIRI company of the inventor Luigi Casale had granted since the 1920s, in the same fields of application. The transfer of chemical technology from abroad, and vice versa, was certainly favoured by the political alliance with Germany, but it would not have been achieved without the enforcement of proper R&D activities. That Italian R&D efforts combined with foreign know how actually led to technical progress is shown by the following examples.

### *2.1 Light metals: the aluminium industry*

In the early 1930s, zinc and lead production were examples of Italy's breach with international cartel restrictions and the build up of national production and technology (thanks to the findings of Livio Cambi). The development of the magnesium industry took a different and rather curious path, in which the chemist-entrepreneur Montesi managed to exclude I.G. Farben by taking advantage of the *Konzern*'s hesitation over

18 ISTAT, *Annuario statistico italiano* (Roma: Istat, 1938 and 1944-48).

19 Alan S. Milward, *War, Economy and Society* 1939-1945 (London: Penguin Books, 1977), p. 206.

20 Rossi and Toniolo, op. cit. (2), p. 552.

the choice of technology and possible alliances (projects had been worked out with Cogne and SAVA). He speedily brought on stream a factory that operated a simple thermic process, which generally was judged backward. However, in the post-1945 period it held first and, after the erection of the Norsk Hydro plant at Herøya, the second position among the European producers. Because of technical improvements and EEC quotas, even in the 1960s the Bolzano plant was one of the most important European suppliers for the car industry.[21] More important, however, was aluminium. From 1927 the Government provided various incentives for the establishment of new alumina and aluminium plants.[22] The previously strong position of ALCOA was weakened by an agreement which Montecatini concluded with the German Vereinigte Aluminium Werke (VAW) for the formation of two companies, subsequently merged into Industria Nazionale d'Alluminio (INA), and a number of industrialists who, together with the Swiss Aluminium Industrie AG (AIAG), incorporated the Società Alluminio Veneto Anonima (SAVA). In the former case, the majority share belonged to the Italian partner, while in the second case the Swiss partner controlled the majority share. After 1935, ALCOA accounted for only between 6 per cent and 7 per cent of primary aluminium production.

The direction of technical change was decided by the competition between INA-Montecatini and SAVA-Alusuisse. Montecatini was at a disadvantage after becoming involved in the VAW's project designed to introduce the Haglund electrothermic system. This offered a less pure alumina that was also less suitable for electrolysis. The lower cost of the Bayer procedure, the problems created by using electrolysis with Haglund alumina, and lastly the difficulty of disposing of the slag from the furnaces led already in 1934 to a temporary closure of the whole of the Montecatini production line. The merger with VAW was dissolved and the Haglund process shelved. When the Bayer process was finally adopted in 1937, SAVA had already managed to carve out a 70 per cent share of Italy's alumina production and 45 per cent of primary metal production. The technological gap in processing alumina was not so wide. In 1938 SAVA introduced the Söderberg system, to be followed two years later by Montecatini, which installed in the Bolzano plant a variant (a single-electrode furnace) modified by its own engineers.[23] Other developments were the establishment of two specialized plants for secondary fusion and alloys by INA and SAVA, an artificial cryolite process, new recycling-methods for scrap-aluminium and extraction of aluminium for chemical

---

[21] For more details see Rolf Petri, "Innovazioni tecnologiche tra uso bellico e mercato civile," in Vera Zamagni, ed., *Come perdere la guerra e vincere la pace*, (Bologna: Il Mulino, 1997), pp. 266-272.

[22] Maurizio Rispoli, "L'industria dell'alluminio in Italia nella fase di introduzione. 1907-1929," *Annali di storia dell'impresa*, **3** (1987), 288-319; Rolf Petri, "L'industrie italienne de l'aluminium à la veille de la seconde guerre mondiale," *Cahiers d'histoire de l'aluminium*, **3** (1988), 15-24.

[23] Mario Mainardis, *I forni elettrici e le industrie elettrosiderugiche, elettrometallurgiche, elettrochimiche, elettrotermiche, elettrolitiche* (Milano: Hoepli, 1953), p.546.

purposes from bauxite waste.[24] Quality standards for marketing the aluminium metal and its alloys became compulsory. It was within the framework of these initiatives that Italy stepped up research into the field of applications, by Montecatini's subsidiary SA Industria Processi Elettrometallurgici, by SAVA's laboratory and by other metallurgical, car manufacturing and aeronautical companies.

The preparations for war hastened the growth of Italian production, which outpaced the international trend in growth. Between 1931 and 1941 Italy's primary aluminium output hovered around 5 per cent of world production. Subsequently, when production grew in many countries as a result of the war effort, Italy's share settled down at 2 per cent. After 1950 the growth rate of Italian aluminium production continued to keep pace with world production, and in the new context of European markets it managed to hold a fairly important position, at least in the sector of alumina. Until the start-up of the Sardinian aluminium pool in the 1970s, this development depended upon the experience and structures that had been created in the inter-war period.

### *2.2 Dyes, explosives and heavy organic chemicals*

Significant technical innovation, which once again was supported by international cooperation, was achieved in the fields of mineral colours, pigments and varnishes. Hundreds of firms contributed to domestic production, but the majority consisted of small factories and works that lacked modern equipment and capital reserves. Innovation, therefore, was promoted by a few big producers, such as the Società del litopone (a joint-venture of Montecatini, I.G. Farben and Sachtleben AG) and DUCO (Dinamite Nobel and Du Pont).

The production of synthetic dyes had to a large extent originated in smooth technical changeover by explosives producers at the end of World War I. The main manufacturers were: the Azienda colori nazionali affini (ACNA), which in 1936 covered 51.6 per cent of Italian synthetic dyestuff output; the Società chimica lombarda A.E.Bianchi in Rho; the Industria chimica Dottor Saronio; and the Industria piemontese colori di anilina (IPCA). ACNA was controlled by Montecatini's 51 per cent share, while I.G. Farben held 49 per cent; 51 per cent of Bianchi's capital was controlled by I.G., whereas Montecatini held the rest. In 1931, when Montecatini had taken over ACNA, I.G. had been 'persuaded' to take part in the affair by a threat of exclusion from the Italian market.[25] The German group, however, restricted, as far as possible, the granting of licences to ACNA and even to Bianchi, in order to defend its exports to Italy and hamper the competition from Italian dyes.[26] The licences granted to Bianchi

---

24 ASV, Camera di Commercio, 1933-III-1, ff. 16, 76; and Giuseppe Agnini, "La criolite e il processo elettrico di fabbricazione dell'alluminio," *La chimica e l'industria*, **5** (1940), 209.

25 AConf, fondo Balella, f Materie coloranti.

26 Anne von Oswald, *Die deutsche Industrie auf dem italienischen Markt 1882 bis 1945* (Frankfurt am Main: Lang, 1996), pp. 101-102.

were linked to the maintenance of the majority shareholding, whereas ACNA had to accept export restrictions on all goods produced with technical assistance from the Germans. Prior to 1940, the technological cooperation with ACNA had, however, been conspicuous.[27] Between 1931 and 1938 the increase of the firm's benzene, toluene and naphthalene consumption was 307 per cent, and new equipment was installed to produce an almost complete range of azo, sulphur, triphenylmethane, indigoid and anthraquinone dyestuffs, as well as various organic chemicals in bulk. ACNA, furthermore, extracted alkenes from the light oil residues of coking using a German procedure, and tripled its phenol capacity thanks to a brand new plant brought on stream in 1941 with the assistance of I.G. Farben engineers. The firm benefited from the cooperation, diversifying into new fields of application, such as synthetic adhesives, glass resins and polyvinyl chloride.[28]

As for organic raw materials, the "scarcity of the national production of coal tar distillates" made it necessary to use "imported benzene, toluene and xylene."[29] In spite of the continuous augmentation of carbonization and tar distillation capacities, and the subsequent spread of benzene rectification plants, the capacities turned out to be unsatisfactory when war needs increased demand for explosives. The aromatic cost problem could not be resolved since the price level of the imported coal influenced both the price levels of its derivatives and of energy. Methanol was derived by the Società Terni from the distillates of lignite using a process developed by Luigi Casale; a minor outlet consisted of unrectified benzene.[30] Other activities in the South Umbrian area fitted in with the warfare effort of the chemical industry. In 1939 Società Terni, Acetati e derivati and SIPE founded SPEA, which from 1940 produced at Narni, on behalf of the navy, heavy chemicals and explosives. The Società italiana ricerche industriali (SIRI) tried to adapt to the industrial scale a new procedure for aromatics based on calcium carbide, via acetylene.[31] Neither the design and functioning of the process (apparently worked by catalytic cyclodehydrogenation), nor its outcome, is well documented, but certainly it could not satisfy the demand for aromatics.

The manifold interlinks between dyes, heavy chemicals and explosives were particularly manifest inside Montecatini. This company controlled three of the six main producers of explosives: Dinamite Nobel, Società generale esplosivi e munizioni, and

[27] HA, OMGUS-Akten, "Activities of I.G. Farbenindustrie AG in the Dystuffs Industry," June 1946, pp. 107-109.

[28] BA, Akten der Direktionsabteilung, 1107, Mikrofilm 72, Italien Farben-und Zwischenprodukte; BASF, Rechtsabteilung, Vertragshauptkartenkartei, 1934.

[29] M. Ragno, *L'industria italiana dei colori e delle vernici* (Roma: USILEA, 1938), p. 47.

[30] ACS-IRI, numerazione rossa, b 419, f Situazione degli impianti nel luglio 1945.

[31] AST, ex-SIRI, num.ord.854, Polimerizzazione termica dell'acetilene.

ACNA (the other three being the Società italiana prodotti esplodenti SIPE, the Bombrini Parodi Delfino BPD, and the Società prodotti esplodenti autarchici SPEA). While the technical performance of the Società generale maintained a rather traditional profile, ACNA, which had mainly produced TNT, now specialized in pentaerythritol tetranitrate. Nobel not only carried on with its traditional productions, but also dedicated "intense research and experimentation in order to find a substitute for trinitrotoluene which possibly might possess the same blasting power as pentaerythritol tetranitrate, and be derived, as well as the latter, from national raw materials," but be less reactive.[32] The company managed to bring on stream the production of the explosive 'T4' (trinitromethylenetriamine, i.e. cyclonite).

Similar innovations, of course, did not confer strategic superiority, but at best improved Italian armaments. Already at that time progress in organic synthesis was however judged from the point of view of industrial change. In 1942 the chemist Livio Cambi argued that "at the end of the war we shall evaluate the contribution of the new procedures to the overall production of explosives, but what is already clear is that they shape the route to further, wide-ranging developments."[33] Most of the more complex processes were introduced by chemical companies. Only in one case did a firm that had specialized in explosives try to diversify towards a wider range of chemical products. The Bombrini Parodi Delfino (BPD) had been among the big powder-mills of World War I. Between 1925 and 1940 the BPD factories began to produce the heavy organic chemicals required for new types of explosives (TNT and pentaerythritol), and subsequently diversified towards the exploitation of byproducts. The firm was capable of producing most of the necessary heavy chemicals and intermediate compounds, such as sulphuric acid, nitric acid, ammonia and methanol (Casale process), formaldehyde, acetaldehyde and phthalic anhydride (by oxidation of naphthalene with a catalyst developed in the firm's laboratories). A new production line was set up in order to obtain the pentaerythritol products called PA and PAN, the latter being a mixture with ammonium nitrate. The persistent shortage of glycerol encouraged the increasing production of the methyltrimethylmethane explosive Metriolo, the synthesis of which was another achievement of the firm's laboratories, directed by Francesco Somlò. The company diversified towards fertilisers, varnishes, etc. Since cartridges and projectiles were not only loaded, but also produced, inside the firm's factories, the connection between chemical, metallurgical and ballistic R&D activities became a distinctive feature of BPD.[34]

The ambitious attempt that the company made in order to penetrate into the fields of

---

32 *Cinquant'anni di storia della Montecatini* (Milano: Montecatini, 1939), pp. 371-372.

33 Livio Cambi, "L'industria chimica dal 1914 ad oggi," *La chimica e l'industria*, **11** (1942), 387.

34 BA, Akten der Direktionsabteilung, 323, Bombrini Parodi Delfino, various documents; *Il gruppo industriale Bombrini Pardodi-Delfino* (Milano: BPD, 1962), pp. 7-28.

plastics, dyes, and pharmaceuticals, was frustrated by the ostracism of the I.G. management, which intended to avoid both Donegani's irritation and further constraints on the Italian export market. The OXO process was granted on the condition that the aldehydes and derivatives not required in the production of explosives had to be sold to the Bianchi, and a request for more general agreements in the field of plastics was rejected.[35] The firm had to wait for peace-time in order to put into practice the desired diversification. In the 1950s and 1960s pesticides, detergents, cosmetics and plastics were added to traditional lines. In 1968, when the fortune of Parodi Delfino's heirs had declined, the production of synthetic fibres had become sufficient to make the company an attractive acquisition for SNIA Viscosa.

***2.3 Pharmaceuticals***

Important progress was achieved by technology transfer for intermediate products, explosives and dyestuffs, apparently because Italy's chemical capabilities in these fields was not that backward, and because these products had political priority. In the field of pharmaceuticals, the situation was more complex. A great number of products and trade-marks placed additional obstacles in the way of potential newcomers. Nonetheless, significant changes took place in this branch too.

The bulk of Italian drug producers consisted of small and medium-sized firms, specializing in the manufacture of natural organic or of inorganic chemical medicinals. Before World War II, only two companies managed to synthesize organic medicinals: Carlo Erba SA and Schiapparelli. The former, founded in 1849 by the pharmacist Carlo Erba, intensified research under the direction of Giovanni Morselli and diversified towards dietetics, artificial syrups and a number of synthetic products, producing vitamins, tranquilizers, insulin, anaesthetics, antibiotics, and chemotherapeutics.[36] The decisive switch in the Schiapparelli firm's history occurred in 1925, when the assets of the company were acquired by Italy's leading gas company Italgas, which at that time was controlled by Rinaldo Panzarasa. Italgas attempted to create a unified chemical division in order to manufacture from the byproducts of coking and tar distillation heavy chemicals, dyes and pharmaceuticals. The ambitious project took Schiapparelli in 1930 across the technological borderline into synthetic pharmaceuticals. However, the financial bankruptcy of Panzarasa placed the Italgas-group in a condition of uncertainty. Like ACNA, Schiapparelli was taken over by Montecatini. A joint-venture with Rhône-Poulenc was established, which in 1936 transformed the firm into Farmitalia.[37] The cooperation with the French shareholder assured the further

[35] BA, Akten der Direktionsabteilung, 323, Bombrini Parodi Delfino, Otto Ambros to Hans Kühne, November 1942.

[36] "Le giornate della chimica italiana alla XXII Fiera campionaria di Milano," *La chimica e l'industria*, **5** (1941), 193; and *La chimica e l'industria*, **3** (1939), 170; "Carlo Erba," in *Artefici del lavoro italiano* (Roma: Istituto Roosevelt, 1956), p.243-248.

[37] Vera Zamagni, "L'industria chimica in Italia dalle origini agli anni '50," in Amatori and Bezza, op.cit (4), p. 97; Ministero per la Costituente, *Rapporto della commissione economica. II Industria. I Relazione*,

development of synthetic pharmaceuticals. But the domestic market for synthetic pharmaceuticals remained to a large extent dependent on imports (estimates range from 50 per cent to 78 per cent).

### *2.4 Oil refining*

Even during the 1930s, in the regime of "economic self-sufficiency," about 75 per cent of the Italian energy requirements had to be satisfied by wood, coal and mineral oil imports. The most urgent problem was the lack of fuels for petrol and kerosene engines, and lubricating oils, due especially to the characteristics of some of the Italian and imported crude oils. The few Italian oils, the Iraqi and Albanian oils, and also some qualities of Mexican crude-oils, did not afford sufficient light fractions to provide, by traditional refining methods, the kerosene and high-octane petrol required by civil and military needs. This is why Italy, after a first wave of technology transfer involving the Dubbs cracking process and vacuum-distillation, looked with great interest at the cartel created by I.G. Farben and Standard Oil to control licensing and information flow in the field of hydrogenation and catalytic refining. The two partners founded Standard-IG and JASCO Inc., which established, jointly with leading chemical and oil companies such as ICI, Shell, Universal Oil and Ruhrchemie, the International Hydrogenation Patents (IHP) and the International Hydro-Engineering and Chemical Company (IHEC).

After cross negotiations between the Italian Government, the Agenzia Generale Italiana Petroli (AGIP), Montecatini, I.G. Farben and IHP/IHEC, AGIP and Montecatini agreed to contribute to the erection of two refineries, one at Leghorn and the other at Bari; AGIP was to provide supplies of petroleum and make available its specific experience in the petroleum field; Montecatini possessed chemical know-how, patents, lignite mines. The plants were to combine the traditional cracking techniques with the new hydrogenation and catalytic refining processes, in order to produce a wide range of products – manufactured petroleum, gas oil, kerosene, petrol of various qualities, paraffin wax, and lubricating oils – deriving them from the most diversified raw materials, ranging from different qualities of mineral oil to lignites and bitumens and rock asphalt oils.[38] In February 1936 AGIP and Montecatini founded the Azienda Nazionale Idrogenazione Combustibili (ANIC). A few months later contracts were drawn up in which IHP and IHEC undertook to grant licences and cooperate in erecting the plants and in bringing them on stream.

The contracts included all the inventions and patents and processes, even future ones,

---

vol.1 (Roma: Istituto poligrafico dello Stato, 1947), p. 132; Bruno Bottiglieri, *Dal periodo fra le due guerre agli sviluppi più recenti*, in Valerio Castronovo, Giuseppe Paletta, Renato Giannetti, and Bruno Bottiglieri, *Dalla luce all'energia. Storia dell'Italgas* (Roma-Bari: Laterza, 1987), p. 217.

[38] Carlo Mazzetti, "L'industria degli olii minerali," in Nicola Parravano, ed., *La chimica in Italia dal 1906 al 1936* (Roma: Tipografia edizioni Italia, 1938), p. 55.

that the Montecatini Company and Giacomo Fauser held, or might hold, in the vast field outlined above. In practice the clause concerned, for the moment, the Fauser patents for the production of hydrogen, which at that time could claim a quite considerable international application, as was the case with Fauser's innovative imitation of the Haber-Bosch ammonia process.[39] The initiative of I.G. Farben had been dictated precisely by the experience gained with processes such as that of Haber-Bosch, the imitations of which quickly spread in many countries and among competing enterprises without any satisfactory return. In the case of hydrogenation, I.G. Farben intended to avoid the repetition of similar developments, and therefore prevent the Italian patents from exercising any threat to its situation. The cooperation seemed to demonstrate the reasonableness of this calculation. The German documents report observations that confirm genuine respect for the research of the Italians. Montecatini transferred its Novara chemical laboratory to ANIC, which, under the direction of Giorgio Kaftal, set up a catalyst factory. The projects of Fauser and Kaftal, especially those on catalysts and on heavy naphtha cracking in order to obtain toluene, interested the BASF head of research Matthias Pier.[40]

From 1938 the ANIC refineries came gradually on stream, and it was owing to them that Italian petrol production rose from 290 thousand tons in 1937 to 520 thousand tons in 1939, providing new high-octane qualities and increasing at the same time the output of other oil derivatives. After the completion of the works, ANIC attained 30.3 per cent of the crude-oil processing capacity the country had available. The company also produced large quantities of iso-octane by polymerization and hydrogenation of the iso-butylene obtained by cracking butane.[41] At Leghorn, dewaxing plants and a Duo-Sol unit for the production of lubricating oils were added. The latter, not embraced by the arrangements of the ANIC-IHP contract, was designed and built, it seems, in 1940, by the American contractor Max B. Miller.[42] But in 1940-41, just when production capacity was increasing, the fuel output dropped sharply: the war had cut off the routes of petroleum supply. In 1941 Bari was forced to suspend production on several

[39] On Giacomo Fauser, see Basil G. Achilladelis, "Process Innovation in the Chemical Industry," Ph.D. Thesis, University of Sussex, September 1973, pp. 148-153. In 1936 three plants for electrolysis of water and 14 synthetic ammonia plants outside Italy used Fauser hydrogen processes. Other Fauser-Montecatini processes were used abroad by 7 nitric acid plants, 7 ammonium sulphate plants and 9 ammonium nitrate plants. At that time, Montecatini's licensing arrangements included firms in France, Germany, the Netherlands, Belgium, Sweden, Canada, Japan, USSR, Czechoslovakia, Poland, Rumania, Yugoslavia, India and South Africa; see *Montecatini Company. A Production and Progress Organization* (Milano: Montecatini, n.d., presumably 1950-51).

[40] BASF, M 1104, Nachlaß Pier, 64, Hydrierung Anic: 3, 6.

[41] "Italian oil development," *The Petroleum Times*, 8 December 1939, p. 40.

[42] BuA, 3, RuK, 1918, Die Raffinierien Italiens; BASF, M 1104, Nachlaß Pier, 64, Hydrierung ANIC, 4, Erfahrungsaustausch, Besprechungen mit den Herren Senator Oreste Jacobini und Direktor Kaftal; Akten-Notiz 21.5.1941.

occasions because of the lack of raw materials. Whereas the Bari refinery survived the war virtually intact, Leghorn was shut down under the German occupation, dismantled and partly transported to the Reich.[43]

After recovery and reconstruction (1948), and renewed cooperation with Standard Oil (1950), the Bari and Leghorn refineries retained an improved and enlarged version of the original design: this was a combination of distillation at atmospheric pressure and under vacuum, with part of the distillates subjected to catalytic hydrogenation, chemical modification of fuels, and the Duo-Sol and dewaxing plants for lubricating oils.[44] Only a few sections, such as that for treating lubricating distillates with phenolic solvent, were added from scratch. The core of the process, hydrogenation and catalytic refining, was recreated along improved and innovative lines: the hydroforming at Leghorn allowed the regeneration of catalysts; and the so-called powerforming installed at Bari employed a platinum-catalyst, apparently similar to the Haensel-system developed by the Universal Oil Company in the post-war period. This was, however, a matter of continuous improvement, based on technical foundations established in the 1930s.

### *2.5 Industrial polymers*

After 1950, the rapid growth of petroleum and methane based production of plastics and synthetic fibres became the most prominent feature of the Italian chemical industry.[45] Montedison, created in 1966 through a merger between the electrical company Edison and Montecatini, in that year produced between 50 and 100 per cent of Italy's aromatics, polyethylene, polypropylene, acrylic fibres, PVC and polyesters.[46] Such a remarkable performance may appear surprising, considering that up to 1950 Montecatini's experience in polymerization had been mainly confined to the more traditional phenol plastics (Bakelite, etc.) and urea-formaldehyde resins (Caurit, etc.). The production of polyvinyl resins and the polyamide fibre nylon had been experimental rather than continuous.[47] The most advanced polymerization procedures

---

[43] ACS-IRI, Numerazione rossa, b 6, ABCD, f Note e memorie degli uffici Iri; BASF, M 1104, Nachlaß Pier, 64, Hydrierung ANIC, 4, Erfahrungsaustausch, Hochdruckversuche, Aktennotiz vom 21.5.1941; BuA, R 3, Beauftragter für den Vierjahresplan, 170.

[44] *L'Anic per l'autarchia dei carburanti. Gli impianti d'idrogenazione di Bari e Livorno* (Roma: ANIC, 1938), Appendix: "Ciclo di lavorazione;" "Stanic Industria petrolifera 1950-1960," *Stanic. Rivista del personale* (Special issue, 1960), 8-9, Appendix.

[45] In 1962 the utilization of hydrocarbon derivatives from oil for petrochemical purposes was 31.3 times higher than in 1954; the extraction of methane gas multiplied by 14.0 between 1950 and 1963; in the same period the volume of plastics output multiplied by 26.2; Zamagni, op.cit. (37), p. 120.

[46] Guido Bodrato et al., *Petrolio, petrolchimica, chimica* (Alessandria: Centro documentazione e ricerche economico sociali, 1971), p. 102.

[47] Luigi Gasperini, *L'industria chimica nella storia italiana* (Messina-Firenze: D'Anna, 1974), p. 87.

had been implemented and operated on industrial scales by other producers.[48] In the face of Montecatini's initial delay in advanced polymerization, the rapid recovery in the 1950s needs to be explained.

The most decisive step took place at the end of the 1940s, when the company acquired the works of the Società anomina italiana gomma sintetica (SAIGS), a manufacturer of synthetic rubber.[49] SAIGS had been founded at the end of the 1930s by IRI and Pirelli in order to minimize the natural rubber imports and was of particular importance for the war economy. The company started collaborating with I.G. Farben and engaged a Russian emigrant, the chemist Alexander Maximoff, who designed the process, and Giulio Natta, later a leading personality in the Italian petrochemical industry and Nobel Prize winner. Butadiene and ethylene production for the years 1941-44 were based on alcohol derived from sugar beet and molasses by fermentation. Catalysis and rectification gave butadiene and ethylene; the latter was transformed by alkylation with benzene into ethylbenzene for manufacture of styrene. While benzene was provided by the state authorities, alcohol came from the sugar refineries. The rubber plant was located at Ferrara, in the central part of an area specializing in beet cultivation and sugar refining, and linked to the sugar refineries by a fairly dense network of roads and navigable canals. It came on stream in October 1941, but, due to the bottleneck created by an inadequate supply of alcohol and price disputes with the sugar syndicate, SAIGS had produced in all not more than 13,027 tons of rubber up to July 1944, when Allied air raids forced the plant to close. Allowing for various subsidies, SAIGS was highly inefficient.[50] It was, at all events, a typical example of the 'distorted' allocation and 'dissipation' of resources during the autarky period. Nonetheless, regarding research, learning, and technological development, the experience of the SAIGS represents the most crucial stage in Italy's path towards petrochemicals.

Production costs were enhanced by the particular conditions of war economy, conditions that would be totally uneconomic in peace-time. Accordingly, in December 1943 the director of the plant, Franco Grottanelli, called together the research staff for a brain storming session that dealt with exploitation of byproducts. Regarding propylene, he suggested polymerization to afford synthetic lubricating oil. Investigations and laboratory experimentation were entrusted to Professor Natta.[51] At that time, Natta had

---

48 An example is Plexiglas: In 1936 Röhm & Haas granted to the FIAT associate Vetrocoke exclusive production and commercialization rights for this glass-resin; ASV, Camera di commercio, 1933-III-1, f 28.

49 Bruno Bottiglieri, "Una grande impresa chimica tra stato e mercato: la Montecatini degli anni '50," in Amatori and Bezza, op.cit. (4), p. 340.

50 Rolf Petri, "Il polo chimico ferrarese," in Pier Paolo D'Attorre and Vera Zamagni, eds, *Distretti, imprese, classe operaia. L'industrializzazione dell'Emilia-Romagna* (Milano: Angeli, 1992), pp. 273-318.

51 ACS-IRI, b 101, f GV1, Relazione sulla riunione del 15.12.1943 per l'esame della situazione dei sottoprodotti di Ferrara.

already published 125 scientific papers and filed 24 industrial patents. Among his numerous studies should be mentioned the separation of hydrocarbon monomers with similar boiling-points, and the fundamental research on the crystalline structure of polymers. His scientific background and experience was eminently suitable for further rapid advancement in the field of polymers.

The continuous technological development was favoured by a relatively low degree of hardware-specialization. In 1947, Natta claimed that the American Distex oil fractionating process, designed to separate hydrocarbon monomers with similar boiling-points, plagiarized his patent. "The procedure is absolutely identical with what I have described by myself in a 1938 patent ... My procedure was tested in 1939-40 at the Bicocca pilot plants and in 1941 adopted on an industrial scale at Ferrara."[52] Although there was a massive cost difference between the raw materials petroleum and sugar beet, Natta confirmed that there were few technical barriers with respect to inputs of raw materials. The shift to oil-based chemical production did not invalidate the accumulated know-how in separating monomers and producing polymers. Technological progress was derived from scientific knowledge, technical skills and experience; it was mainly the human capital that made ready the plants and their processes for future developments.

While the production of synthetic rubber was resumed by ANIC in the 1950s on an improved technological basis, Montecatini acquired, together with the SAIGS's plants and patents, the cooperation of Natta. Petroleum was now transported to Ferrara by pipelines to be cracked for chemical manufacture. Stimulated by the low pressure-low temperature polymerization discovered by Karl Waldemar Ziegler in 1952-53, Natta convinced the company to invest in research on catalytic polymerization and managed to obtain polypropylene plastics that were distinguished by excellent mechanical properties. When the new processes came on stream gradually after 1955, Montecatini gained conspicuous commercial success. Thanks to Natta's research on crystallization and stereoregularity it was possible to extend the results to a wider range of applications, producing new copolymers, and adjust the crystalline structure in order to change the properties of plastics and fibres. According to Saviotti "the development of catalysts has mainly to be ascribed to Ziegler ... whereas the elucidation of the crystalline structure of polypropylene and the explanation of stereoregularity in terms of a general phenomenon of macro-molecular chemistry was to the credit of Natta's team. This is why in 1963 the Nobel Prize in chemistry was given jointly to Natta and Ziegler."[53]

---

[52] Giulio Natta, "Sul processo 'Distex' per il frazionamento di miscele di idrocarburi," *La chimica e l'industria*, **1** (1947), 10.

[53] Pier Paolo Saviotti, "Il ruolo della ricerca e della tecnologia nello sviluppo della Montecatini," in Amatori and Bezza, op. cit. (4), p.391.

### 3. Conclusion

Was Italy a case of state-sponsored industrial catching-up? The developments in the chemical industry suggest that state intervention was critical at least for the rise of heavy industrial sectors. Forces behind industrialization, however, cannot be limited to Gerschenkron's triad of market, state and banks.[54] For example, non-state and non-market institutions, such as family relations, associations and informal networks, played essential rôles in the recent development of Italy's small and medium-sized industry.[55] What distinguishes a national course of industrialization is not so much the presence and weight of individual components, but the more or less accidental convergence of its phases with (and its dynamic response to) international cycles of production and technical change.

Reinert has recently argued that the explanation for divergent national growth performances should avoid "the neoclassical notion that 'all economic activities are alike.'"[56] In different periods and changing environments some branches, processes, and techniques perform better than others in terms of long-term growth and structural change. This idea, already common to Schumpeter, as well as to List, seems to be more cogent to the historical process of industrialization. What is best, of course, becomes certain only in retrospect. Innovation depends on risk, even if prompted by potentially self-fulfilling prediction. In the case of the Italian chemical industry the state was ready at the right moment to carry part of the risk by giving up some short-term advantages in favour of technology transfer and innovation. But it was not due to the autarky policy as an isolated theoretical phenomenon that a decisive change in the sector's development occurred. It was a matter of timing and linking to international developments at a particular moment and by the right means. Free-trade policies in the 1930s might well have blocked innovation no less than autarky policies might have blocked innovation in the 1980s.

Some of the major chemical innovations of the interwar period led Italy from the position of outsider to that of a participant in the diffusion era of key technologies, even if in the subordinate rôle of an imitating newcomer. Considering that most of these technologies were new at the international level, it may be assumed that the timing of the lock-in was favourable. Quite often for a firm or national industry that has acquired prior scientific know-how and technical expertise, entry during the initial phase is advantageous because of savings in R&D and other costs, and because sufficient space

---

54 Alexander Gerschenkron, *Economic Backwardness in Historical Perspective* (Cambridge, Mass.: Harvard UP, 1962).

55 Giacomo Becattini, *Italy*, in Werner Sengenberger, Gary W. Loveman, and Michael J. Piore, eds, *The Re-emergence of Small Enterprises: Industrial Restructuring in Industrial Countries* (Geneva: ILO Publications, 1990), pp. 144-172.

56 Erik S. Reinert, "Catching-up from way behind," in Jan Fagerberg, Bart Verspagen, and Nick von Tunzelmann, eds, *The Dynamics of Technology, Trade and Growth* (Aldershot: Elgar, 1994), p.174.

is left for adaptation and autonomous research.[57] This sometimes opens up unexpected development prospects, as was the case for industrial polymers. For the economy as a whole it was significant that successful catching-up occurred in a field of novel and technologically sophisticated activities, rather than in textile or foodstuffs production.[58] Driven by a Listian industrial policy, the Italian chemical industry enjoyed extraordinary growth-rates before and during the "economic miracle" and contributed to changes in economic structure and productivity that enabled Italy to join the group of highly industrialized nations.

By kind permission Bulzoni Editore, and of FORMIT, Rome, sponsor of the Italian version of this paper published in Daniela Brignone (ed.), *Innovazione tecnologica ed industria. Studi sui settori alimentare, chimico ed aeronautico tra Ottocento e Novecento,* Roma, Bulzoni Editore, 1998, pp. 161-224.

57 Carlota Perez and Luc Soete, "Catching up in technology: entry barriers and windows of opportunity," in Giovanni Dosi, *et al.*, eds, *Technical Change and Economic Theory* (London/New York: Pinter, 1988), pp. 471-475.

58 Reinert, op. cit. (56), p.185.

# THE FRUSTRATED RISE OF SPANISH CHEMICAL INDUSTRY BETWEEN THE WARS

NÚRIA PUIG
*Facultad de Ciencias Económicas y Empresariales*
*Universidad Complutense de Madrid*
*28223 Madrid, Spain*

## 1. Introduction

In many ways World War I marked the take off of the Spanish chemical industry. In common with several neutral countries, Spain enjoyed the benefits of a natural protection that encouraged numerous entrepreneurs, investors and adventurers to found or to enlarge industrial plants to satisfy the growing demand from both domestic and European markets. Within the chemical industry, 'old' sectors such as explosives, sulphuric acid and alcohol profited from the European collapse without having to undergo dramatic changes. The 'new' sectors, dyes and nitrogen compounds, emerged in those years, in spite of the poor natural and scientific resources of the country. Phosphate fertilisers, a field in which Spain had a comparative advantage, went through a boom that left a multitude of factories scattered all around the Peninsula. Not so impressive, though quite remarkable, was the development of perfumes, pharmaceuticals, and several other chemicals.

The end of the war, however, was to reveal that not all these industries were capable of competing with the firms recovering in northern Europe. Neither the protectionist laws of 1917, 1924 and 1926, which explicitly defended the chemical industry as a "national interest," nor the high tariffs, that made some imports extremely expensive, could prevent the shutdown or decline of many local firms in the years, or even months, that followed the peace treaty. Not just the international situation, but the well known reasons for Spain's industrial backwardness in the previous century were now put forward to explain the collapse of some war-time industrial business. There were, on the one hand, the lack of raw materials, cheap power, capital, modern transport facilities, and adequate scientific and commercial knowledge, and, on the other, a narrow domestic market, due to slow growth of population and agriculture. Apart from these factors, and as some well qualified voices were to report a few years later, the most important reason for the failure of chemical manufacturing during the golden years was



*A.S. Travis et al. (eds.),*
***Determinants in the Evolution of the European Chemical Industry, 1900–1939,*** 301–320.

bad management.[1] According to those voices, the entrepreneurial and scientific attitudes of most industrialists had been, and remained, short-sighted or lacked self-confidence. A great opportunity had been lost anyway.

Nevertheless, the war had positive long-term effects. Capital and technical know-how accumulated during the conflict were ready to be invested in industrial activities afterwards. Besides this, the chemical industry was greatly assisted by public institutions in Spain during the interwar period. A highly interventionist and corporatist framework emerged and was consolidated by successive political regimes (liberal monarchy, dictatorship and republic) until the outbreak of the Civil War in 1936. The Spanish economy underwent a remarkable growth, especially in the industrial sector, which was strongly stimulated by public works in the 1920s, prior to stagnating in the 1930s.[2] And, even if Spain remained an agrarian country, the long delayed structural change began, and for the first time a modern capitalist society emerged.[3]

Accumulated capital and know-how, protection and increasing domestic demand seemed ideal conditions for the growth of chemical industry in an era of normalization. Did the industry take full advantage of them ? If not, then why ? It is the aim of this contribution to approach these questions by describing the state and development of the chemical sector, by reviewing existing research, and by analyzing the main factors of growth, and retardation, that were at work during the first third of the 20th century in the Spanish chemical industry. My study is based on the strategies of five representative sectors, sulphuric acid and soda, explosives, fertilisers, dyestuffs and alcohol, and looks at the supply and demand sides and the institutional framework. The Italian case, examined in this volume by Rolf Petri, has inspired most of my comparative remarks.

## 2. The Spanish chemical industry

In common with many industrially less advanced countries on the eve of World War I, the Spanish chemical industry had been an auxiliary of agriculture and textile production. As in many of those countries as well, the slow development of its main clients and the unavailability of cheap power and technological skills accounted for the

---

1 The chemist, professor and entrepreneur Josep Agell in *Química e Industria*, **17** (1925), 147-153.

2 According to Carreras, Spanish industry grew at a rate of 2.01 per cent between 1913 and 1935. In the same period, the United Kingdom grew at 1.28; France at 0.25; Germany at 0.46; Italy at 1.94; and Sweden at 2.81 per cent. Albert Carreras, ed., *Estadísticas históricas de España. Siglos XIX y XX* (Madrid: Fundación Banco Exterior, 1989), p.174. The relatively small impact of the Depression on Spain is usually interpreted as arising from lack of integration with world markets.

3 See Santiago Roldán, José Luis García Delgado and Juan Muñoz, *La formación de la sociedad capitalista en España, 1914-1920* (Madrid: Confederación Española de Cajas de Ahorro, 1973), a pioneering study of the economic and social transformation of Spain. See also the more recent works of Jordi Nadal, Albert Carreras and Carles Sudrià, eds, *La economía española en el siglo XX. Una perspectiva histórica* (Barcelona: Ariel, 1987), and Gabriel Tortella, *El desarrollo de la España contemporánea* (Madrid: Alianza, 1994).

industry's lack of dynamism. Even in 1920, the leading German industrialist Carl Duisberg described Spain as a country "not suitable for the chemical industry."[4]

The extraordinary conditions created by the European war converged with an acceleration of structural changes and were a big push for local chemical industrialists. Even if there was a lot of speculation in the boom years, chemicals became a relevant and dynamic sector of the growing Spanish industrial economy between the end of World War I and the Civil War. The performance of the chemical industry was quite unequaled: some older sectors were revived; many others arose and developed more or less successfully; while some very important sectors, such as petrochemicals and synthetic textiles, were almost completely absent before 1939. Unfortunately there are no reliable production data before 1958. This "statistical misery" which was bitterly regretted by contemporary corporations and journals, makes it difficult for the historian to reconstruct the evolution of the chemical industry and to draw comparisons with other countries.[5] Some efforts have been made, however, to overcome these difficulties. The reconstructed estimates of output of sulphuric acid, the classical indicator of the chemical industry, suggest that the industry made a remarkable contribution to general industrial growth in so far as it grew at a faster rate than most other sectors. This has been actually interpreted as responsible for the take-off of chemicals in Spain.[6] According to some estimates, the chemical sector employed 5 per cent of the labour force and represented more than 10 per cent of industrial added value.[7] The available data on investment by both public and private banking show that the chemical industry had certainly become a strategic sector for the development of Spanish industry: chemicals drew 10 per cent of the official credits through the Banco de Crédito Industrial (founded 1921), and were the second sector in which the emerging industrial banks participated.[8]

---

4 Carl Duisberg, in a private letter to Arthur Gwinner, president of Deutsche Bank. Gwinner had been commercial attaché in Madrid before the war and was eager to promote joint ventures and German industrial investments in Spain. Bayer Archiv Leverkusen 084-030. More optimistic was the American report of 1932. See U.S. Department of Commerce, *The European Chemical Industry in 1932* (Washington: U.S. Government Printing Office, 1933).

5 As Agell, in *Química e Industria*, or Péres Unzueta, in *Anuario de Industrias Químicas* (Madrid: Cámara Nacional de Industrias Químicas, 1932), pp. I-IV, the first (and last) industrial yearbook of the chemical industry published before 1936. They complained that Spanish publications only included data on foreign industries and blamed suspicious entrepreneurs for that.

6 Albert Carreras, op. cit. (2), pp. 173-187.

7 See Jürgen B. Donges, *La industrialización en España* (Barcelona: Oikos Tau, 1977), pp. 151ff.

8 On the Banco de Crédito Industrial, see Gabriel Tortella and Juan Carlos Jiménez, *Historia del Banco de Crédito Industrial* (Madrid: Alianza, 1986), pp. 49 ff. Between 1921 and 1931, the chemical industry was third, behind the metal and metallurgical industries and hydroelectricity. On private industrial banking, see Roldán et al., op. cit. (3), pp. 250-251. Around 35 per cent of the chemical sector was controlled by the

For a first approximation, in order to get a description of the chemical industry before the Civil War, we can rely on the business yearbooks, first published in 1916.[9] According to the information available in the 1934/35 directory, 311 enterprises (out of a total number of 4,793) could be considered as belonging to the chemical industry. These 311 firms had a subscribed capital of 658,406 thousand pesetas, less than 5 per cent of the total corporate capital. However, if we exclude the non industrial capital, then the chemical industry accounted for 9 per cent of the total activity and was third behind the mining and textile industries.

If we look now at the size of firms we find a quite dualistic structure, shared by countries such as Italy: a few large enterprises, on the one side, and a multitude of small firms, on the other. The concentration rate was indeed high: the ten largest firms represented 47.34 per cent of the total issued capital, while the 20 largest subscribed 59.43 per cent. These 'big' businesses were to be found in several branches: heavy chemicals, explosives, wood derivatives, fertilisers, perfumery, alcohol and chemical intermediates. It is still more interesting to look at the dates of foundation. Out of the whole sample, only twenty-seven firms had been set up before World War I. This demonstrates how decisive the war had been for neutral Spain, even though twelve of the twenty biggest firms had actually been founded by the turn of the century, with considerable foreign technology and foreign financial capital. Thus the dominant characteristics of the Spanish chemical industry arose during and after World War I. This becomes even more apparent if we consider that, besides corporate firms, many traditional factories and small works were engaged in the manufacture of chemical products. Further, a close look at the financial and technological connections of corporate enterprises reveals that foreign capital and know how together with Spanish partners, as national law required after 1917, was often present in such branches as organic dyes, fertilisers, and pharmaceuticals, the most technically advanced branches. As for the geographical distribution of the chemical industry, Barcelona, in Catalonia, surpassed any other province in Spain, with one third of the chemical firms concentrated there. The demand from traditional industries such as textiles, along with higher income and good transport, commercial and educational systems around Barcelona, explains this concentration. Madrid, where many firms used to have their head offices for political and administrative reasons, and Vizcaya, in the Basque

---

banks in 1921.

9 The *Anuario Financiero y de Sociedades Anónimas* (Madrid, 1934/35) included all kind of corporate firms existing in 1933. This publication, with the usual limitations of business yearbooks, provides information about a firm's trade name, head office, subscribed capital, industrial activities, and – only in the case of large enterprises – board of directors and balance sheet. For this exercise I have taken the chemical, pharmaceutical and distilling enterprises and have left out the oil (vegetable and mineral) industry. A systematic analysis of this source has been undertaken by Albert Carreras and Xavier Tafunell, "La gran empresa en España (1917-1974)," *Revista de Historia Industrial*, **3** (1993), 127-147, who studied the rise of "big business" in Spain within a Chandlerian framework.

Country – with the powerful Unión Española de Explosivos (UEE) and Unión Resinera Española – came next, with 18.85 per cent and 23.68 per cent of corporate capital, respectively.

In order to analyze the strategies of the chemical businesses we have to turn to private archives. A good deal has been done in this field, and, in spite of the many problems that surround company records in Spain (as elsewhere), some of the big enterprises have already been or are being studied. As will be shown in the next section, this research has primarily focused on the entrepreneurial side, that is, the economic and the political decision-making processes. Little attention has been paid to the technical and scientific and technological dimensions of chemical industry. Three main reasons explain, in my view, this peculiarity of Spanish historiography: the interests of the authors, most of them business and economic historians; the kind of documents that have been kept, such as accounting books and records of the board of directors' meetings; and, last but not least, the fact that no Spanish company had a R&D department that could have left valuable records and justified a scientific and technological history. This is in fact the big difference that arises between Spain and most of the European countries reviewed in this volume. The comparison with Italy, a country facing similar difficulties, while developing modern and creative enterprises in the chemical field after the war, is undoubtedly the most reasonable, and the most puzzling. We will find no Montecatini, and of course no Fauser in Spain.

There is still a second printed source we can rely on to draw up the general outlines of the Spanish chemical sector: foreign trade statistics. According to 1935 statistics, imports exceeded exports tenfold.[10] Nitrogenous fertilisers made up half of total imported chemical manufactures, while other traditionally imported products, such as pharmaceuticals (10 per cent), general chemicals (10 per cent) and dyes (5 per cent), represented much less than they had done in the previous decade. Except for the nitrogenous fertilisers, therefore, a remarkable change in imports had taken place. The traditionally most important supplier, England, had been replaced by Germany and Holland, that were responsible for massive imports of nitrogenous fertilisers. Chemical intermediates, pharmaceuticals and dyes were also important products imported from Germany. Even if Chile and the United States contributed as exporters of saltpetre and several chemicals, Spanish chemical imports remained essentially European (over 80 per cent).[11] As for exports, they were dominated by tartaric acid, carbide, perfumes and natural essences. Europe took only 46 per cent of Spanish exports, the American markets attracting 3 per cent. International comparisons suggest that the catching-up

---

10 *Estadística del Comercio Exterior de España* (1935). 130,239,000 pesetas *versus* 11,098,000 pesetas. There are no substantial differences between 1935 and the two previous years.

11 Germany, 30 per cent; the Netherlands, 13 per cent; France, 8 per cent; Italy, 8 per cent; Belgium, 7 per cent; United Kingdom, 5 per cent; Switzerland, 5 per cent. The shift from England to Germany had accelerated in 1931. United States, 10 per cent; and Chile, 8 per cent.

process for Spain had been much more modest than in most European countries. During this period, Italy partially left the southern periphery, but Spain remained fully part of it.[12]

The lack of suitable education and research institutions were one of the most frequently stated reasons for Spanish industrial backwardness. During the first third of the century some initiatives and reforms were undertaken in attempts to overcome Spain's scientific isolation. The future of existing French-style, theory-oriented technical schools were discussed. However, no consistent scientific policy was ever implemented, and the two most remarkable institutions of the time were actually private or regional. Their stories, even if modest, reveal many of the problems that hindered the development of an appropriate scientific and technical framework in Spain. The Instituto Químico de Sarrià, Barcelona, was founded by the Jesuits in 1916.[13] Its artificer, the distinguished chemist Eduardo Vitoria, educated in Leuven, was aware of the deficiencies of the Spanish scientific and industrial system and the great opportunities of World War I. He designed this relatively modest institution after the German model, and it played an important role in the industrial region of Barcelona, by taking up applied research for small industries and training chemists for larger ones. The Instituto de Sarrià remained – and remains – essentially flexible, and its publication, *Afinidad*, reflected its adaptive vocation.[14] A second institute emerged in the same year and in the same town, the Institut de Química Aplicada.[15] This was a civil institution conceived and directed by another distinguished chemist, Josep Agell, who also intended to introduce the industry-oriented educational model of the German-speaking countries. Agell's institute included a school for managers of chemical industry, several laboratories available to nearby companies and, from 1924 on, a specialized journal, *Química e Industria*. As with Vitoria, Agell and his collaborators wanted the Spanish chemical industry to catch up with the rest of Europe and to keep entrepreneurs up to date. The school and the laboratories suffered from the anti-regionalism and scientific indifference of the military regime of general Primo de Rivera and were shut down in 1925. A business association, the Cámara Nacional de Industrias Químicas, founded in Barcelona in 1919, took over the more informative tasks of the institute and the journal, that appeared until 1937. This journal is the only useful publication for following trends in

---

12 As the editors of *Química e Industria* constantly stated in the statistical section of the journal.

13 See Núria Puig Raposo and Santiago López, *Ciencia e Industria en España. El Instituto Químico de Sarrià, 1916-1992* (Barcelona: Institut Químic de Sarrià, 1992) and *idem*, "Chemists, engineers, and entrepreneurs. The Chemical Institute of Sarrià's impact on Spanish industry (1916-1992)," *History and Technology*, **11** (1994), 345-359.

14 The institute was only briefly closed, since the Society of Jesus was nominally expelled from Spain in 1932.

15 The history of this institute was studied by Alexandre Gali, *Història de les institucions i del moviment cultural a Catalunya 1900-1936* (Barcelona, 1981), pp. 129-134 and 203-214.

Spanish chemical industry before the war. It confirms that Spain was a second-rank country, satisfied to be informed about the main technological innovations from abroad and to have the means to apply them. When analyzing the state of the chemical industry in Spain, the Cámara, an official corporation in 1931, praised the progress made from the beginning of the century and showed a deep confidence in the opportunities proferred by the rich raw materials of the country. However it complained, as had the journal since 1924, about the main obstacles to the future of the industry: the Spanish character (!), inadequate training, education and research, and lack of a stable commercial protection.[16]

Whether the general industrial policy implemented in the first third of the century encouraged or hindered the economic development of Spain is passionately debated among economic historians. Research has traditionally focused on the legislative and administrative dimensions of regulation and has shown the quite cumulative and contradictory nature of it.[17] It seems, in fact, as if the law-makers were unable to be selective about the specific industries to be protected. Only in recent years have scholars dealt with the real effects of that regulative and protectionist framework on the performance of specific industries and firms; they have looked as well at the strategies of particular pressure groups. Their often opposite results have intensified a debate which stands at the core of Spanish economic history, and plays a role in explaining successes and failures. However, nothing similar has yet been undertaken for the chemical industry, which was nominally very protected after 1917. The records of the Comisión Protectora de la Industria Nacional and other bodies in charge of the application process for individual companies show that requests were dealt with slowly and usually required a display of personal pressure.[18] More important than that, however, was the fact that the continuous revision of laws and commercial agreements often invalidated the privileges previously granted and therefore made business decision-making very difficult. These perverse effects of over-regulation were pointed out by both the entrepreneurs – through the Cámara and several employers' associations – and the civil servants and consultants of the official corporations. They reacted angrily when the League of Nations published in 1927 a well-known report on the

---

[16] See, for the complaints, the chemist and editor Uhthoff in *Química e Industria*, **2** (1924), 39-41. For an optimist perspective, see the chemist and businessman A.M. Llopis, president of the corporation during this period, "La industria química en España," *Anuario de Industrias Químicas* (1932), pp. XIV-XXIV.

[17] For an exhaustive description of the interventionist and corporatist framework, see the classical studies of Román Perpiñá, *De economía hispana, infraestructura, historia* (Barcelona: Ariel, 1972), pp. 64-89, and Juan Velarde, *Política económica de la dictadura* (Madrid: Guadiana, 1973), pp. 137-173. An interesting survey and research agenda is given in Francisco Comín in Nadal, *et al.*, op. cit. (3), pp. 105-149, and Gabriel Tortella, op. cit. (3), pp. 255-267.

[18] See the self-deprecatory description of the Comisión in: *Comisión Protectora de la Producción Nacional, Anexos* (Madrid: S.E. de Artes Gráficas, 1919).

chemical industry and described Spain as the most protectionist country in the world.[19] Over-regulation helped also to understand the division of labour that soon appeared in Spanish companies between 'politicians' and 'technicians,' the first taking care of relations with the administration and the second of daily business. In those sectors closely linked to defense interests, the companies had to intensify their political relations. The profit they took from the military plans of "industrial mobilization" in 1917 and the Civil War is a topic that needs to be specifically addressed.[20]

In retrospect, and in relation to developments in other European countries, it seems that in spite of the repeated efforts undertaken by successive governments to listen to the chemical industrialists and to establish a stable framework for the emerging national industry, in Spain no adequate framework arose. It is tempting to blame political instability for the lack of convergence in this respect. The Italian case, once more, shows that both Mussolini's fascism and World War II stimulated the catching-up process that so successfully contributed to the postwar Italian 'miracle.' Spanish fascism under Primo de Rivera was too weak – and unwilling – to create such a 'convergent' climate, and the outbreak of the Civil War brought Spain, and most of her chemical industries, far away from technologically creative Europe until the 1950s.

## 3. The development of five specific sectors

We will now address the performance of five representative branches. Our analysis focuses on relatively large companies for which the years between the wars were crucial. Finally, we shall pay attention to their backgrounds and to the branch as a whole.

### *3.1 Sulphuric acid and soda*

Up to World War I most industrial countries experienced a rapid increase in the production of sulphuric acid.[21] This was a basic input product for the manufacture of traditional industries (such as textiles, iron and steel), as well as the newer heavy chemicals (oil refining, artificial fertilisers, and explosives) and the emerging fine chemicals (such as synthetic dyes). The strongest reasons for the early rise of this industry in Spain are the abundance of raw materials (sea salt and pyrites) and the existence of a consumer industry (textiles). The development of the acid industry was, however, difficult and slow in the previous century. The story of the pioneer enterprise, Cros, founded by a native of southern France in the surroundings of Barcelona in the

19 League of Nations, *The Chemical Industry* (Geneva: League of Nations, 1927), p.56.

20 On the industrial mobilization see Elena San Román, "La movilización industrial (1914-1918). Apuntes sobre el caso español," M.A. Thesis, Universidad Complutense de Madrid, 1993. It seems that the main short-term impact of this undertaking was the introduction of industrial statistics in the Spanish army.

21 For a brief international review see Gary Goertz, *The World Chemical Industry Around 1910. A Comparative Analysis by Branch and Country* (Geneva: Centre of International Economic History, 1990), pp. 15-30.

1820s, helps to understand this apparent paradox. Cros was initially able to get some privileges from the Spanish crown, so that imported sulphur and saltpetre were obtained by his factories at good prices.[22] Unfortunately, these privileges were short-lived, and he soon had to compete with British and French products and with Spanish natural vitriol (produced in Alcañiz, Aragon). Meanwhile, many local imitators had opened up, encouraged by the expansion of the textile industry in Catalonia. Consequently, the situation became difficult for Cros.

At the same time the European chemical industry was undergoing substantial changes through the implementation of the Leblanc process for artificial soda, which would soon become a natural ally of sulphuric acid in much larger plants. Spain was again especially well endowed with salt, sulphur, pyrites and manganese for producing Leblanc soda and its chlorine derivatives, though it had to deal with the coal problem, and Spanish participation in the fabulous Chilean saltpetre mines had dramatically declined.[23] But in spite of these extraordinary conditions, the Leblanc process did not take root in Spain; in fact they served to consolidate this industry in northern Europe. The Cros story again gives us the clue.[24] Besides the comparatively high prices of salt (controlled by a state monopoly) and coal, all consumer industries other than textiles (glass, soap and paper) performed very poorly at the time and frustrated the development of a domestic market. So the existing manufacturers of sulphuric acid did not, as in Britain or France, combine it with other new specialities, and their businesses remained modest and old fashioned. Artificial soda, when required, was imported from Liverpool or Marseilles at low prices.

Actually, Spanish industrial consumers became sufficiently mature when the ammonia soda (Solvay) and electrolytic processes were established in Europe at the end of the century. At that point, market conditions had improved, but the production conditions, especially the problem of energy, were still problematic. However, several factories opened in the first decade of the century, most of them made possible with foreign capital.[25] World War I gave them additional impetus, thanks to the development of new or renovated chemical factories. That explains the increasing production of sulphuric acid in the following decades: it rose from 33,000 metric tons in 1914 to 334,000 metric

---

[22] See Jordi Nadal, "La debilidad de la industria química española durante el siglo XIX. Un problema de demanda," *Moneda y Crédito*, **186** (1986), pp. 41-44. This is an excellent analysis of Spanish chemical industry during the 19th century.

[23] According to the Ungewitter report of 1927, by 1913 Spain had more than 55 per cent of the world pyrite reserves, but less than 1 per cent of natural sulphur. League of Nations, op. cit. (19), p. 32.

[24] Nadal, op. cit. (22), pp. 50-59.

[25] Solvay, Griesheim and Schuckert and Thomas & Houston were the main investors and technology providers. Cros was to remain the largest investor.

tons in 1935.[26] In fact this essential input is a useful index not only of the chemical industry, but of Spanish industry in general. It was not until the 1960s that Spain became a major sulphuric acid producer and at the same time an industrialized country.

### *3.2 Explosives*

The dynamite industry constituted one of the most solid branches of Spanish chemical industry from the 1870s through the 1930s. The main explanation for this is to be found in consistent demand from the expansive – and mainly foreign-controlled – mining sector. The story of the trust, the Unión Española de Explosivos (UEE), which between 1897 and 1917 held a legal monopoly on the production and distribution of explosives, exemplifies the development of this industry in Spain.

The first significant enterprise was the Sociedad Española de la Dinamita (SED), the Nobel subsidiary founded during 1872 in Bilbao. In common with many other firms which emerged in the following years, with national or foreign capital, the SED set its sights on the rich mines of iron, lead, copper and mercury spread over the Iberian Peninsula.[27] The aggressive price war over explosives that broke out in the 1880s and 1890s all over the world gave rise in Spain to a succession of agreements among the competing firms. The outcome was Unión Española de Explosivos, a trust that integrated nine enterprises: the Spanish Nobel branch, SED, two German-linked firms, and other small companies related to the mining business in northern Spain. Of course the SED, as the hegemonic company, engineered the trust-making process that took place in 1896 under the inevitable inspiration of the European Nobel Trust. The rationalization of the explosives industry was soon achieved, in spite of high costs. Manufacturing was concentrated in the pioneer plant of Gualdácano, in the Basque Country; the other factories became exclusively marketing establishments. Each of them kept its traditional market. Besides this, the Unión conducted a very restrictive credit policy, and acted as a financial intermediary among the integrated firms; its links to several Spanish and French banks were remarkable.

In contrast to the general European trust pattern, however, another factor emerged in Spain: the government. Its budgetary difficulties, made worse by the American war in

---

26 Carreras, op. cit. (2), pp. 206-207. Italy had reached that production by 1905.

27 They had been liberalized after 1869. The economic historian of the SED and the UEE, Tortella, has showed the high correlation between the mining exports and the SED sales, on the one hand, and the fluctuations of British industrial production, on the other. He states therefore that it was the impressive development of Spanish mining as a foreign enclave that allowed the SED, and later the UEE, to become the first big Spanish chemical enterprise. See Gabriel Tortella, "La primera gran empresa química española: la Sociedad Española de la Dinamita (1872-1896)," *Historia económica y pensamiento social. Homenaje a Diego Mateo del Peral* (Madrid: Alianza, 1983); *idem*, "La implantación del monopolio de explosivos en España," *Hacienda Pública Española* (Homenaje a Don Ramón Carande), **108-109** (1987), 393-410; and *idem*, "La integración vertical de una gran empresa española durante la dictadura de Primo de Rivera. La Unión Española de Explosivos, 1917-1929," in José Luis García Delgado, coord., *Economía española, cultura y sociedad. Homenaje a Juan Velarde Fuertes* (Madrid: EUDEMA, 1992), pp. 359-393.

the 1890s, explain the early interest of the Spanish state in encouraging the syndication of this very important and promising industrial activity. According to Tortella, The Unión was born with the aim of holding a monopoly.[28] And this is what happened just one year later. For the Treasury, the monopoly meant more revenue (without having to increase other taxes) and more regularity in the provision of war explosives. For the Unión, the deal, even if risky in the short run, offered, in the long run, the benefits of any monopoly. For both of them, the war-bound prospects were greater explosives consumption, more revenue, and more exports as the natural result of the peseta's devaluation. In fact, criticism came from the main clients of the explosives business, the mineowners, who quickly organized a joint campaign against the legality of the monopoly's constitution, as well as against the higher prices, declining quality and deteriorating distribution of networks for explosives.[29] The customers' grievances were actually valid, so the Unión (among whose partners there were prominent mineowners) had to make some concessions. This did not disturb either the fabulous profits nor the image of the UEE in Madrid. In the first year, as an example, the Unión earned profits of 67 per cent on sales and 37 per cent on capital. The Treasury, on its side, received 3 million pesetas, exactly 5 per cent of the UEE issued capital. So the monopoly helped to transfer benefits from the mining companies to the Spanish government and to the trust.

Under the twenty-year long monopoly, the Unión experienced some changes. The most outstanding one was the 'nationalization' that took place during the first years of this century. As the whole business lost its appeal for foreign capitalists (because of a revalued peseta and decreasing dividends), Spaniards progressively took over, and the UEE became more independent of the Nobel Trust. Later on, the European war created great opportunity for an increase in the scale and scope of the explosives business. The Unión actually took on the manufacturing of fertilisers and tried to get into the business of dyes and nitrogen. These promising conditions made the expiration of the monopoly in 1917 quite irrelevant. Until the Civil War, the extremely cautious board of directors of the Unión pushed forward a smooth process of both vertical integration and diversification: manufacture of navy gunpowder and other explosives under British patents; acquisition of phosphate mines in Morocco, potash mines in Catalonia and pyrite mines in Huelva; and investments in waterpower in Castile. In the short run, the strategy was very successful in the explosives and potash fields. The Unión enjoyed the benefits of a monopoly *de facto* in the Spanish explosives market, and fertilisers generated profits in a still agrarian country.[30]

---

[28] Tortella, op. cit. (27), pp. 397ff.

[29] Tortella has estimated that the most common types of dynamite became around 80 per cent more expensive. As for the bad quality, the material was tested, and this was confirmed by technicians. The distribution problems can be put down to the reorganization process. Tortella, op. cit. (27), pp. 400-404.

[30] Tortella states that this conservative strategy (if compared with those of Du Pont or Nobel) was reasonable for the 'narrow' horizon of the Spanish economy of the 1920s and 1930s. Tortella, op. cit. (27), p. 381.

### *3.3 Dyestuffs*

As elsewhere outside Germany, World War I marked the start of an indigenous coal tar dye industry in Spain. In the postwar years the development of this young industry required – as in many other countries also – strong protection and support from the government because production costs remained high and German competition had to be faced again. There are probably few such clear examples of the benefits and limitations of a national industrial policy in the economic history of Spain as that of dyestuffs. But there are also few cases where the joint efforts and political pressures of the manufacturers combined with foreign technological and financial cooperation brought better results. The development of this industry in Spain will be demonstrated through that of a Catalan firm, Fabricación Nacional de Colorantes y Explosivos (FNCE), partially controlled by I.G. Farben from 1926.[31] Between 1922 and 1936, this company practically dominated the Spanish domestic market.

Before the war, Spanish textile manufacturers – the most important consumers of synthetic dyes – had to rely on the many German salesmen and agents that were established in the country. Bayer, BASF, Cassella, Hoechst and AGFA held the largest market shares. [32] The few Spanish companies that had undertaken the manufacture of synthetic dyes were not serious competition; they merely produced small amounts of direct and cheap dyes for their local clients. Through higher productivity and lower export prices, German companies could easily overcome increasing Spanish tariffs. The war, however, created the ideal conditions for replacing imports by Spanish products, as well as a chance for Swiss dyemakers to exploit the Spanish market. Most of the Spanish corporate companies were founded between 1914 and 1918, during which time a hasty introduction of technology and qualified labour took place.[33]

The evolution of the young Spanish dyestuffs industry after the war was closely observed by the German firms.[34] The latter bitterly complained that the Spanish market

---

[31] Its development was not as "truly marvelous" as the Italian case, but quite remarkable. See the enthusiastic Italian report in: League of Nations, op. cit. (19), p.120. References to the Spanish case are unfortunately scarce in such basic studies as Harm Schröter, "Kartelle als Form industrieller Konzentration: Das Beispiel des internationalen Farbstoffkartells von 1927 bis 1939," *Vierteljahrschrift für Sozial- und Wirtschaftsgeschichte*, **74** (1987), 479-453, and Gottfried Plumpe, *Die I.G. Farbenindustrie A.G. Wirtschaft, Technik, Politik 1904-1945* (Berlin: Dunker & Humblot, 1990). This section is based on ongoing research into the Fabricación Nacional de Colorantes y Explosivos, 1922-1965. Archival files from Bayer Barcelona (AFNCE) and Bayer Leverkusen (BAL) are cited.

[32] With 24.17 per cent, 23.38 per cent, 19.59 per cent, 10.93 per cent and 10.49 per cent, respectively in 1913. Firmenarchiv Hoechst AG (FAH), C/3/1/b, Filiale Barcelona.

[33] Vero y Vidal, the oldest and largest manufacturer (1882), for example, hired a chemist from AGFA, who would later be responsible for the only high quality (black) dye produced in Spain, according to a German report (BAL 084-030). Vero y Vidal (later Sagnier) was the major promoter of the FNCE. I have not been able to find out its origins.

[34] See the I.G. Farben reports on Spain between 1922 and 1926 in: BAL 126/6.

was protected by a double barrier: tariffs of up to 492 per cent, and currency devaluation of 80 per cent.[35] Two of the most protectionist laws of Spanish industrial history were enacted in 1924 and 1926, the second prohibiting the imports of several dyestuffs. Within this framework, the dyestuffs business went through an important reorganization. On the commercial front, a succession of mergers gave rise to the Unión Química y Lluch, which from 1926 on represented the German firms in Spain. On the industrial front, the five most important dye manufacturers merged at the end of 1922 to found the FNCE.[36] The first years of FNCE were certainly successful, and rationalization, re-investment, and attractive dividends went hand in hand. The decisive step to dominate the Spanish market was made in 1926, through the acquisition by I.G. Farben of 50 per cent of the shares. The Spanish partners had previously obtained special support from the government.[37] The German conglomerate was doing nothing but "coming to terms with the Spanish manufacturers," as the I.G. Direction had been advised to do, "if the tariff war is kept on."[38] The I.G., which controlled the Spanish firm through a Technical Committee, introduced further rationalization of production.[39] Several agreements were concluded immediately with the Unión Química y Lluch, which obtained the exclusive right to sell the FNCE products, and with the electrochemical plant of Electroquímica de Flix (Griesheim), a potentially competing firm that leased its dyestuffs plant to FNCE.[40] In the following years, FNCE came to further agreements with the Unión Española de Explosivos and other firms, in a double policy of competition suppression and diversification. The firm's records show smooth and prosperous operation, with high profits and continuous expansion.[41] At the same time, it expressed an ongoing dissatisfaction with the tariff policy, which was considered too unstable to be efficient. As for the technological attitude of the managers, they did not hide their basic lack of self-confidence. However, they were deeply interested in the introduction of several non-dye I.G. innovations in Spain, which became apparent with the Civil War, when the FNCE started to produce

---

35 *Ad valorem*. The average tariff for dyes rose from 38.2 to 90.3 per cent in 1913 to 128.3 to 266 per cent in 1926. See League of Nations, op. cit. (19), p.56.

36 See the official history of Unicolor, *Historia de Unicolor* (Barcelona, 1967). A detailed description of the merger conditions can be found in the first accounting diary and record of FNCE (AFNCE). The firm was led from the beginning by the families of several chemists and entrepreneurs.

37 R.D. 28.9.26. The highest foreign participation was set at 25 per cent. No wonder that the influential industrialist and senator Sedó was one of FNCE partners.

38 BAL, 126/6.

39 See the conditions of the merger in the meetings of 22 December 1926 and 27 December 1926.

40 FNCE bought 15 per cent of Unión Química y LLuch and gave 25 per cent of its benefits to Flix, which represented the firm Griesheim. On the leasing contract see BAL 700-0374.

41 By 1936, FNCE had 280 workers and around 12,000 clients.

ammonia with the Stickstoff Syndikat [nitrogen cartel] and ICI, and explosives, in Franco's Spain. Plans for the manufacture of synthetic fuel by I.G.'s high pressure process had to be abandoned.[42]

The bitter side of the FNCE story was that it had to face harsh criticism from the public. The German-speaking economic press condemned the foundation and the development of the Spanish synthetic dye industry and drew attention to the "hothouse policy" of the Spanish government and the unfair habits of the local manufacturers.[43] The domestic economic press, for its part, suggested that the FNCE was actually a cover for the sale of German dyestuffs. The end of the 1920s, with this campaign underway and several social problems, was the most hostile time for the company.[44] According to I.G. Farben records, however, the bulk of the Spanish sales (5 per cent of I.G. total sales) was actually produced in Spain.[45] On the eve of the Civil War, the production had risen to 1,800 tons, which covered almost 85 per cent of the national demand. The production capacity of FNCE (that is, of Spain) was 1.8 per cent of world capacity.[46]

### *3.4 Fertilisers*

The development of synthetic fertilisers in Spain during the first third of the 20th century had two completely different aspects. Whereas the production of superphosphates rose spectacularly (from 76,000 metric tons in 1905 to 1,076,000 metric tons in 1935) satisfying domestic demand, nitrogen fertilisers had to be imported on a massive scale. Nitrogen fertiliser production rose from practically nothing by the turn of the century to 3,800 metric tons in 1935.[47]

The fortunate evolution of superphosphate production was due to an equally fortunate combination of natural resources and geographic and political circumstances from the

---

42 In 1927, some of the FNCE managers participated in an official commission that visited the I.G. Farben plants in Germany and produced a study on Spain's possibilities. According to the commission, the FNCE was the company most suited to produce synthetic fuel. But the project was not accepted, something that the Spanish company interpreted as "public incompetence." See the official report in: César Serrano and Antonio Mayorga, *Síntesis de los compuestos nitrogenados. Hidrogenación de los carbones. Informe sobre la obtención del nitrógeno y sus derivados* (Madrid: Compañía General de Artes Gráficas, 1929).

43 BAL 034-030.

44 See the meeting of 31 March 1930 (AFNCE).

45 The 'sendings' were less than 25 per cent. Spain was the ninth country (behind Germany, the United States, China, Japan, India, Czechoslovakia, Italy and the United Kingdom) within the I.G. Farben business. BAL 15D.3A.1.

46 According to Dr. Struss' estimates. BAL 4/B 14.3.3.

47 Carreras, op. cit. (3), pp. 206-207.

end of the 19th century.[48] These were well exploited by Spanish and foreign entrepreneurs. The availability of the rich phosphates of neighbouring Tunis and Algeria, together with the availability of sulphuric acid in the Peninsula, made possible the growth of this industry in Spain. It was in the 1910-20 period when most of the new factories emerged, strategically located and financed by both Spanish and foreign capital. British and French investors, who considered this business a natural extension of the exploitation of phosphates and pyrites, soon felt unhappy with the more modest profits of the industry. Spanish investors were, on the contrary, encouraged and so a 'nationalization' of the superphosphates business took place in the late 1920s. The economic euphoria of the war years, together with the increase of domestic production, as well as the availability of electric power, pushed new and old industrialists into the business. Among the last was the Cros firm, which started a remarkable vertical integration process that made it a major fertiliser producer and distributor (approximately 50 per cent of the Spanish output).[49] The results were impressive for the whole country, which became the fourth world producer behind the United States, France and Italy. The abundance of superphosphates was not only an example of a successful process of replacement of imports, but had very positive effects on the growing Iberian agriculture.

The poor performance of the nitrogen fertiliser industry represents an apparently puzzling chapter of the nationalistic period we are dealing with. No protectionism, actually no policy at all, was put forward for the development of this strategic industry before 1940.[50] The ultimate reason for this can be seen in the pressure exerted by the agrarian lobbies, which were understandably interested in cheap fertilisers. As is well known, international dumping made nitrogen fertilisers widespread throughout the 1920s and 1930s.[51] At the same time, however, because nitrogen was considered essential for the national defence, the government somehow had to promote this industry. It was a difficult task. Industrialists complained of the lack of *ad-hoc* protection; they also feared nationalization. The result was a Joint Nitrogen Commission in 1932 and a decree against dumping enacted two years later. It was only

---

[48] See the short but brilliant analysis of Jordi Nadal, "La consolidació pel biaix dels adobs. 1914-1939," in Jordi Nadal, et al., *Història econòmica de la Catalunya Contemporània* (Barcelona: Enciclopèdia Catalana, 1993), pp. 151-155.

[49] In 1929, 83 per cent of turnover came from superphosphates. Nadal underscores the impact of electric power on this new industry that had been hampered by the lack of coal for steam power. Nadal, op. cit. (48), p.153.

[50] Francisco Bustelo, "Notas y comentarios sobre los orígenes de la industria española del nitrógeno," *Moneda y Crédito*, **63** (1957), 23-40, offers a very interesting and deep analysis of this matter.

[51] On markets and prices, see Josep Pujol, "Los condicionamientos de la oferta y la demanda en la difusión de los abonos químicos en España durante el primer tercio del siglo XX," *VII Simposio de Historia Económica* (Barcelona: Universitat Autònoma de Barcelona, 1994), pp. 109-129.

in 1952 that Spain produced sufficient nitrogen. Until then, this deficit constituted a major obstacle to the modernization of Spanish agriculture and therefore to the structural change of the Spanish economy.

In spite of a missing adequate framework, several initiatives were taken to obtain nitrogen by any of the new processes.[52] Significantly, Spanish gas works were too small to produce ammonium sulphate. On the eve of World War I, Spain heavily relied on British imports of ammonium sulphate. The new processes soon attracted many investors in the second and third decades of this century. First were processes for electrochemical capture of atmospheric nitrogen. A joint venture of the Sociedad Ibérica del Azoe and the Norsk Hydro Elektrisk established a plant in Catalonia that never operated, because of the outbreak of World War I and the subsequent difficulties created in the supply of electric power at an economic price. A second attempt was made also in Catalonia by the Société des Produits Azoés and the Sociedad Española de Carburos Metálicos, which had been founded in 1897 to produce acetylene gas and was interested in diversification. Again, the process required too much electric power; cyanamide was imported by the French partner. Nevertheless, the head office built or participated in several plants all around Spain. The Sociedad Española de Industrias Aragonesas, an originally French company founded in 1918, focused on the alternative and ultimately more successful high pressure synthesis of ammonia. Sites for hydroelectric power in the Pyrenees were bought and, in the 1920s and 1930s, the Aragonesas not only became the first producer of nitrogenous fertilisers (synthetic cyanamide, ammonia and ammonium sulfate), but succeeded in diversifying. Its only serious rival was founded in the mining region of Asturias by the Sociedad Española del Nitrógeno (SIN), actually a group of Spanish banks, the Société Centrale des Industries de l'Air Liquide and the Société Chimique de la Grande Paroisse that also operated the Haber-Bosch synthetic ammonia process. The SIN had a difficult birth in the 1930s, because of social conflict in the region and tough international competition. The last major project of the SIN with Saltos del Duero, UEE and ICI had to wait until 1940: This was Nitratos de Castilla, one of the distinguished firms of the period of autarky.

### *3.5 Alcohol*

Distillation was one of the most successful businesses in 18th and 19th century Spain. A combination of clear comparative advantages on the supply side (wine production, traditional know how and sea transportation) with increasing demand from the American and northern European markets had made this possible. The major technological and commercial modernization that took place in this industry from the 1880s onwards, however, favoured the spread of manufacture of alcohol among the consumer countries. Alternative raw materials, such as sugar beet, grain or potatoes, fed the new, large-scale distilleries of northern Europe. These distilleries, called 'industrial,' were usually highly capitalized and able to benefit from good communications, and

52 For a description of the four main projects see Bustelo, op. cit. (50), pp. 29-34. Bustelo was one of the partners of the SIN, of which see later.

constitute a real Chandlerian example of the second industrial revolution.[53]

In Spain, the coming of industrial distillation was strongly encouraged by the sugar factories, a flourishing business, following the loss of the American colonies, in which many prominent investors participated. Costs of production were much lower here than in the small-scale, traditional wine distilleries. Besides this, the modern processes of continuous distillation and rectification produced a quite homogeneous, high-grade alcohol suitable not only for the manufacturing of alcoholic beverages, but also for several chemical industries. It was soon clear to the distillers that diversification was necessary for their survival.

The development of this industry in a rapidly internationalizing market was strongly influenced by public regulation and fiscal necessities. French alcohol policy actually inspired the Spanish law-makers, who were sensitive to the more than 2,000 wine distilleries that cried out for government protection against a few powerful industrial distillers. The shaping of an alcohol policy became a major social issue in Spain between 1888 and 1936 and a multitude of pressure groups arose. Four issues were at the core of the public debate: the fight between wine and industrial alcohol; the loss of foreign markets; stagnating domestic consumption; and the low tax revenue on alcohol. In the 1930s, Spain was a second-rank producer, far behind France and Germany, and even behind Britain.[54] Official consumption was strikingly low: half a million hectoliters, between 50 and 75 per cent of the total production. And alcohol made up less than 2 per cent of the total state revenue. The result of the passionate alcohol debate was unsteady and intricate legislation that yearly tried to regulate the alcohol market with mainly fiscal criteria. Industrial alcohol was heavily taxed and prohibited for alcoholic beverages, while some industrial uses were fiscally encouraged. The problem was, however, that output – not the raw material – was taxed, and so the system did not promote technological progress. The alternative of a public or private monopoly was discussed, as elsewhere in Europe, especially after World War I, but it could not take place under the social pressure exerted by the wine sector. This activity remained – with a few exceptions – technologically and commercially backward and geographically scattered, and the state became its major representative and defender.[55] The dictatorial and republican regimes promoted the institutionalization of interest groups, which gave

---

[53] As his splendid analysis of whisky manufacturing shows. See Alfred D. Chandler, *Scale and Scope. The Dynamics of Industrial Capitalism* (Cambridge, Mass.: Belknap Harvard, 1990), pp. 149-170 and 262-266.

[54] See the international study of Tapio Voionmaa, "An international survey of the production and consumption of alcoholic beverages," *Twenty-second International Congress Against Alcoholism* (Geneva: United Nations, 1941), with data for 1920-1941.

[55] This issue has been studied by Núria Puig Raposo, "La modernización de la industria del alcohol en España," *Revista de Historia Industrial*, **4** (1993), 91-109, and Juan Pan-Montojo and Núria Puig Raposo, "Los grupos de interés y la regulación pública del mercado de alcoholes en España (1887-1936)," *Revista de Historia Económica*, **XIII** (1995), 251-280.

rise in 1933 to an Institute of Wine, where all of them were represented. It was the aim of this institute to permanently regulate the alcohol market by reducing wine production and encouraging technological modernization, concentration and diversification of distilleries. It met with a lot of resistance from the agrarian groups, until the outbreak of the Civil War.

Nevertheless, if we look at the firms, we see that there was a remarkable progress in the first third of this century. Among the industrial alcohol producers, three big and influential companies (linked to the sugar giants) were dominant. They were technically advanced (by acquiring French and German technology), pursued a strategy of concentration and diversification, and finally established a *de facto* cartel in 1935. The creation of the public monopoly CAMPSA in 1928 constituted a golden opportunity for them, in so far as the monopoly was obliged to yearly buy important stocks of industrial alcohol to produce "national fuel." Only a few wine distillers, on the contrary, grew and diversified in a similar way. The most successful experiences were those of vertical integration in the alcoholic beverages field, and diversification towards tartaric derivatives, such as tartaric acid. The tartaric acid industry was actually born in the late 1920s, but its development was fast and its contribution to chemical exports remarkable. On the whole, the so-called industrial uses of alcohol, however, advanced much less than in other European countries: industry consumed just one third of the distilled alcohol, and most of it at the expense of the Treasury.[56]

**4. Conclusion**

The years 1914-1936 are usually stated to be those in which the Spanish chemical industry emerged. Conditions were indeed good. The European war provided markets, capital and technology for already established branches and for new ones. During and after the war, the chemical industry claimed, and obtained, specific protection. Successive laws set tariffs, fiscal exemptions and other privileges for the chemical plants, and created specific bodies to look after this industry in the interwar years. At the same time, some modest initiatives from inside the industry led to the foundation of industrial schools and laboratories, journals and associations, to promote the development of chemical industry on a sound basis. Along with this, Spain went through a structural change that increased domestic demand for chemicals and contributed to the greatest prosperity of the country until that time.

The five cases examined in this chapter show that the chemical industry, on the whole, took advantage of the favourable atmosphere through investment, acquiring foreign technology and, eventually, partners, and expansion, in spite of fierce international competition. That was clear for heavy chemicals and fertilisers, which greatly benefited from the growing domestic market. The nitrogen and alcohol examples reveal that the

[56] The average consumption during 1926-1935 was: beverages, 66.8 per cent; combustible (denatured alcohol), 12.1 per cent; perfumery, 5.9 per cent; sanitary uses, 5.2 per cent; gunpowder, explosives and ether, 4.8 per cent; fuel (dehydrated), 4.8 per cent; various industries, 4.0 per cent.

protection nominally granted to the chemical industry had to be reconciled with the interests of agriculture, and that international competition was tough. The explosives sector also experienced the constraints of international trusts, and the vagaries of intervention at home. Dyestuffs showed that there was no consistent and stable protection, and that the domestic market itself could not guarantee the progress of this new industry. The availability of some raw materials and the spread of electric power encouraged the introduction of new processes, though there were no improvements. In the five cases, the local entrepreneurs tried to influence the law-makers in Madrid, while establishing versatile cooperation with foreign firms and displaying successful integration and diversification strategies, but did not innovate at all.

If this was the state of the large, for Spanish standards, enterprises, then the chemical sector on the whole, made up basically of small, semi-industrial establishments, surely looked worse. Technological dependence, small size and poor competitiveness, the main structural weaknesses of Spanish chemical industry until the 1980s, when the time for full integration arrived, were deep-rooted. Throughout the 20th century, chemical manufacturing kept pace with new technologies arriving from abroad, but both the public and the private institutions have shown an essentially adaptive rather than innovative attitude. This was so even in the dark period of Franco's dictatorship, when the national watchword was autarky and therefore domestic industrial creativity. The poor outcome of such ideology became obvious with the implementation of a liberal policy from the late 1950s. The Spanish entrepreneurs then had to face competition from a technologically and commercially dynamic international industry, and only few of them could meet the challenge. The main body of the surviving chemical companies had to come to terms with foreign firms, which showed little or no confidence in Spanish technological capabilities. So it very often meant cessation of manufacturing and downgrading to sales branch office.

But let us return to our chronological setting, the interwar years. In spite of the good prospects created by World War I, there was no structural change within the chemical sector that could have favoured, as in Italy, an accumulation of scientific, technical and managerial capacities. Spanish and foreign contemporary observers blamed the political powers, unable or unwilling to create an adequate framework for the development of a national chemical industry, and the old-fashioned scientific and technological institutions, which often involved politicians and public servants as well as scientists and entrepreneurs. Can we make, from our historical, privileged perspective, the same statement ?

The five cases summarized in this contribution suggest that those observers were not very far from wrong. If we draw comparisons with advanced industrial countries, then we should consider firstly the relatively narrow horizon that a still agrarian country such as Spain offered for the chemical industry until the 1930s. It is true that both agriculture and the industrial sector made progress and that this benefited some chemical branches, but economic progress was, for several reasons, more modest than in Italy. We come, secondly, to the question of industrial policy. The extraordinary political instability following the Civil War of the 1930s did not help to foster a national industrial project with the agreement of public authorities and entrepreneurs. Instead, over-regulation and erratic administrative behaviour, from the state, and continuous political pressure and short-term plans, from industry, became the main features of the chemical sector. Two further, essential, differences between Italy and Spain arise when comparing this very important field of industrial policy, namely the absence in Spain of government faith in the public sector and of a powerful company comparable to Montecatini. Voices of nationalization were very rare in interwar Spain, which in spite of all its corporatist display remained a capitalist country.[57] Actually, what happened in Spain was that the state promoted the stimulation and development of small, often not very competitive companies, as the business yearbooks suggest, by spasmodically granting them the privileges based on legislation. The industrial history of Spain after World War II confirms the lack of consistency in industrial policy; not even a long military dictatorship was able to force that necessary convergence from above.

A comparative analysis of the interwar scientific and technological institutions related to the chemical industry also confirms the pessimistic diagnosis of contemporary observers. The two modest initiatives, the Instituto Químico de Sarrià and the Institut de Química Aplicada, reveal the backwardness of Spanish institutions. One cannot make the private sector alone responsible for Spain's low level. It seems much more as if the merely adaptive attitude of the companies was a rational response to the absence of a scientific and technological policy. There was not in Spain any institute like the Milano Politecnico, and there were no outstanding personalities. Looking at this key aspect of industrial progress, it is hard to believe that, had the Civil War not taken place, and had Spain entered World War II, the Spanish chemical industry would have emulated that of Italy. The scientific and technical deficit of Spain remains one of the more puzzling questions of Spanish economic history.

It is then not surprising that Spain is absent from L.F. Haber's splendid work on the structural changes of the international, and particularly, interwar chemical industry.[58] The story of Spanish chemical industry between the wars is that of a promising rise, a meritorious development, and a many-sided frustration that political history does not, by itself, explain.

---

57 A very interesting exception was Suárez Inclán, a prominent member of the earlier mentioned Comisión Protectora de la Producción Nacional and expert on the chemical industry. He played an important role in post-war industrial policy. See his project on nationalization in: Félix Suárez Inclán, *Industrias esenciales. Organización económica del Estado* (Madrid: Imprenta R. Velasco, 1922), where he proposed a complete reformation (Germanization) of the educational system and the nationalization of strategic industries such as explosives, dyestuffs and pharmaceuticals.

58 L.F. Haber, *The Chemical Industry 1900-1939. International Growth and Technological Change* (Oxford: Clarendon Press, 1971).

# THE TAKE-OFF PHASE OF DANISH CHEMICAL INDUSTRY, ca. 1910-1940

HELGE KRAGH
*History of Science Department*
*Aarhus University*
*Ny Munkegade*
*8000 Aarhus, Denmark*

## 1. Introduction

Scientific knowledge may be international, but industrial and other applications of that knowledge are not. This is particularly true of chemistry and chemical industry. Thus, although particular chemical processes or production plants can be inaugurated in any nation, a variety of political, economic, geographic and historical circumstances tend to give the structure and development of the chemical industry a distinctly national character. Perhaps the most important of these circumstances are, simply, a nation's size and geography, including its natural resources. These factors are constraints of great importance not only to national chemical industries, but also to the development of the local chemical communities upon which the industries depend.[1] This generalization applies to Denmark's chemical industry, which developed under constraints characteristic for a small country with no natural resources apart from its fertile soil. From this standpoint it is not surprising that the development of Danish chemical industry has been dominated by the biotechnical and agrochemical sectors.

The period between 1910 and 1940 witnessed not only a growth in these crucial sectors, but also a proliferation of new chemical enterprises unrelated to agricultural products. Though most of the new companies were small and internationally insignificant, together they made up an important and progressive part of Danish industry that would greatly expand after the country had been transformed into an industrial, mass-consumption society in the 1950s. To use the terminology of Walt W. Rostow, chemical industry in Denmark (and several other minor countries) experienced its take-off phase,

[1] The importance of demographic and material conditions for the development of science is argued in Harold Dorn, *The Geography of Science* (Baltimore: Johns Hopkins University Press, 1991). See also Norman J. G. Pounds, *An Historical Geography of Europe 1800-1914* (Cambridge: Cambridge University Press, 1988), which includes much information relevant to chemical industry.



*A.S. Travis et al. (eds.),*
***Determinants in the Evolution of the European Chemical Industry, 1900–1939,*** 321–339.

and then developed to the stage of maturity characterized by the capacity to move beyond the original activities that powered its take-off.[2] What follows is a brief account of some pertinent aspects of this development as it took place prior to 1940. From an international perspective, Danish chemical industry was neither particularly innovative nor of great economic significance.[3] All the same, the historical development of the country's chemical industry is of some interest in a wider, international context. This is so for comparative reasons, and also because the development included a few remarkable examples of chemical industries that were specific to the Danish situation.

In this paper I adopt a liberal definition of the term "chemical industry" under which label is included medico-pharmaceutical products as well as various biotechnical products related to agriculture, such as sugar refining, brewing and other fermentation technologies. I realize that this is at variance with the customary definition but find it reasonable – especially in the Danish case – to include sectors which are after all chemical in nature, make extensive use of chemical knowledge, and employ chemically trained personnel. It is no accident that the director of Copenhagen's Polytechnical College, the chemist and industrialist Gustav Hagemann, in 1904 described agriculture as "Denmark's largest chemical industry."[4] The standard definition identifies chemical industry with the production of chemicals, including basic chemicals (acids, alkalis etc.), chemical products to be used in further manufacture (plastics, vegetable oils, etc.), and finished products (drugs, fertilisers, etc.). However, "the chemical industry, which cannot be defined with scientific accuracy, means many different things depending on how it is classified by the statisticians."[5] Not only are definitions matters of convenience, they also tend to be ahistorical and neglect national differences.

## 2. Economic and political developments

The development of a national chemical industry is evidently influenced by the general industrial developments, that are in turn determined by, or at least constrained by, the broader economic and political life of a nation. As a general background, a brief

---

[2] Walt W. Rostow, *The Stages of Economic Growth*, 2nd ed. (Cambridge: Cambridge University Press, 1971).

[3] The lack of international significance is reflected in the standard histories of chemical industry. Thus no Danish companies or developments are mentioned in either L.F. Haber, *The Chemical Industry, 1900-1930* (Oxford: Clarendon Press, 1971) or F.S. Taylor, *A History of Chemical Industry* (New York: Arno Press, 1972). Fred Aftalion, *A History of the International Chemical Industry* (Philadelphia: University of Pennsylvania Press, 1991) briefly treats Denmark and Norway as appendices to Sweden and singles out Nordisk Insulin and Novo, the two pioneering companies in insulin production, as noteworthy Danish enterprises in the interwar period (on p. 313).

[4] Quoted in Helge Kragh and Hans J. Styhr Petersen, *En Nyttig Videnskab: Episoder fra den Tekniske Kemis Historie i Danmark* (Copenhagen: Gyldendal, 1995), p. 189.

[5] Charles A. Thomas, "Chemical industry," in *Encyclopædia Britannica*, vol. 5 (Chicago: Encyclopædia Britannica Inc., 1966), pp. 378-382, on p. 378.

introduction to Denmark's economic and political history during the first four decades of the 20th century will therefore be useful.[6] Among the most decisive determinants are the geographical and demographic factors. Denmark (excluding Greenland) is a small country of about 43,000 $km^2$ with few natural resources in the form of minerals, oil, coal or extensive forests; hence the natural conditions for chemical industry are limited and tend to exclude the establishment of a heavy industry. Around the turn of the century the population numbered 2.4 million, which increased steadily through the following decades and reached about 3.8 million in 1940, that is, an increase of 58 per cent. The growth in the workforce, reflecting the beginning of industrialization, was 68 per cent during the same period (see table 1). Part of the increase was the result of Germany's defeat in World War I; in 1920 Southern Jutland was reunified with the kingdom after fifty-six years of separation, which increased the population by 165,000 inhabitants, of which 40,000 worked in agriculture.

Although the first phase of industrialization had occurred in the 1870s, Denmark was still predominantly an agricultural society around 1920. At that time about 600,000 people worked in the agricultural sector, or more than three times the workforce employed by industry. The workforce in agriculture reached its peak in 1929 and only then began the shift to industry and service jobs, a trend which has continued to the present.

Table 1. Danish population and workforce 1900-1940. All figures in millions.

| **Year** | 1900 | 1910 | 1921 | 1930 | 1940 |
|---|---|---|---|---|---|
| Population | 2.43 | 2.74 | 3.29 | 3.54 | 3.83 |
| Total workforce | 1.19 | 1.32 | 1.61 | 1.83 | 2.00 |
| Workforce in agriculture | 0.51 | 0.53 | 0.60 | 0.60 | 0.56 |

Danish industry experienced a marked expansion during the interwar years, with an increase in the industrial workforce between 1914 and 1935 from about 134,000 to 190,000. In the same period mechanization accelerated, and the energy requirements of industry increasing from around 146 khp to 430 khp. However, even by the late 1930s agriculture remained dominant and the century-long preparation for industrialization had still not resulted in a noticeable modernization process.[7] As an illustration, by 1939

[6] For information about Danish economic and industrial history in the early part of the 20th century, see Svend Aa. Hansen, *Økonomisk Vækst i Danmark*, 2 vols. (Copenhagen: Akademisk Forlag, 1977); S. Aa. Hansen, *Early Industrialization in Denmark* (Copenhagen: Gad, 1970); and Hans Chr. Johansen, *Industriens Vækst og Vilkår 1870-1973* (Odense: Odense Universitetsforlag, 1988). The 19th century industrial history is covered in Axel Nielsen, *Industriens Historie i Danmark*, 2 vols. (Copenhagen: Gad, 1944).

[7] Ole Markussen, "Danish industry, 1920-39: Technology, rationalization and modernization," *Scandinavian Journal of History*, **13** (1988), 233-256. For a general survey of technological development in Denmark, see Henry Nielsen and Michael Wagner, "Technology in Denmark," in Jan Hult and Bengt

the number of horses exceeded the number of motor vehicles of all kinds. (Only in 1960 did the total export value of Danish industrial products exceed that of the country's farming products.) The limited degree of industrialization is roughly illustrated by the figures in table 2. They indicate that by 1930 Denmark was at about the same stage of industrial development as Norway, but clearly behind countries such as Switzerland.

Moreover, Danish industry was not technologically advanced and was dominated by many small firms that mainly produced consumer goods for the home market. Although technological inventiveness flourished in the period, few of the domestic inventions were developed into industrial innovations. The interwar years were a period of transition, a preparation for modern industrial society, but the result of the transition became clearly visible only in the 1950s. Yet the period was marked by continual technological progress and a gradual transformation of the old and protected industrial structure that existed before 1914. For example, notwithstanding periodic restrictions on foreign trade, industry became more competitive and open to the world market. At the same time, the home market expanded and created a need for new industrial products and a demand for higher quality. The economic crises of 1921 and 1932 had a less serious impact in Denmark than in most of the larger European nations. It has been argued that the crises had a healthy effect on the modernization process by helping to weed out enterprises of low productivity and competitiveness and gave rise to new and more technologically advanced forms of production.[8]

Table 2. Percentage of workforce employed in agriculture (including fishing and forestry) and industry (including crafts and mining) about 1930. For smaller European countries, the ratio between industry and agriculture may be considered as a crude measure of the level of industrialization. Source: Topsøe (note 26), p. 41.

| | families (millions) | per cent in agriculture (a) | per cent in industry (b) | ratio (a):(b) |
|---|---|---|---|---|
| Norway | 1.2 | 35 | 26 | 0.74 |
| Denmark | 1.6 | 35 | 27 | 0.77 |
| Sweden | 2.9 | 36 | 32 | 0.89 |
| Netherlands | 3.2 | 21 | 38 | 1.81 |
| Switzerland | 1.9 | 21 | 45 | 2.14 |
| Belgium | 3.7 | 17 | 48 | 2.82 |

Nyström, eds, *Technology & Industry: A Nordic Heritage* (Canton, Mass.: Science History Publications, 1992), pp. 1-27.

[8] Markussen, op. cit. (7), p. 235. See also Francis Sejersted, ed., *Vekst Gjennom Krise* (Oslo: Universitetsforlaget, 1982), which documents the process in the case of Norway. Most of the points made in relation to Norway are valid also for Denmark, and for Scandinavia in general.

The dominant position of agriculture in Danish politics and economy was reflected in the structure of the research and development system, where priority was given to agricultural rather than industrial research. Technical research was not given high priority by the state until the 1930s and only very few private companies were large and progressive enough to form their own research units. As a result, even though applied research had a strong tradition within the agricultural sector, technical and industrial research was limited in scope and less advanced than in other small countries such as Switzerland, the Netherlands, and Sweden. It was only in the 1930s that the importance of technical research and development was fully realized and new institutions and funds were established to promote applied research in the industrial sector.[9] The most important was the Academy of the Technical Sciences (Akademiet for de Tekniske Videnskaber), founded in 1937. However, the efforts to modernize Danish industry by means of applied research and development were scattered and did not give immediate results. Here, too, the 1930s was a decade of preparation for the transition toward modernization that would take place only in the 1950s.

Before 1914, Danish economic politics was liberal and there was very little state regulation of trade and industry. The situation during World War I forced the government to abandon its liberal politics and the state's regulatory role continued after 1920. Although most of the commodity restrictions were abolished, the state intervened frequently in order to maintain control of prices and the import of goods. During the interwar years the government was either liberal-conservative or social democrat, but in spite of their very different political ideologies there was a broad consensus about the economic politics and the kind of limited role that state intervention should play.[10] The social democrats, who were in power throughout the 1930s, had state ownership of the means of production on their agenda, but in practice they refrained from direct state engagement in industrial companies and projects. Several workers' cooperative industries were, however, established by the social-democratic trade unions. These included chemically-oriented industries, such as breweries and margarine factories, but they were not particularly successful and were of limited economic importance. By and large, in Denmark the chemical industry developed under ordinary market conditions; state intervention was not a major determinant of the development. This was contrary to the situation in some other European countries, such as Spain, where protectionism and

---

[9] On the development of education in the technical sciences in Denmark, see Henrik Harnow, *Den Danske Ingeniørs Historie, 1850-1920* (Herning: Systime, 1997). Aspects of higher technical education in Europe are treated in Robert Fox and Anna Guagnini, eds, *Education, Technology and Industrial Performance in Europe, 1850-1939* (Cambridge: Cambridge University Press, 1993), which includes a chapter on Sweden (by Göran Ahlström), but none on Denmark or Norway.

[10] From 1901 to 1924 and from 1926 to 1929 the governments were liberal and with strong bonds to agriculture. The first social democratic government was in office from 1924 to 1926, and from 1929 to 1940 the government was formed by a coalition between the social democratic party and a smaller liberal party ("the radicals").

state intervention were of crucial importance.[11]

The following example illustrates the cautious attitude of the social democratic government with regard to intervention in market mechanisms. By the mid-1930s motorized transport had grown considerably and Denmark imported large amounts of petrol for the rapidly increasing number of cars. At the same time, unemployment was a record high and the economic situation difficult. Jacob Ellehammer, a Danish inventor, therefore, suggested large-scale production of ethanol from potatoes, and the passing of a law that made it obligatory to add ethanol to the petrol. He obtained support from sectors of Danish industry, including one of the largest chemical companies, De Danske Spritfabrikker (a distillery with expertise and experience in alcohol production from potatoes), which made a detailed plan for the project. Although the ethanol would be more expensive than the imported petrol, it was argued that as a whole the project would have a beneficial effect on Danish economy, not least because it would create a large number of new jobs. However, the project required state support, which was not forthcoming. The reason was that the government considered it too risky from an economic point of view, fearing retaliation from foreign trade partners. With the government's decision, the ambitious ethanol from potatoes programme was dead.

**3. Main trends in chemical industry**

Until about 1920, the chemical industry in Denmark was small, scattered and of limited economic importance. It had developed from pioneering industries established between 1850 and 1890, several of which – such as the Carlsberg brewery (Carlsberg Bryggeri) and the cryolite industry – were highly successful and made use of scientific innovations and advanced technological processes. However, the few large and successful enterprises were not representative of the chemical industry as a whole. A study of the chemical industry in the period 1870-1914 shows that the growth in Denmark was disappointingly low and that Danish chemical industry did not partake in the expansive development that characterized the large European nations.[12] The growth was modest not only compared with foreign countries, but also compared with other sectors of domestic industry. For example, whereas the average power consumption per company in Danish industry rose from 17 hp to 74 hp during the period 1897-1914, the increase in the chemical sector was from 14 hp to 52 hp. As another indication of the relative backwardness, consider the size of the companies in the chemical sector. In 1914, there were only nine companies employing more than 100 workers; in total 1,440 workers, or 28 per cent of the total workforce in chemical industry.

---

11 See the essay by Núria Puig in this volume.

12 Richard Willerslev, "Den teknisk-kemiske industri 1870-1914," *Erhvershistorisk Årbog*, **12** (1960), 27-68. A more recent and comprehensive analysis of the history of Danish technical chemistry from about 1760 to 1990 is presented in Kragh and Styhr Petersen, op. cit. (4). The present article is based in part on material more fully discussed in this book. Although Kragh and Styhr Petersen do not question the modest significance of chemical industry in economic growth before 1914, they emphasize the innovative nature of Danish chemical industry and hence its importance for economic growth in the interwar years.

All the same, the sector did experience a growth. While in 1897 the gross output of the entire chemical industry was 37 million kroner, in 1905 it was 52 million kroner; and in 1913 it had increased to 89 million kroner, corresponding to about 37 per cent of the output of the total industry.[13] However, the growth was paralleled by an increasing dependency on foreign chemical industry, especially a large import of raw materials and semi-manufactured goods from Germany and England. In most sectors the imports grew steadily and it was only in a few areas of chemical industry that domestic production was sufficient to satisfy the country's need for chemical products. The most important feature of Danish chemical industry was perhaps the absence of synthetic organic chemical manufacture based on coal tar (and later oil). This important area, the backbone of the German chemical industry, was never established in Denmark.[14] The strongest part of Danish chemical industry in the early part of the century was clearly associated with agriculture, whereas sectors such as sulphuric acid, dyes and pigments, metal working, and soap were much less significant. At the outbreak of World War I, the chemical economy was dominated by sugar refineries, margarine factories, breweries, distilleries, and oil mills, which together made up about 60 per cent of the industry's gross income.

The healthy development of the agri- and bio-technical sectors had its roots in the last three decades of the 19th century when chemists, pharmacists and bacteriologists assisted the new dairy cooperatives in order to turn the dairies into small chemical industries. One of the most successful of the new generation of applied scientists was Sigurd Orla-Jensen, a specialist in the chemistry and bacteriology of cheeses and the country's (and indeed the world's) first professor of "biotechnical chemistry."[15] There is a nice parallel between the development in production of dairy products in Denmark and Finland, where Virtanen played a role not unlike that of Orla-Jensen in Denmark.[16] However, whereas the Finnish cooperatives established their own research laboratory,

---

[13] H. J. Styhr Petersen, "The emergence of the Danish chemical industry and the role of the chemists" (unpublished manuscript, 1995).

[14] There were, however, works which distilled coal tar from the gas works. Danske Gasværkers Tjærekompagni (today 'Tarco') was founded in 1919 and produced a variety of tar products, especially for roads. With the exception of a limited use of tar derivatives for pharmaceutical products during World War II, there was no synthesis of organic compounds in connection with the Danish tar industry.

[15] His chair was originally, in 1908, in "fermentation physiology and agricultural chemistry," and changed to "biotechnical chemistry" in 1913. On Orla-Jensen as a pioneer in biotechnology, see Robert Bud, "The zymotechnic roots of biotechnology," *British Journal for the History of Science*, **25** (1993), 127-144, and *idem*, *The Uses of Life: A History of Biotechnology* (Cambridge: Cambridge University Press, 1994), pp. 5-10. For the 19th century background, see Helge Kragh, "Out of the shadow of medicine: Themes in the development of chemistry in Denmark and Norway," in David M. Knight and H. Kragh, eds, *The Making of the Chemist: The Social History of Chemistry in Europe 1789-1914* (Cambridge: Cambridge University Press, forthcoming).

[16] On Virtanen and Finnish dairy production, see the essay by Timo Myllyntaus in this volume.

research in Denmark was mostly based at the Polytechnical College.

Table 3. Development of Danish industry between 1914 and 1935. Only firms with six or more employed workers are included. The "chemical-related industries" include dairies, sugar refineries, margarine factories, breweries, cement works, glassworks and works producing stone and claywares. Source: Johansen (note 6), pp. 93 and 144.

| | 1914 | 1935 |
|---|---|---|
| Industrial firms | 4,476 | 6,050 |
| chemical industry | 95 | 178 |
| chemical-related industries | 707 | 866 |
| Industrial workforce | 133,630 | 190,290 |
| chemical industry | 2,890 | 6,143 |
| chemical-related industries | 21,558 | 27,142 |
| Average power consumption per worker (hp) | 1.1 | 2.3 |
| chemical industry | 1.6 | 3.9 |
| chemical-related industries | 1.9 | 5.0 |

After 1920, Danish chemical industry began to expand and a number of new manufactures were established, most notably in the young medical-pharmaceutical sector. Whereas the number of people employed in the chemical sector increased by 57 per cent between 1897 and 1914, in the period 1914-1935 the growth was 120 per cent (table 4).

Table 4. Workforce, in thousands, employed in a select group of industries. The category "other chemical industry" include cryolite, matches, asphalt and tar products, dyes and pigments, rubber, sulphuric acid and superphosphate. Source: Topsøe (note 26), p. 33.

| Year | 1897 | 1906 | 1914 | 1925 | 1935 | growth 1897-1935 |
|---|---|---|---|---|---|---|
| agricultural industry | 38.9 | 44.9 | 49.0 | 55.0 | 57.2 | 47 per cent |
| other food industry | 9.4 | 10.3 | 12.3 | 13.8 | 15.8 | 68 per cent |
| stone, clay, and glass | 16.0 | 17.4 | 19.5 | 20.0 | 22.4 | 40 per cent |
| other chemical industry | 4.4 | 5.9 | 6.9 | 10.1 | 15.2 | 245 per cent |

According to a 1932 survey, "chemical and related industries" had in 1930 a sales value of about 1.9 billion kroner, although this figure included products from dairies and bacon factories; if these sectors are omitted, the value becomes 800 million kroner.[17] The two largest sectors were classified as "general technical chemistry" (311 m. kr.) and "biotechnical chemistry" (303 m. kr.), whereas metallurgical industries (115 m. kr.) and "mortar, glass and ceramics" (73 m. kr.) were less important. Among the chemical industries most important to the economy were gasworks, sugar refineries, oil mills, breweries, distilleries, dairy products and margarine factories, again indicating the dominance of activities related to the agricultural sector. Tables 5 and 6 give further data for Danish chemical industry in 1930.

One of the factors that obviously influenced growth in the chemical industry was the education and training of chemical engineers and other technical personnel needed for the industry. Geert Somsen (this volume) describes how academic chemistry in the Netherlands was divorced from chemical industry during 1890-1910, the very same period when Dutch chemistry experienced its remarkable "second golden age." A somewhat similar situation occurred in Denmark twenty years later. In the 1920s Denmark became an important centre of chemical research, primarily through the works of George von Hevesy, Niels Bjerrum, and Johannes Brønsted. To speak of yet another "golden age" may seem trite, but it is nonetheless appropriate.[18] But brilliant scientific results were not followed by any contacts between academic chemists and industry. By and large the two groups ignored one another and, contrary to the Dutch case, there was no rapprochement in the 1930s either. However, this does not mean that chemists played no role in industry, for in Denmark chemical engineers were no less 'chemical' than university-trained chemists; contrary to the situation in the Netherlands, Danish chemical engineers had an academic status and a harmonious relationship to their colleagues at the university.

Chemical engineers had been educated at the Polytechnical College (Polyteknisk Læreanstalt, now Denmark's Technical University) in Copenhagen since its foundation in 1829, and the first engineers graduated in 1832. During the 19th century the annual average number of graduates in chemical engineering was a modest 3.8, but after 1900 the number grew rapidly, from 10.1 in 1900-1909, over 17.8 in 1910-1919, to 29.5 in the decade 1920-1929.[19] The education was generally of a high quality, not least scientifically. There was, however, constant discussions about the high priority given to scientific aspects of chemistry, and many industrialists complained that the young

---

[17] Peter E. Raaschou, "Fabrikingeniørens indsats i samfundets økonomi," *Kemisk Maanedsblad*, **13** (1932), 29-32, and Kragh and Styhr Petersen, op. cit. (4), pp. 242-244. In what follows, "m. kr." stands for millions of kroner.

[18] Anita Kildebæk Nielsen and Helge Kragh, "An institute for dollars: Physical chemistry in Copenhagen between the wars," *Centaurus* (forthcoming).

[19] H. C. Nielsen, "Fortid, nutid, fremtid," *Kemisk Maanedsblad*, **10** (1929), 1-14, which includes a complete list of chemical engineers graduated from the Polytechnical College 1829-1929. A larger historical perspective is provided in Harnow, op. cit. (9).

engineers lacked insight into practical production processes and commercial aspects of technical chemistry.

Table 5. Main sectors of Danish chemical industry in 1930. Source: Raaschou (note 17).

| | chemical engineers | sales value (m. kr.) | workers |
|---|---|---|---|
| technical chemistry, general | 175 | 311 | 13,259 |
| biotechnical industries | 70 | 303 | 11,953 |
| dairies, export bacon factories | — | 1066 | — |
| metallurgical industry | 25 | 115 | 10,821 |
| ceramics, glass, cement | 60 | 73 | 9,865 |

Table 6. Biotechnical industries in 1930. Source: Raaschou (note 17).

| | number of companies | workforce | sales value (m. kr.) |
|---|---|---|---|
| breweries | 212 | 4,043 | 84 |
| distilleries, yeast | 4 | 410 | 12 |
| mineral water | 50 | 550 | 9 |
| milk condensates | 11 | 641 | 18 |
| margarine | 135 | 1,284 | 78 |
| chocolate | 17 | 1,696 | 20 |
| tanneries | 31 | 641 | 16 |

The Danish philosophy in the education of chemical engineers was modelled on German experiences, where the descriptive study of process technologies (Verfahrenstechnik) was the core element. When the concept of unit operations and the American tradition of chemical engineering were developed in the 1920s, the new trends were ignored in Denmark, as in most other European nations with education systems modelled on German or French traditions. (It was only in the 1950s that unit operations became a central part of the curriculum for Danish students of chemical engineering.) In spite of this element of conservatism, Danish chemical engineers were generally highly qualified and played a significant role in the expansion of the industry that took place in the 1930s. According to the 1932 survey mentioned above, chemical industry in Denmark employed 294 chemical engineers, all of them from the Polytechnical College. This may seem a small number, but it reflected an increasing role for the chemical engineer. Whereas the main source of employment for chemical engineers had earlier been in the public sector, in the 1920s there was a change in the pattern toward employment in industry and in this the chemical industry led. Between 1920 and 1929, 32 per cent of new engineering graduates were employed in industry, as compared with 22 per cent in the first decade of the century.[20]

Yet another indicator of the general development of chemical industry is the data gained from patent statistics. A modern patent law of the type existing in other countries such as Germany and England was introduced only in 1894, when it replaced the old, and, at the time, antiquated system of royal monopolies. According to the new patent law, processes for producing pharmaceutical products could be patented, but not the products themselves; and foodstuffs, whether the product or the process, could not be patented at all. Danish chemical industry, which to a large extent focused on just such products, felt that these clauses hampered its productivity and competitiveness and for this reason there was during the interwar period a constant debate between the Danish patent authorities and a highly frustrated chemical and pharmaceutical industry.[21]

In general, as measured by the patenting activity, Danish industry was not very innovative and the chemical industry was no exception. The relative failure of Danish chemical industry to innovate and market its inventions is illustrated by two examples: first, very few of the applications for Danish patents came from Danish companies; second, the proportion of successful patent applications (i.e., those actually granted a patent) was smaller among Danish applicants than among foreign companies seeking a Danish patent. Between 1894 and 1942 a total of 113,008 patent applications were filed, of which 38 per cent were Danish and 62 per cent foreign; the proportion of patents granted to Danes was somewhat smaller, namely 34 per cent. If we look at inventions related to the chemical sector, the picture is considerably darker. Only 19 per cent of the applications and 12 per cent of those granted came from Danish companies or individual inventors. In the most important of the chemical classification groups (no. 12 in the German classification system, "Chemical methods, processes and apparatus") only 105 patents were granted to Danish companies in 1894-1939, or 6.7 per cent of the total number of granted patents.[22]

Patent statistics should of course be handled with care and it would be wrong to equate low patent frequency with failure in productivity and economic development. Yet it is

---

20 "Ingeniørens beskæftigelse," *Ingeniøren*, **41** (1932), 70-73.

21 S. Schønning, "Nogle hovedspørgsmål vedrörende patentering af kemiske opfindelser i Danmark," *Kemisk Maanedsblad*, **16** (1935), 145-156.

22 For more data and details, see Kragh and Styhr Petersen, op. cit. (4), pp. 231-235.

clear that in many areas Danish chemical industry was remarkably slow in turning inventions into innovations and often left the initiative to large foreign companies. As an example, consider the important pH concept, of Danish origins, and methods for determination of pH in industrial processes. The definition of pH was introduced in 1909 by S. P. L. Sørensen, the director of the department of chemistry at the Carlsberg laboratory, who also developed practical colorimetric methods for the determination of pH. Other chemists in Copenhagen took up the topic and pH measurements became a Danish speciality in the 1920s; the developments included Einar Biilmann's invention of the quinhydrone electrode, for a decade the favourite method of practical pH determinations. The quinhydrone electrode was not patented, but was widely used in both agriculture and industry. Thus, in the late 1930s the Danish state plant growth laboratory (Statens Planteavlslaboratorium) made more than 200,000 pH determinations annually with Biilmann's electrode. At first the interest in pH was primarily directed toward biochemical and medical uses, but pH measurements were also applied in industrial processes at an early stage.[23] For example, they were used extensively at the Carlsberg breweries from 1920 and were also important in bakeries and tanneries. Radiometer, a Danish electronic company founded in 1935, specialized in pH-meters for medical use, but the development of automatic pH-meters for industrial purposes was neglected. Instead, American instrument manufacturers successfully developed industrial pH-meters with a wide variety of applications, such as in breweries, paper mills, alum works, and water treatment systems.[24]

The discovery of the element hafnium provides a further example of a missed opportunity to turn scientific achievement into practical application. Hafnium was discovered at Niels Bohr's institute in Copenhagen in 1923 and quickly turned out to be fairly abundant and of potential industrial interest. However, no Danish chemical company seems to have taken an interest in the discovery. The two discoverers, Hevesy and Dirk Coster, sold the patent rights to the Philips firm in the Netherlands. Philips at once took out several patents on the use of hafnium compounds and developed methods for applying the metal in electrodes, filaments and fireproof enamels.[25]

The most serious weakness of Danish chemical industry between the wars was structural. The industry was comprised of many small sectors and companies, but the sectors were not integrated into a larger system and the industry therefore lacked the organized nature that characterized the chemical industries of larger European nations.

---

[23] On early pH measurements and their industrial use, see Holger Jørgensen, *Wasserstoffionen-konzentration (pH) und deren Bedeutung für Technik und Landwirtschaft* (Dresden: Th. Steinkopff, 1935) and Kragh and Styhr Petersen, op. cit. (4), pp. 249-261.

[24] John T. Stock, "Early industrial pH measurements and control," *Bulletin for the History of Chemistry*, **10** (1991), 31-34.

[25] Danish patents nos. 37,002, 37,266 and 37,687, all of 1923. See also Helge Kragh and Peter Robertson, "On the discovery of element 72," *Journal of Chemical Education*, **56** (1979), 456-459.

This was again a consequence of the absence of works that processed basic raw materials such as coal, oil, wood, sulphur and lime. The need of such "basic works" as a starting point for organic and inorganic industries was emphasized by the engineer (and later successful industrialist) Haldor Topsøe in a critical and perceptive analysis of Danish chemical industry published in 1943. According to Topsøe, the structure of the industry was "unsatisfactory and accidental" and production was located in "limited, mutually isolated and apparently arbitrarily selected areas with no technical connection". Lacking was the large-scale planning that might give the industry coherence and flexibility: "A really fruitful development of the chemical industry can only be achieved by basing the industry on a limited number of raw materials ... In this way a perfect flexibility can be achieved and changing needs can be satisfied merely by changes in the domestic production enterprises based on existing raw materials."[26] Topsøe's plan was to engage the state and private business in a system of oil refineries and coke works, and integrate these works with a synthetic organic chemical industry. However, it was unrealistic in the early 1940s and also, it turned out, in the postwar era.

Attempts to establish chemical industries of the kind suggested by Topsøe had occasionally been made, but the results were invariably disappointing. One area of strategic importance was the supply of nitrogen fertilisers which from about 1905 was mainly satisfied by calcium nitrate imported from Norway. The supply was of crucial importance for Denmark, as an agricultural country, and for this reason there were several plans to start domestic production of nitrogen fertilisers. In 1918, when the Norwegian supply was threatened because of war, the government sponsored an investigation which led to the recommendation of a pilot plant based on the Haber-Bosch process. It hoped that this would lead to Haber-Bosch works capable of producing about 8,300 tons of nitrogen per year in the form of synthetic ammonia. However, the project remained on the drawing board and it was only in the 1930s that another attempt to establish domestic production of artificial nitrogen compounds took place. This time it was based on an ingenious modification of the Birkeland-Eyde arc process invented by the engineer Johan Nyrop.[27] This appeared promising, and in 1933 test trials with the so-called arc centrifuge resulted in a yield of 310 grammes of nitric acid per kWh of electric energy (as compared with the conventional Birkeland-Eyde method, where the ratio was about 62 grammes per kWh). However, although Nyrop received support from leading Danish chemical companies, the invention was never transformed into an industrial innovation. One reason, but probably not the only one, was political: the government was eager to maintain ordinary trade relations with Norway, which included a profitable export of agricultural products, and for this reason it declined to support the project. Other factors were the strong opposition to this project by the international nitrogen cartel, in which major firms such as Norsk Hydro,

---

[26] H. Topsøe, *Danmarks Produktionsliv* (Copenhagen: Reitzel, 1943), pp. 247-249.

[27] G. Sørensen, *Initiativets Pris: En Bog om Johan Ernst Nyrop* (Copenhagen: Niro Atomizer, 1983).

ICI, and I.G. Farben were organized, and diplomatic intervention from Germany.[28]

Although poor in mineral resources, Danish territory (including Greenland) does include some valuable deposits of coal, iron, etc., which to a limited extent had been utilized since the early 18th century. At the turn of the century mining had almost stopped completely, but there were frequent attempts to revive the industry. During World War I the price of coal soared and it became difficult to obtain sufficient amounts of coal from the traditional suppliers, England and Germany. In this situation the government decided to reopen the small coal fields on Bornholm (in the Baltic), but production was limited and kept alive as much as a work programme as a supplement to the coal supply. When market conditions were normalized in the early 1920s it was too costly to subsidize locally produced coal and production ceased. The difficult supply situation from 1916 to 1920 was also the background for an ambitious plan to establish Danish blast-furnace works for the production of iron from the deposits of bog iron in Jutland. The bog iron ores were poor in iron content and quality (30-50 per cent iron), but a state commission concluded that the project was realistic and that locally produced iron would satisfy the needs of Danish industry. The amount of bog iron was estimated to correspond to 125,000 tons of pig iron, but the economics of the project was questionable, to say the least. The war ended before the planned blast furnaces were established and it was soon realized that under normal market conditions a local production of iron could not compete with foreign supplies, even if heavily subsidized.[29]

The situation in Greenland looked more promising. However, in spite of several attempts to establish a mining industry there it remained only of local importance. The large deposits of coal and graphite in western Greenland were exploited to some extent, but the low quality of the coal and the high cost of transportation made it unfeasible to use the coal outside Greenland. Graphite mining in Greenland ended in the late 1920s. Even less successful were attempts to utilize the deposits of copper. Copper mining was abandoned in the 1860s, but revived in 1907 by Greenland Mining Incorporated. After seven years of difficulties and financial loss the company gave up in 1914. By and large, the mining industry in Greenland never took off.

The exception was the cryolite industry, which in the second part of the 19th century

---

[28] See Harm G. Schröter, *Aussenpolitik und Wirtschaftsinteresse. Skandinavien im Aussenwirtschaftlichen Kalkül Deutschlands und Grossbritanniens 1918-1939* (Frankfurt a. M.: Peter Lang, 1983), p. 296, and H.G. Schröter, "Privatwirtschaftliche Marktregulierung und Staatliche Interessenpolitik: Das Internationale Stickstoffkartell 1929-1933," in H.G. Schröter and Clemens Wurm, eds, *Politik, Wirtschaft und Internationale Beziehungen* (Mainz: Ph. v. Zabern, 1991), pp. 117-137, on p. 128.

[29] Somewhat similar iron and steel projects, but on a larger scale, were initiated in the Netherlands during the war. In contrast to the Danish project, the Dutch factories became an important part of Dutch industry. See Geert Somsen's essay in this volume.

produced large quantities of the mineral cryolite for the manufacture of soda.[30] However, by the 1890s the cryolite soda adventure was over and during the following decades the cryolite factory in Copenhagen strived to find new applications for the Greenlandic mineral. Some of the raw cryolite was sold to soap factories, some was used for the manufacture of enamels, and some used as insecticide (under the name 'cryocide'). It was increasingly used as flux in the expanding international aluminium industry. Whereas 25 per cent of the cryolite was sold to the aluminium factories in 1904, at the end of the 1930s the figure had increased to more than 80 per cent.

From about 1930 the natural cryolite met tough competition from "synthetic cryolite," that is, compounds with the same stoichiometric composition as cryolite but produced from soda, hydrofluoric acid and alum earth. Such products, made by German chemical firms since the beginning of the century, were increasingly used in the aluminium industry as a substitute for natural cryolite. As a result of the competition and the gradual depletion of the deposit in Greenland the cryolite industry was constantly faced with difficulties. However, it remained one of the country's largest and most profitable chemical industries during the interwar period. Sales and income varied with market trends, but generally the last era of the cryolite company in Copenhagen (Øresunds Chemiske Fabriker) was marked by economic progress. Table 7 indicates the increase in economic value and the increasing importance of the aluminium industry in the period from 1914 to 1939. The Greenlandic mineral became of strategic value during World War II, when the entire production of cryolite was shipped to the American and Canadian aluminium plants that worked flat out to satisfy the demands of the aeroplane industry.

Table 7. Sales and net profits of the cryolite company. The second and third columns list the cryolite sold for aluminium production, in weight and in per cent of total sale of cryolite. Source: Topp (note 30).

| | sale of cryolite (tons) | aluminium (tons) | per cent of aluminium | profit (m. kr.) |
|---|---|---|---|---|
| 1904 | 2,110 | 543 | 25 % | 0.43 |
| 1914 | 5,760 | 2,950 | 51 % | 0.85 |
| 1925 | 16,760 | 10,760 | 64 % | 6.60 |
| 1939 | 30,050 | 24,610 | 82 % | 13.65 |

30 On the early history of the cryolite industry, see Helge Kragh, "From curiosity to industry: The early history of cryolite soda manufacture," *Annals of Science*, **52** (1995), 285-301. The later phase is treated in Kragh and Styhr Petersen, op. cit. (4) and in Niels-Henrik Topp, *Kryolitindustriens Historie 1847-1990*, 2 vols. (Copenhagen: Kryolitselskabet Øresund, 1990), vol. 2.

**4. Some examples of successful industries**

The cryolite industry was not representative of Danish chemical industry in the interwar period and neither were the failed projects in, for example, mining and fertilisers. In spite of many shortcomings and failures the period was generally one of progress. Examples of relatively successful industries include the margarine factories and the pharmaceutical (biomedical) industry – the one in the traditional sector with its roots in the 19th century and the other in a progressive 20th century sector.

The first Danish factory for production of margarine was established in Aarhus during 1883. The enterprise, Aarhus Oil Factory Ltd (Aarhus Oliefabrik), experimented in the 1890s with using coconut oil as raw material, and thanks to the work of the engineer Malthe C. Holst it succeeded in developing a vacuum process for refining and deodorizing the coconut oil.[31] The process was inspired by a German patent of 1895, but Holst developed his own version on which the factory based its production of plant margarine. Holst was also a pioneer in the introduction and modification of the hydrogenation process originally invented by Wilhelm Normann in Germany in 1902. Hydrogenation was introduced at Aarhus Oil in 1911, the first of its kind in Scandinavia. As a result of the innovations and the general demand for margarine, the Aarhus factory expanded and concentrated on products based on coconut oil. The growth is illustrated in table 8.

Table 8. Growth of Aarhus Oil. Source: Kragh and Styhr Petersen (note 4), p. 168

| Year | 1896 | 1931 |
|---|---|---|
| margarine and plant oil (tons per day) | 14 | 700 |
| engine power (hp) | 70 | 4,000 |
| employees | 100 | 1,000 |
| technical personnel (chemists, engineers) | 1 | 20 |

The Aarhus factory was soon followed by many other margarine factories. In 1908, the sixteen Danish factories had an annual production of 30,000 tons of margarine; twenty years later the number had risen to 141 factories, mostly small and serving local areas. In the late 1920s the total annual production was 84,000 tons. The boom in plant margarine was of great economic value for the Danish nation, both directly and indirectly. This might suggest that margarine competed with butter and thus constituted a threat to the single most profitable Danish export product. In a way it did, but the growing use of margarine in Danish households turned out to benefit the butter export

[31] E. Cohn, *Margarineindustrien i Danmark 1883-1933* (Copenhagen: Otto Mønsted, 1933); H. J. Styhr Petersen, "M. C. Holst, spiseolie og margarine," *Dansk Kemi*, **74** (1993), 26-30.

market: during the 1920s and 1930s the dairy industry sold all the butter it could produce, and because of Danish housewives' partial conversion to margarine a large amount of butter was kept free for the export market. It was estimated that in this way the margarine industry resulted in an annual currency gain of around 60 million kroner in the mid-1930s. Aarhus Oil was one of the largest and most internationally oriented of the Danish chemical companies. In the 1880s it had established a large oil mill in Lipaja (then Lipau) in Latvia, and until the outbreak of World War I the Latvian factory was a highly profitable business that served part of the Russian market. About 1910 the Russian activities expanded with a new oil mill in Odessa and a margarine factory in Warsaw. Aarhus Oil also had business interests in Morocco and Syria, where refineries and oil presses were established in the 1930s.

As for the pharmaceutical industry, it is worth noting that this development started in the late 1940s, but the foundation was laid before the war. The first Danish company to enter the medical market was Løvens Kemiske Fabrik which started production in 1908 and by the mid-1920s had expanded to become the main supplier for the home market of many medical products. The 1920s witnessed a proliferation of medical companies, although the sector was still small and had to rely on imported raw materials. In the early 1940s, the pharmaceutical industry employed about 1,200 workers and exported about 36 per cent of its production.[32] The young industry was hampered by the absence of a local organic chemical industry to supply the necessary raw materials and for this reason the production of synthetic drugs was limited. In a 1944 analysis of the Danish pharmaceutical industry, the chemist Hans Baggesgaard Rasmussen concluded that the main problem was the lack of supplementary industries able to supply the necessary organic materials. "In this respect we are far behind our neighbour Sweden, not to mention Germany and Switzerland. ... [We need to] establish a cooperation between our pharmaceutical factories and other chemical industries and to develop it considerably ... It is however questionable if this alone is enough to create a sufficiently versatile pharmaceutical industry. We will probably have to initiate cooperation with foreign factories which are able to supply important groups of intermediary products."[33]

Only in a few cases was reliance on organic materials unnecessary, as was the case with the inorganic drug sanocrysin, developed by the company Ferrosan in the mid-1920s. Ferrosan, incorporated as a Danish-Swedish concern in 1920, specialized in ferrous drugs of which Idozan, an iron(III) citrate and tartrate, was a considerable success both on the home market and abroad. Sanocrysin is a gold coordination compound with the composition $Na_3[Au(S_2O_3)_2]$, $2H_2O$ and was used therapeutically against tuberculosis. Between 1925 and 1931, 200 kg of gold was used for the new drug, which was believed

---

32 H. Baggesgaard Rasmussen, "Medicinfabrikation i Danmark," *Ingeniørvidenskabelige Skrifter 1939*, no. 2.

33 H. Baggesgaard Rasmussen, "Om den danske medicinalindustri," *Ingeniøren*, **53** (1944), 1-6, 9-13, on p. 5.

to be a promising cure against tuberculosis. However, the effects were disputed and by the late 1930s treatment with gold compounds ceased. (Sanocrysin and similar compounds are still used for chronic polyarthritis.)

The most spectacular success in the Danish pharmaceutical industry was undoubtedly the production of insulin based on the method that the Nobel Prize laureates Frederick Banting and Charles Best developed in 1921 and which almost instantly proved its efficiency in the fight against diabetes.[34] In 1922 the physiologist August Krogh (Nobel prize,.1920) received authorization to introduce insulin into Scandinavia, and with Hans C. Hagedorn he quickly began the organization of Danish insulin manufacture based on supplies of pork pancreas from the bacon factories. The company, Nordisk Insulin Laboratory (Nordisk Insulinlaboratorium), was an instant success and developed the original method into a more precise and reliable manufacture of insulin. Among other improvements, Hagedorn and his colleagues developed in 1936 a new insulin product (protamin insulin) which was absorbed more slowly than the original insulin and therefore did not give rise to so-called insulin shock. After a few years of production, Nordisk Insulin became a large firm with 90 per cent of output exported. Another early player on the insulin market was Novo (Terapeutisk Laboratorium Novo), founded in 1924 by two former employees of Nordisk Insulin, the brothers Harald and Thorvald Pedersen. In competition with Nordisk Insulin, Novo became in the 1930s one of the largest companies in the market for insulin and related products.[35] The growth and competition continued after World War II. (In 1989 the two companies fused to form Novo Nordisk, which is presently the largest chemical company in Denmark and the world's second-largest manufacturer of insulin products.[36])

There were of course other Danish chemical industries of domestic, and sometimes also of international, significance. I mention three examples.[37] Beet sugar refineries were started in the 1870s, and from about 1880 this branch of organic chemical industry expanded under the leadership of the firm De Danske Sukkerfabrikker, which soon

---

34 Michael Bliss, *The Discovery of Insulin* (Chicago: University of Chicago Press, 1982).

35 H. Richter-Friis, *Livet på Novo* (Copenhagen: Gyldendal, 1991).

36 The largest manufacturer is Eli Lilly in the United States. With an export value of 13 billions of kroner and an export surplus of about 8 billions of kroner (1995 figures), the medical-pharmaceutical sector is presently the largest and most successful part of Danish chemical industry. As for export of medical products per inhabitant, the country ranks as number three in the world (Switzerland is number one and Ireland number two).

37 On sugar refineries, see P. Drachmann and L. Estrup, *De Danske Sukkerfabrikker 1872-1922* (Copenhagen: DDS, 1922). The production of sulphuric acid is the subject of H. J. Styhr Petersen, *Svovlsyre i Danmark* (Copenhagen: Dansk Selskab for Historisk Kemi, 1992), and the chlorine-alkali plant is analyzed in Kragh and Styhr Petersen, op. cit. (4), pp. 296-305 and, from a contemporary perspective, A. Larsen, "Produkt-industriernes expansion: Nogle nyanlæg indenfor dansk teknisk-kemisk industri," *Kemisk Maanedsblad*, **17** (1936), 77-101.

obtained an almost monopolistic position in the home market. In the late 1930s the production of sugar amounted to about 240 million kg per year, almost seventeen times as much as the production in 1880. Another branch of considerable domestic importance was sulphuric acid and superphosphate, which from 1903 was a monopoly of DSSF (Dansk Svovlsyre- & Superfosfatfabrik, later Superfos Ltd.). Between 1907 and 1930 DSSF established five new factories for sulphuric acid and superphosphate with the result that the domestic need for superphosphate was fully satisfied by domestic production. Dansk Sojakagefabrik, founded in 1909 by the trade company ØK, established in 1935 a technically advanced chlorine-alkali plant in Copenhagen. The production of hydrogen and chlorine products was based on the Castner-Kellner process; this process, which makes use of mercury cathode cells in the electrolysis of sodium and potassium hydroxides, was at the time unusual in the European chemical industry and the Copenhagen plant was in many respects technically innovative. It was the country's first advanced inorganic chemical plant and the single largest industrial consumer of electrical power. However, in the 1930s its main product was hydrogen for the margarine industry, whereas the chlorine it produced was considered less interesting and was sold to Swedish and Norwegian cellulose plants.

## 5. Conclusion

During the period between 1910 and 1940 the Danish chemical industry experienced steady growth and developed into a sector of great importance to the national economy. Although the major part of the chemical industry continued to be related to agricultural products, there was considerable diversification and many new chemical companies were established within the inorganic and biomedical branches. The chemical industry in Denmark was severely limited by the absence of basic organic and inorganic works, such as for the production of ammonia, mineral oil products and metals. In particular, the absence of a synthetic organic chemical industry meant that the country was to a large extent dependent on foreign chemical products, especially from Germany. In spite of these and other limitations, progress in Danish chemical industry was considerable. By 1940 a sound foundation for further development had been established and the more rapid development after World War II relied substantially on this foundation.

In the history of science and technology it is often the case that prior developments and events can only be properly evaluated by reference to what happened at a later date. Such evaluation is clearly 'whiggish,' but it may nonetheless furnish a better and more reliable account of what happened in the past than purely historicist or 'anti-whiggish' accounts that rely solely on the records of the period in question.[38] The case of the Danish chemical industry seems to be one among many that brings this point home.

[38] The crucial question of whig contra anti-whig historiography has been discussed more in the history of science than in the history of technology and industry. For general discussions, see Helge Kragh, *An Introduction to the Historiography of Science* (Cambridge: Cambridge University Press, 1987), pp. 89-107, and Stephen G. Brush, "Scientists as historians," *Osiris*, **10** (1995), 215-232.

# NEGLECTED POTENTIAL? THE EMERGENCE OF THE FINNISH CHEMICAL INDUSTRY, 1900-1939

TIMO MYLLYNTAUS
*Department of Economic and Social History*
*University of Helsinki*
*POB 54, Snellmaninkatu 14*
*00 014 Helsinki, Finland*

**1. Perspectives on the past**

It has often been suggested that industrialization proceeded in three revolutionary waves. The first one started in Britain in the 1760s and was characterized by the introduction of mechanized, power-driven textile machines. Between 1800 and 1830, the British industrial revolution was fuelled by the development of railways, steamships, the factory system and a liberal economic policy. At the same time, these processes of industrialization spread to some other European countries and North America.

The second revolution, from the 1870s to the 1930s, was international from the very beginning, although the leading group was composed of half a dozen west European countries and the USA. Its main feature was the new application of science to industry. The chemical industries and electrical engineering were the flagships of this phase.

After 1945, the third revolution emerged, bringing with it large technological systems and the "information society." Industrial technology relied more than ever on science, while automation and computerization penetrated not only the industrial processes, but also almost every corner of the "post-modern" society.

The first and second industrialization phases spread to the European periphery at almost the same time and overlapped in various sectors. By 1900, problems related to the phenomena of the first industrial revolution were mainly resolved. Steam engines, steamships, railways, textile machinery and blast furnaces relied on what had become standard technology, the transfer of which could easily be achieved in appropriate economic circumstances. The introduction of new science-based technologies, such as hydroelectric power plants and chemical processes, was much more challenging to peripheral, late-industrializing countries such as Finland. In these cases, raising funds



*A.S. Travis et al. (eds.),*
***Determinants in the Evolution of the European Chemical Industry, 1900–1939,*** 341–360.

for investment and importing the required production hardware were not the main problems. Acquiring the critical technological and commercial know-how created the pivotal bottleneck. Difficulties encountered around 1900 in adopting science-based chemical production processes were more complex than those met in any earlier industry. The special problems associated with transferring technology and developing new industries may be understood by examining late-industrializing countries. Case studies of countries on the European periphery exhibit the characteristics of the second industrial revolution that was in full swing during the years 1900 - 1939.

**2. Prerequisites for industrialization**

Finland resembles several other small countries on the European periphery that only emerged as independent nations in the early 20th century. In common with Ireland, Norway, Estonia and the former Czechoslovakia, for example, Finland detached itself from a large empire and gained its political sovereignty, cultural autonomy and the freedom of economic self-determination.

For about 700 years, Finns had been the inhabitants of four provinces in the Swedish Kingdom until, by means of a war in 1808-09, Russia conquered these provinces, east of the Gulf of Bothnia, and annexed them to the territories ruled by its tsar. However, this was not a catastrophe. On the contrary, the ruler, Alexander I, "elevated the Finns among the group of nations" by transforming the conquered territory into an autonomous grand duchy.[1] In the short run, this 'elevation,' however, hardly changed the life of the majority, the peasantry. Agriculture remained the dominant sector up to the mid-20th century.

In the second quarter of the 19th century, modern industry began to grow practically from scratch. The contribution of manufacturing to the gross national product (GNP) had risen to nearly 20 per cent by 1917, when – just after the October Revolution in Russia – Finland gained its political independence. In the years 1920-38, the growth rate of Finnish industry was extraordinarily high, 7.9 per cent annually, whereas between 1860 and 1913, the annual growth rate had been 5 per cent.[2] In the interwar years, the economy of the other sectors also developed briskly, and on the eve of World War II, the living standards of 3.7 million Finns were catching up with the average West-European level.

Prior to 1940, the development of the Finnish chemical industry was paradoxical. Undoubtedly, chemical production had an impressive history in Finland. In the late 18th

[1] Possibly, the model for this arrangement was taken from the Grand Duchy of Warsaw founded by Napoleon after the peace negotiations at Tilsit in 1807.

[2] Riitta Hjerppe, *Suomen talous 1860-1985. Kasvu ja rakennemuutos* (Helsinki: Suomen Pankki, 1988), p. 65; Riitta Hjerppe, *Finland's Historical National Accounts 1860-1994* (Jyväskylä: University of Jyväskylä, 1996), pp. 109-129.

and early 19th centuries, there was hardly any other European country in which chemical products accounted for such a large proportion of the total exports. Tar, produced by burning resinous pine billets in the traditional soil-covered tar pits, was the country's most important export product. Occasionally, pitch, saltpetre and potash were included among the leading items shipped to foreign markets.[3] However, these products were not manufactured industrially. Because industrialization had not even started at the time, at best we can call the production of tar, pitch, saltpetre, and potash, cottage industries, or proto-industrial activities.

Finnish chemistry had a firm foothold in the early 20th century, with a very long tradition and notable achievements. Thus Pehr Adrian Gadd became the first Finnish professor in chemistry at Åbo Akademi, the University of Turku, as early as 1761. After the beginning of industrialization in the mid-19th century, and acceleration before and after World War I, chemistry became the field of science in Finland with the highest international recognition. At the turn of the century, the Finnish professor Ossian Aschan (1860-1939), who made significant contributions to research on terpenes, the technological use of naphthenic acids and the production of synthetic rubber, was several times a candidate for the Nobel prize.[4] In 1945, his student Artturi Ilmari Virtanen was awarded the Nobel prize for his discoveries in biochemistry; and Virtanen still remains the only Finnish Nobel laureate in science.[5]

Though Finland lacked deposits of coal, oil and many minerals, it did possess natural resources suitable for chemical production. In the 1910s, a new, rich copper mine at Outokumpu was opened; its ore contained sulphur pyrites, among other things. However, the most valuable resources for the chemical industry were timber and water power. Processing staple wood products gave the forest industries opportunities to extract various chemicals as byproducts. Hydroelectricity was a new energy source for the electrochemical industries at the time.

Although Finland seemed to have traditions in chemical production, natural resources and know-how, its chemical industry, nevertheless, remained fairly small up to the eve of World War II. In 1938, the chemical industry contributed only two or three per cent

---

3 Kustaa Hautala, *European and American Tar in the English Market during the Eighteenth and Early 19th Centuries* (Helsinki: Suomalaisen tiedeakatemian toimituksia B 130, 1963); Sven-Erik Åström, *From Tar to Timber, Studies in Northeast European Forest Exploitation and Foreign Trade, 1660-1860*, (Ekenäs: Commentationes Humanarum Litterarum 85, 1988).

4 J. R., "Adolf Ossian Aschan," in *Finlandssvenka tekniker II*, Utg. av Jonatan Reuter, (Helsingfors, 1924), p. 55; G. K. Bergman, "Ossian Aschan, Minnesteckning," *Svenska tekniska vetenskapsakademien i Finland, Förhandlingar*, no. 10, 6 (Helsingfors, 1941); Terje Enkvist, *The History of Chemistry in Finland, 1828-1918* (Helsinki: Societas Scientiarum Fennica 1972), pp. 83-89.

5 Matti Heikonen, *AIV-keksintöjen aika*, (Helsinki, 1990).

to Finnish industrial output (see table 1 and figure 2).[6] It has traditionally been a home market industry with limited diversification in terms of products. Before World War II, production was not very science-based, and modern, industrial chemical products were exported only in modest amounts.

This paper focuses on the question of why the modern chemical industry developed in Finland so slowly and so late, and attempts to explain the difference between apparent potential and unpretentious achievements.

**3. The Production of Traditional Chemicals**

The considerable proto-industrial production of tar, pitch, saltpetre and potash declined during the 19th century. The change was attributed to at least three phenomena. First, demand for these products decreased abroad due to technological and economic changes. The transition from wooden to iron ships substantially reduced the consumption of tar. Second, synthetic products replaced traditional ones. Coal tar as a cheaper substitute for wood tar is a case in point. Saltpetre production methods used by Finnish peasants were not competitive with the more developed industrial processes applied abroad. Third, the Finns failed to modernize their proto-industrial processes and to industrialize their chemical production. However, some Finnish firms made attempts to distill tar from timber industrially in retort ovens. In 1900, sixteen factories were manufacturing tar, but their output was just a fraction of the quantity that traditional pits had been producing half a century earlier.

In the mid-19th century, mills for pressing oil from linen seeds constituted an important branch of the Finnish chemical industry. By the turn of the century they, too, had lost their significance. During the years 1900-39, of all the traditional chemical branches, only paint and printing ink factories, candle and soap factories, and match factories produced substantial outputs, as indicated in table 1. These branches, as well as the manufacture of bone meal and other fertilizers, explosives, bleaching chemicals, perfumes and pharmaceuticals, worked for the home market. They aimed to substitute imports, but due to insufficient domestic supply were forced to use a lot of foreign raw materials. The scale of their production was fairly limited because foreign products were often competitive in the small Finnish market. However, ten to fifteen Finnish match factories controlled the home market. There were hardly any imports, and exports also remained modest.[7]

6 *Industristatistik år 1938*, Finlands officiella statistik [Official Finnish Statistics], Serie 18, (Helsingfors, 1940).

7 *Industristatistik åren 1900-1913* (Helsingfors, 1902-15); *Album öfver den inhemska industrin* (Helsingfors, 1913), pp. 215-217.

Table 1. Employment and Output in Industries Related to Chemistry in Finland, 1900, 1913 and 1938.

| | **1900** | | **1913** | | **1938** | |
|---|---|---|---|---|---|---|
| Industrial products | Work-force | Output, in constant 1913 prices FIM 1,000 | Work-force | Output in constant 1913 prices FIM 1,000 | Work-force | Output in constant 1913 prices FIM 1,000 |
| Paint and smoke black | 46 | 330 | 94 | 1,184 | 404 | 10,728 |
| Lubricants, candles and soap | 111 | 2,116 | 242 | 5,987 | 386 | 10,412 |
| Fertilizers | 59 | 199 | 68 | 351 | 158 | 4,474 |
| Terpenes, tar and charcoal | 622 | 2,936 | 114 | 512 | 149 | 1,447 |
| Matches and explosives | 664 | 1,266 | 656 | 1,492 | 678 | 4,114 |
| Calcium carbide | 135 | 310 | | | | |
| Chlorate | | | 62 | 824 | 50 | 851 |
| Chlorine | | | | | 118 | 2,719 |
| Carbonic acid and soda | | | 23 | 228 | 23 | 368 |
| Hydrochloric acid | | | | | 17 | 298 |
| Sulphuric acid | | | | | 108 | 1,526 |
| Water glass | | | | | 27 | 202 |
| Pharmaceuticals | | | | | 658 | 6,904 |
| Other chemical products | 74 | 400 | 187 | 1,289 | 368 | 1,386 |
| **Main chemical industry** | **1,711** | **7,558** | **1,446** | **11,867** | **3,144** | **45,430** |
| *% of total manufacturing* | *2.1* | *2.3* | *1.2* | *1.7* | *1.5* | *2.5* |
| Dyeing | 254 | 1,167 | 73 | 182 | 301 | 947 |
| Rubber | 95 | 506 | 92 | 1,016 | 2,571 | 18,904 |
| Chemical pulp | 684 | 3,267 | 2,869 | 24,479 | 8,236 | 226,930 |
| Wine, juice and mineral water | 469 | 3,383 | 771 | 6,123 | 451 | 3,053 |
| Beer and stout | 1,559 | 9,199 | 949 | 5,929 | 1,669 | 15,474 |
| Spirit,* yeast and vinegar | 703 | 5,964 | 567 | 18,707 | 387 | 24,746 |
| **Sum** | **5,475** | **31,044** | **6,767** | **68,304** | **16,759** | **335,482** |
| *% of total manufacturing* | *6.8* | *9.4* | *5.8* | *10.0* | *7.8* | *18.1* |
| **Manufacturing, grand total** | **80,556** | **329,973** | **115,795** | **682,926** | **214,387** | **1,850,184** |

* Including sulphite spirit in the interwar years
Source: *Industristatistik åren 1900 - 1938* (Helsingfors 1902 - 1940).

## 4. The breakthrough of new industrial branches

In the latter half of the 19th century, the new, science-based chemical industry greatly advanced in various European countries. In Finland, such new processes obtained a foothold only from the 1890s onwards. In 1864 the Swedish inventor Alfred Nobel set up a dynamite factory near Stockholm and in 1865 his brother Robert founded a similar factory in Finland. This subsidiary, just outside Helsinki, was closed only a year later when Robert Nobel moved back to Sweden and became the chief executive of the family firm, Nitroglycerin Aktiebolag. It was only in 1893 that dynamite manufacture began properly in Finland. Then a Belgian firm, Compagnie de la Forcite-Dynamite, established a factory in Hanko. The number of its workers rose from 31 in 1913 to 129 in 1938, when it was still the only firm producing explosives in the country.[8]

In 1898 the first rubber factory based on the natural product, Suomen Gummitehdas Oy (later Oy Nokia Ab rubber factory), was set up to produce galoshes, i.e. overshoes. The firm specialized in rubber footwear and flourished. In 1926 it began to manufacture tyres for bicycles, and in 1934 tyres for cars. Two years later it introduced its own innovation, winter tyres for cars. The growth of the rubber industry was outstanding. In 1913, it employed over 90 employees and in 1938 as many as 2,570, or 82 per cent of the total workforce of the main chemical industry and 42 per cent of its output, as shown in table 1. Before the Winter War of 1939-40, rubber manufacturing was one of the three biggest chemistry-related branches of Finnish industry.[9]

Because Finland's largest rivers have numerous rapids, there were great hopes at the turn of the century for large-scale electrochemical industries. The river Vuoksi in south-eastern Finland was the biggest and provided the best building sites. Prior to World War I, various plans for hydroelectric power plants were worked out. The most ambitious of them contained a proposal to excavate a new riverbed and build a gigantic power plant – more efficient than that constructed at the Niagara Falls in the early 1890s. All the plans to build huge hydroelectric power plants on the river Vuoksi included a power transmission line to St. Petersburg, which is why the Finnish government and mass media opposed them. Together with other factors, this opposition was sufficient to prevent the construction of any major power plant along the river Vuoksi right up to the fall of the Russian Empire.[10] In contrast, at least some leading chemists took a more reconciliatory stance among the conflicting interests, and strongly supported hydro-electric power plants and the setting up of electrochemical factories in the south-eastern part of Finland.[11]

The Finnish authorities allowed the construction of smaller hydroelectric power plants along the river Vuoksi for supply to local factories. Consequently, in the 1890s two factories were set up to produce calcium carbide and another to make chlorate, mainly

---

[8] *Industristatistik åren 1913-1938,* (Helsingfors, 1915-1940); Ahti Haapaniemi, "Kemian teollisuutemme historiaa, Tervanpoltosta teollisuudeksi," *Kemia,* **10**, no. 6 (1983), 503; Vilh. Westman, *Suomen Forsiitti-Dynamiitti Osakeyhtiö 1893-1943* (Hanko, 1943).

[9] There are historical surveys that include rubber manufacturing in the main chemical industry: Yrjö Talvitie, *Suomen kemiallisen teollisuuden kehitys,* Kemian keskusliiton julkaisuja 5, No. 1, 1945 (Helsinki, 1945), pp. 104-105; Viktor Hoving, *Finska gummifabriks aktiebolaget 1898-1948* (Helsingfors, 1948).

[10] Sven Hirn, *Imatran tarina, Matkailuhistoriamme valtaväylältä,* (Imatra, 1978, 168-173); Timo Myllyntaus, "Samhällelig kontroll av teknikövertöring: Teoretiska anteckningar över erfarenheter från Finland," *Polhem, Tidskrift för teknikhistoria*, **8**, no. 3 (1990), 207-211.

[11] See for example, Ossian Aschan, "Synpunkter i fråga om vår kemiska industri," *Tekniska Föreningen i Finland, Förhandlingar*, **10** (1910).

for Russian match manufacturers. A sharp decline in prices on the world market from the year 1900 forced one of the carbide factories, Hämekoski, to convert into a pulp-grinding mill; the other one, the Linnankoski plant, was destroyed by fire. The latter was not rebuilt and a great scheme for two hydroelectric power plants surrounded by carbide and phosphorus factories with transmission lines to St. Petersburg was abandoned by the three founders of the company, a Danish businessman, Theodor Höffding, a Swedish engineer, William Waern, and a Major P. Laurell from Stockholm.[12] In 1908, there still remained an excess capacity of 50 per cent in the world's calcium carbide production.[13]

On the eve of World War I, only one electrochemical factory was still working in Finland. This was the chlorate factory, Finska Elektrokemiska Ab, which was founded in 1897 at Tainionkoski by the trading house Hackman & Co. in collaboration with its partners. The factory bought its electric power, 2,200 horse power, from a nearby wood-processing mill owned by Oy Tornator, and used imported raw material (Strassfurt salts), and applied the electrolytic process developed by the Swedish engineer Cassel. The annual output of the factory was limited to 1,100 - 1,200 tons of chlorate, as dictated by the European cartel.[14] After losing the Russian market because of the October Revolution in 1917, the Finnish owners sold the factory to a Scandinavian transnational firm, Swedish Match, in the early 1920s.

Before World War I, Finland clearly underutilized its water power resources for generating electricity. The number of electrochemical and electrometallurgical plants increased from 3 to 8 between 1900 and 1917, while in Sweden the number grew from 8 to 75 during the period 1904-17.[15] The profitability of these industries depended on carefully-chosen raw materials, highly technical know-how, and cheap energy. Finland did not have many appropriate raw material resources for this kind of production, but neither did Norway or Switzerland, which still managed to develop thriving electrochemical industries mainly on the basis of imported raw materials. Finland also lacked the technological expertise and experience; the diffusion of foreign electrochemical know-how was obstructed by many patents, expensive licenses and the cartels of transnational companies.

---

[12] *Prospect der Siitola-Actien-Gesellschaft. Gemeinde Ruokolahti, Gouv. Wiborg in Finnland,* (Riga, 1898), pp. 6-9; *Statut der Siitola Actien-Gesellschaft* (Riga, 1899); Kalevi Heitto, "Kronikka Siitolasta," *Vanhaa Imatraa sanoin ja kuvin*, Kanta-Imatran Seuran julkaisu no. 2 (Imatra, 1973), pp. 12-34.

[13] Of the total output (200,000 tons), Norway and Sweden produced together about 25 per cent and the USA roughly 20 per cent. *Suomen teollisuuslehti*, **27** no. 15-16, (1909), 100; **28** no. 3 (1910), 32.

[14] Julius Hirn, ed., *Suomen kauppa, meriliike ja teollisuus. Viipuri* (Turku, 1915), pp. 84-86; *Album öfver den inhemska industrin* (Helsingfors, 1913), p. 216.

[15] *Berättelse öfver Finska elektrokemiska aktiebolagets verksamhet under år 1914* (Wiborg, 1915); *Teknikern*, **28**, no. 1094 (1918), 95.

Another critical aspect was the shortage of cheap electricity. Before the Imatra power plant was commissioned in 1929, there was not enough inexpensive energy available for a larger chemical factory. In the early 1920s, the lowest electricity prices for the Norwegian electrochemical industry were just under 3 current Finnish pennies (FIP) per kWh, and for the corresponding Swedish industry FIP 4 - 6. In Finland, some electrochemical factories closed after World War I because they were required to pay FIP 12 - 18 per kWh for their energy.[16] More favourable geographical and hydrological conditions made hydroelectricity cheaper in Sweden as well as in mountainous Norway and Switzerland than in flat Finland. Furthermore, in those countries, abundant waterpower resources stimulated the construction of power plants specifically for producing electrochemicals. In the early 1920s, the electrochemical industries consumed nearly half of the hydroelectric power generated in Norway, 30 per cent in Sweden, but only 2 per cent in Finland.[17]

It has been a Finnish peculiarity that the large pulp and paper industry has had such a remarkable effect on the development of the electrochemical industry. In some cases, pulping and paper manufacturing competed with electrochemical production for waterpower. Generally, the big wood-processing companies had the upper hand and limited the opportunities for growth of the chemical industry. By contrast, wood-processing also needed various chemicals, which stimulated the manufacture of some of them in the country. The extensive requirement for bleaching chemicals is an example. The small-scale production of chlorine by the Siemens process was begun at the Kuusankoski paper mill in 1901.[18] Other mills continued importing their bleaching chemicals. However, four chlorine factories were put in operation by 1939. All these factories applied the electrolytic process in production. The most important and largest of them was Finnish Chemicals, a joint venture of three big transnationals, Britain's Imperial Chemical Industries Ltd. (ICI), the German I.G. Farbenindustrie AG, and the Belgian Solvay & Cie.[19]

**5. The rise of the chemical pulping industry**

Finns have been producing paper from rags since 1667. In 1841, the first paper-making machine was imported into Finland, but the lack of rags became an ever more acute bottleneck in paper making. The wood grinding machine, a German invention, was first introduced in 1859. However, this mechanical pulping provided only a partial solution

[16] V. Sihvonen, "Suomen sähkökemiallinen teollisuus," *Teknillinen aikakauslehti* (1926), 149-155; Arnold Almqvist, "Suomen sähkökemiallinen teollisuus," *Teknillinen aikakauslehti* (1926), 427.

[17] Ibid; Bror Sjögren, "Användbarhet av Finlands vattenkraft för elektrokemisk industri," *Tekniska Föreningen i Finland. Förhandlingar*, no. 2, (1929), 21-22.

[18] Talvitie, op. cit. (1945), p. 85.

[19] Haapaniemi, op. cit. (1983), pp. 510, 527; Karl-Erik Michelsen, *Sähköstä ja suolasta syntynyt. Finnish Chemicals Oy, Nokia Chemicals 1937-1987* (Jyväskylä, 1989), pp. 120-127.

to the permanent shortage of raw material, because some amount of rag pulp was still needed to keep wood-based fibres together in paper. Finally, chemical pulping provided the opportunity to utilize the huge standing stock of Finnish forests for paper making, and this freed the industry from the shortage of fibres, the key raw material for several decades.[20] In the late 19th century, pulping and paper making became the fastest-growing industry. At the same time, it almost succeeded in superseding the products of foreign paper mills in the huge Russian newsprint market.

Historically, chemical pulping is included in the paper industry because pulping often takes place in the same mills along with the rest of paper making. Chemical pulping, however, has its roots firmly in chemistry. Furthermore, two different chemical techniques, the sulphate and sulphite processes, were developed. The first one, sulphate pulping, was developed from the soda method, invented in the USA in the 1850s. The process involved boiling wood chips in caustic soda. In 1879, the German C.F. Dahl replaced soda with sodium sulphate and that opened up the way to successful industrial applications. Mixed with mechanically grounded pulp, it made good paper.

In the 1860s, the American Benjamin C. Tilghman managed to break up wooden chips into fibres using sulphuric acid. This invention was turned into a commercial innovation in the following decade by the German Alexander Mitscherlich and the Swede Carl Daniel Ekman, who independently developed their own industrial cooking methods for manufacturing bright sulphite pulp. In Ekman's method, spruce chips were boiled with magnesium bisulphite. Two other Swedes, Waldemar Flodquist and Victor Folin, also made significant contributions to developing industrial sulphite pulping processes.[21]

The soda method was also attempted in Finland in the 1870s and early 1880s. Theodor Höffding and another Danish engineer were involved in setting up the first two pulp mills close to Viipuri, near the eastern border.[22] However, commercial production first started in 1886 in the Walkiakoski paper mill, a Finnish company in the western part of the country that employed the sulphate process. Due to technical setbacks, the output remained modest for many years. Paper mills in Nokia and Kuusankoski adopted the other process, sulphite pulping, in 1885, and succeeded much better. Imitators followed, and for several decades that process dominated chemical pulping in Finland.[23]

---

20 G. Nikander and I. Sourander, *Lumppappersbruken i Finland* (Helsingfors, 1955), pp. 20-47.

21 Talvitie, op. cit. (1945), p. 9.

22 Ingwald Sourander, "Finlands första cellulosafabrik och dess tekniker," *Suomen Paperi- ja Puutavaralehti*, **10** (1928), 225-234; Per Schybergson, "Två danska cellulosapionjärer i Finland," *Folk og erhverv tilegnet Hans Chr. Johansen*, Festschrift (Odense, 1995), pp. 235-242.

23 Karl-Erik Michelsen, "Paperiteollisuus," in Timo Myllyntaus, Karl-Erik Michelsen and Timo Herranen, *Teknologinen muutos Suomen teollisuudessa 1885-1920* (Helsinki, 1986), pp. 116-130.

Technological problems constituted the main obstacles in adopting chemical pulping. At the very beginning, Finnish chemists and engineers could not master the new processes. Factory owners even had difficulties in obtaining knowledge of the various options, buying cooking vessels and other equipment, as well as in purchasing licences to use the production processes. Devices were ordered from British, Swedish and German engineering workshops, and foreign experts were recruited to install new machinery and supervise pulping. Nevertheless, poor quality and low productivity were frequent problems.

Some knowledge of chemistry and its application was a key factor in establishing a pulp or paper mill. This is supported by the fact that among the most successful entrepreneurs in wood-processing there were several pharmacists. For example, in the 1870s the Walkiakoski groundwood and paper mill was set up by the apothecary E.J. Granberg. His colleague, also from Tampere (Tammerfors), Gustaf Adolf Serlachius, founded first a wood grinding mill and then a chemical pulp and paper mill at Mänttä. Pharmacists were entrepreneurs in various other industries, ranging from tobacco manufacturing to mining and engineering. The 'gründerzeit' of industrialization in the late 19th century owes much to pharmacists as technical experts, merchants as financiers and salesmen, and officials and former army officers as negotiators with Finnish and Russian authorities.[24]

Nevertheless, the transfer of the latest foreign technology to Finland met with various setbacks. The best processes were often kept secret; foreign innovators and entrepreneurs were reluctant to reveal or even sell them to Finnish firms. Because Finnish mills lacked technological competence for their own R & D, they sometimes used rather underhand methods to obtain details of processes in greatest demand. A case in point was an invention by the Swedish engineer Alvar Müntzing. In the late 1880s, Müntzing had succeeded in making an important breakthrough with his durable and strong "kraft pulp."[25] Through industrial espionage, the Finnish Walkiakoski paper mill managed to locate Müntzing's associate who, in 1895, was enticed to sell the recipe for the secret chemical pulping process for FIM 50,000, a sum corresponding to the annual salaries of over 50 Finnish papermill workers at the time. From Walkiakoski, the process spread all over Finland; sawmill companies were especially anxious to refine their wastes into valuable "kraft pulp." Consequently, eight sulphate pulp mills had been set up by 1920. In the same year, twenty sulphite mills were working in Finland.[26]

---

[24] K. O. Alho, *Suomen uudenaikaisen teollisuuden synty ja kehitys 1860-1914* (Helsinki: Suomen pankki, 1949); Kerttu Peldan, *Suomen farmasian historia* (Helsinki, 1967).

[25] Anders Lundgren, "The development of the chemical industry in Sweden and the contribution of academic chemistry after 1900," in this volume.

[26] Matti Autio and Toivo Nordberg, *Vuosisata paperiteollisuutta I. Valkeakosken, Simpeleen, Myllykosken ja Jämsänkosken paperiteollisuuden vaiheet Yhtyneet Paperitehtaat osakeyhtiön perustamiseen saakka* (Valkeakoski, 1972), p. 125.

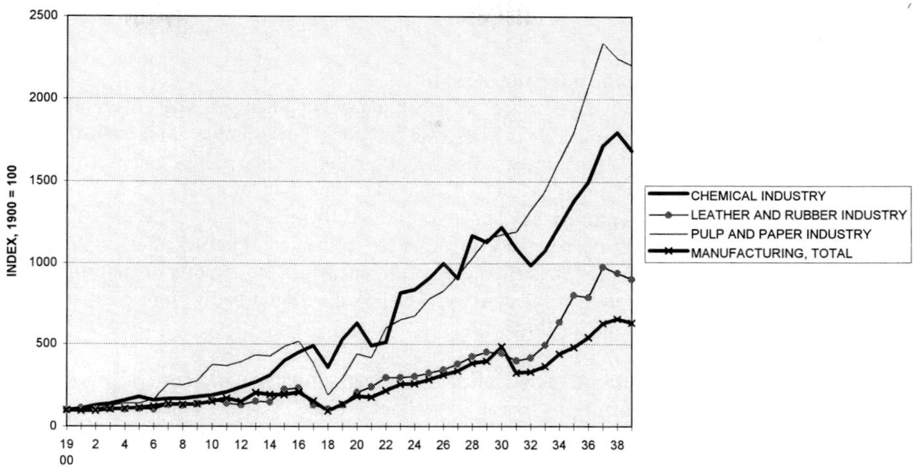


Figure 1. The volume index of production in Finnish large and medium-scale industry, 1900 - 1939.

Compared to many other countries, the interwar period was very successful for the Finnish economy. In fact, a number of industrial breakthroughs took place during that epoch. The pulping and engineering industries grew fastest due to large investments. In terms of production value, they both surpassed sawmilling, formerly the biggest industry in the country. In 1938, there were 26 sulphite pulp mills, 13 sulphate pulp mills and 26 paper mills. Between 1913 and 1938, the workforce engaged in chemical pulping rose from 2,900 to 8,200, while its output grew by a factor of ten, to 1.47 million tons. Chemical pulping contributed as much as 12 per cent to industrial output.[27]

Paper making consumed several thousand tons of various chemicals annually, and this demand stimulated the development of certain chemical branches, such as the production of sulphur, chlorine, and caustic and calcium salts. Nevertheless, in 1938 more than half of the chlorine, chloride of lime, sulphur and Glauber salt required was imported.[28] Therefore, even in this field not all potential for growth was used.

During the years 1900-38, Finnish industrial output increased by a factor of six, as shown in figure 1. The pulp and paper industry was the fastest growing manufacturing

27 See Table 1; *Industristatistik åren 1913-1938* (Helsingfors, 1915 - 1940); Michelsen, op. cit. (1986), pp. 255-257.

28 *Industristatistik år 1938* (Helsingfors, 1940).

sector. During the interwar period, its output, in fact, soared. From the early 1910s, the growth rate of the chemical industry also clearly surpassed that of all other manufacturing sectors.

**6. State intervention in the interwar period**

After Finland gained its independence, its economic policy changed. The government began to take a more vigorous role in economic activities. For example, it intervened in mining and industry by founding new firms and buying existing mines and factories from foreign owners who wished to sell out. The state purchased the Outokumpu copper mines, which supplied sulphur pyrites as a byproduct. In 1922 the state established a manufacturing company to produce sulphur for industry and sulphate fertilisers for agriculture. The main purpose was to substitute the imports of foodstuffs and increase self-sufficiency in the country, which had suffered badly from economic blockade and famine during World War I.[29]

The state-owned hydroelectric power plant at the Imatra rapids in eastern Finland was constructed in 1929. Because its transmission lines crossed southern Finland from the eastern border to the western coast, the plant significantly promoted the electrification of the country. When its capacity was increased from 56 MW to 125 MW by 1937, the plant was also able to supply electric power for various electrometallurgical and electrochemical industries, as well as for electric boilers at some pulp mills.[30]

Despite this favourable development, the Finnish chemical industry employed only 3,140 people in 1938, and the output was little more than one tenth of that in the pulping and paper industry.[31] As a whole, the chemical industry failed to achieve a breakthrough in terms of increased output in the hectic interwar years.

As depicted in figure 2, on the eve of World War II, chemical production contributed roughly 2 per cent to industrial output, and rubber manufacturing about 1 per cent. In the early 20th century, the rather one-sided structure of Finnish industry, with its strong emphasis on engineering, as well as sawmilling and other wood-processing, did not sufficiently support the development of chemical production. For instance, the Finnish chemical industry was not boosted by wine making and spirit distilling to the same extent as in many other countries.

Spain is a case in point. Núria Puig includes the production of alcohol among the five

---

29 The present transnational company, Kemira, developed from this state-owned firm. *30 vuotta. Suomen rikkihappo- ja superfosfaattiteollisuutta. Rikkihappo-ja superfosfaattitehtaat Oy 1922-1952* (Helsinki, 1952); Markku Kuisma, *Kuparikaivoksesta suuryhtiöksi, Outokumpu Oy 1910-1985* (Forssa, 1985).

30 Jaakko Auer and Niilo Teerimäki, *Puoli vuosisataa Imatran Voimaa* (Helsinki, 1982), pp. 44-66.

31 *Industristatistik år 1938.*

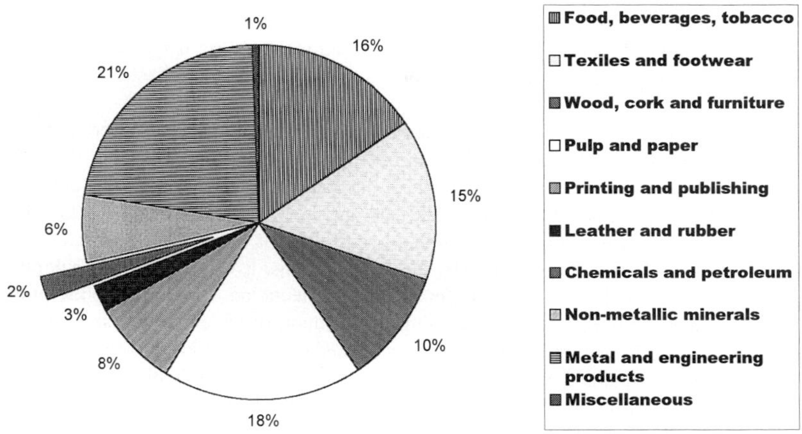


Figure 2. The contribution of Finnish industry to GDP, at factor cost, 1938. (Breakdown by manufacturing industry.) Source: *Industristatistik år 1938* (Helsingfors, 1940).

largest sectors of the Spanish chemical industry and states that the wine sector contributed ten percent of GNP.[32] In the 'long' 19th century, neither brewing, wine making nor spirit distilling was a notable industry in Finland. Whatever the activities, they supplied the local market; practically no alcohol was exported. Despite quite moderate consumption per inhabitant, the temperance movement was active and vocal.[33] In the first years of independent Finland, the movement succeeded in achieving one of its main goals: the Prohibition Act outlawed sales of alcohol from June 1919 to April 1932, when, following a referendum, production and sales were legalized again. As in Norway and the USA, which had corresponding laws almost concomitantly, prohibition did not prevent citizens from consuming alcohol.[34] Nevertheless, it brought

32 Núria Puig, "The rise of the Spanish chemical industry between the wars," in this volume.

33 Prior to World War I, the annual per capita consumption of alcohol in Finland (about 1.5 litres) was probably the lowest in Europe. Väinö Voinomaa, *Die Abstinenzbewegung in Finnland*, (Helsingfors: Raittiuden ystävät, 1907), pp. 10-18; Tapio Voionmaa, *Kansainvälinen yleiskatsaus alkoholijuomien kulutukseen vv. 1925-1936* (Helsinki, 1939), p. 14; Tapio Voionmaa, *An International Survey of the Production and Consumption of Alcoholic Beverages,* Reprint from the Volume of Proceedings of the Twenty-second International Congress Against Alcoholism (Helsinki, 1941), pp. 36-37.

34 According to some people, the Prohibition Act actually promoted the consumption of alcohol. "Owing to smuggling and clandestine distillation, spirits had never been so cheap and easily accessible in Finland as they were during the prohibition era, 1919-1932." Terje Enkvist, *The History of Chemistry in Finland, 1828-1918* (Helsinki: Societas Scientiarum Fennica, 1972), p. 143.

the legal production of alcoholic drinks down to a minimum and also hampered the development of beverage-related industries, such as the production of yeast and industrial alcohol.[35] Furthermore, Finnish exports as a whole suffered, since the prohibition policy encouraged some wine-producing countries of southern Europe to levy higher import duties on Finnish products than on merchandise from countries with more liberal alcohol policies.

The severe shortage of liquid fossil fuels during World War I forced the government of Finland to consider substitutes for petrol and kerosene. During the Civil War of 1918, two wood-processing companies suggested to the Vaasa Senate that it should support the start of alcohol production from the spent liquor of sulphite pulp processes. The senate accepted the proposal and bought the Swedish patent from Gösta Ekström, who had developed the process around 1916. Long before this, it was known that sulphite spent liquor contained sugars suitable for alcohol fermentation. Cooking sulphite pulp produced a huge amount of waste, including spent liquor, of which fermentable sugars accounted for 2-2.5 per cent. Initially it was calculated that about 40 litres of 95 per cent alcohol could be obtained from liquor left after producing one ton of sulphite pulp. However, fermentation into alcohol using traditional methods was not possible because the sugar content was so low in the large volume of spent liquor.[36]

The Prohibition Act caused the sulphite alcohol programme to be scaled down. At first only one of the six alcohol plants under construction was completed, and over 2 million litres of ethyl alcohol had to be imported annually. The army and the Economic Defence Council demanded national self-sufficiency in ethyl alcohol and the supply of liquid fuels in case of crisis, and the production of sulphite alcohol was substantially expanded between 1932 and 1941. Eventually the annual total capacity of the seventeen factories rose to over 31,000 tonnes of sulphite alcohol, expressed in terms of 100 per cent alcohol.[37]

### 7. Education and research

After the fire in Turku (Åbo) in 1827, the university was moved to the new capital, Helsinki (Helsingfors), and the teaching of chemistry continued there. In mid-19th century Finland, a great shift took place from inorganic and mineralogical chemistry to

---

35 However, after the repeal of the Prohibition Act, the legal production of alcohol did not soar in Finland as it did in the USA, which also had a similar act in force during the years 1920-32. Voionmaa, op. cit. (1941), p. 12; Olli Vehviläinen, *The Story of Rajamäki* (Helsinki, 1967), pp. 33-35.

36 Enkvist, op. cit. (1972), p. 143; Antti Kaukoranta, *Sulfiittispriiteollisuus Suomessa vuosina 1918-1978* (Helsinki, 1981), pp. 7-9.

37 In 1941-45, due to various problems, the total production of sulphite alcohol was only 61,852 tonnes, and that was 30 per cent of petrol imports. Thus sulphite alcohol decisively mitigated the chronic shortage of liquid fuels and secured the operation of motor traffic during World War II. Kaukoranta, op. cit., pp. 7-9, 17-112

organic chemistry, and this gave the discipline a new boost. In the 1860s, the Polytechnic Institute in Helsinki started teaching in applied chemistry. Terje Enkvist has claimed that the competition between the two chemistry departments in the university and at the Polytechnic Institute was as a "noble contest" between Swedish- and Finnish-speaking chemists, and stimulated Finnish chemistry to achieve its "golden age" during the years 1900-20.[38]

The appointment of Adolf E. Arppe to the chair of chemistry in 1847 meant a decisive turn in Finnish chemistry. Arppe was the first pioneer of organic chemistry and, unlike his predecessors, he had travelled much abroad and studied for long periods in Germany. In addition, he replaced Latin with German in his publications and followed up-to-date German models in organizing his education and new laboratory. Since he enjoyed the trust of the Russian authorities, in 1858 he became the Rector of Helsinki University, following student protests. In his new post, his task was to warn students about Scandinavism. Later he received more political nominations, such as for the chairmanship of the press directorate, which censored all publications in Finland.[39]

The next professor of chemistry, Johan J. Chydenius, was even more deeply interested in organic chemistry than Arppe. Chydenius, unlike Arppe, had studied and worked in Paris. Through his efforts to improve education, organic chemistry was at the time emphasized more in Finland than in the Scandinavian countries.[40] After holding the post for eleven years (1871-82), Chydenius was forced to retire due to poor health. His successor, Edward Hjelt, was appointed to the chair at the age of only 27. In the 1870s and 1880s, he had studied in Dresden, Würzburg, Munich and Strasbourg. In Germany Hjelt started his research on the amides of camphoronic acid. This heralded an epochal turn in Finnish chemistry, and for the next forty years research on camphor and the terpenes was its main focus. Moreover, Hjelt substantially strengthened the links of Finnish chemists with their German colleagues and with German chemistry. At the turn of the century, lively and close contacts with the leading continental universities kept Finnish chemists at the forefront of developments.[41]

Hjelt became well-known as an industrious author of textbooks that were used for decades not only in Finland but also abroad. With Ossian Aschan he wrote a popular textbook in Swedish, and four new volumes, totalling 4,000 pages, in the series edited by Henry Roscoe and Carl Schorlemmer, *Ausfürliches Lehrbuch der organischen*

---

[38] Enqvist, op. cit., (1972), pp. 22, 145.

[39] Enkvist, op. cit., (1972), pp. 50-62.

[40] Anders Lundgren, "Den svenska kemin runt sekelskiftet," *Kemia - Kemi*, **18,** no. 11-12 (1991), 1019-1023.

[41] George B. Kauffman, "Edvard Hjelt - kemisti ja valtiomies," *Kemia - Kemi*, **18** (1991), 1010-1013.

*Chemie*.[42] Hjelt gave up his career as a scientist when he became the rector of Helsinki University. Later he was nominated to the post of vice-chancellor and chancellor of the university, as well as a senator of the Grand Duchy. He continued his academic career in independent Finland, too, and was the first Finnish ambassador to Germany. After the turn of the century, his work as an administrator forced him to cease his experimental work in chemistry, and he switched to research into the history of science. His main monograph in that field was a comprehensive history of organic chemistry, published in Germany in 1916.[43]

Hjelt's successor, Ossian Aschan, had also studied in Germany under the same supervisors, including Johannes Wislicenus, as many other Finnish chemists did at the time. Aschan produced an enormous scientific output of articles and monographs. He published some 250 works, most of them dealing with camphors and terpenes.[44] His book on alicyclic compounds is regarded as his most significant work.[45]

In the 1920s, Ossian Aschan wrote that, as a young lecturer, he had believed in the possibilities of the chemical industry in Finland but had become thoroughly disillusioned. He and Gustav Komppa had attempted to produce synthetic camphor, but this failed commercially. At the beginning of the 20th century, there was a great demand for camphor as a solvent for nitrocellulose in the manufacture of celluloid. The Finnish producers were not able to obtain the necessary raw materials at competitive prices because the Japanese cartels for the marketing of natural camphor from Formosa were successful in counteracting synthetic camphor producers.[46]

In 1903, Gustaf Komppa, the professor of chemistry at the Helsinki Polytechnic, later the University of Technology, invented a synthetic method for manufacturing camphor, but this was not enough for a commercial breakthrough. For forty years he attempted to achieve the synthesis of pinene, a key raw material in producing synthetic camphor. The long delay in doing this, together with falling world-market prices for camphor,

---

42 Edward Hjelt and Ossian Aschan, *Lärobok i organisk kemi*, 1st edition (Helsingfors, G.W. Edlund, 1893), 4th edition (Stockholm: Samson & Wallin, 1922); H. Roscoe and C. Schorlemmer, eds, *Ausführliches Lehrbuch der organischen Chemie*, Teil 3-7 (1896-1901).

43 Edward Hjelt, *Geschichte der organischen Chemie von ältester Zeit bis zur Gegenwart* (Braunschweig: Vieweg & Sohn, 1916), 556 pages.

44 Enkvist, op. cit. (1972), pp. 83-88; Pentti Mälkönen, "Suomen orgaanisen kemian kulta-aika 1800- ja 1900-lukujen taitteessa," *Kemia-Kemi*, **18,** no. 11-12, (1991), 1016-1018.

45 Ossian Aschan, *Chemie der alicyklischen Verbindungen* (Braunschweig: Vieweg & Sohn, 1905), 1163 pages.

46 Ossian Aschan, "Finska kemistsamfundets möte (6.3.1926)," *Finska kemistsamfundets meddelanden*, **35** (Helsingfors, 1926), p. 6; Enkvist, op. cit. (1972), pp. 90, 121-126.

destroyed the possibilities of producing synthetic camphor in the Grand Duchy of Finland. Finally, in the interwar years, Komppa succeeded in developing a commercial process to synthesize pinene. This kind of production clearly had potential in the country because pinene was later manufactured cheaply in Finland as a byproduct of chemical pulping.[47]

The Finnish chemists of the "golden age" were very prolific, and their articles were published in the leading German chemistry journals. They were appointed members of various Scandinavian and German academic societies, received honorary doctorates from universities of the same countries and were awarded major scientific prizes. For example, Gustaf Komppa was the first Nordic scholar to receive the silver von Hofmann-medal of the Gesellschaft Deutscher Chemiker. Nevertheless, the factories founded by them were, as a rule, unsuccessful. Komppa, however, was a co-founder of three sound companies, the pharmaceutical factory Orion, the state-owned powder factory at Vihtavuori, and another state-owned company for the manufacture of superphosphate and sulphuric acid.[48]

Practically the only commercial successful innovations made by Finnish chemists in the early 20th century were related to tall oil soap and some other byproducts of wood-pulping. The Finnish scholar Alfons Hellström discovered two more natural terpenes from juniper resin, invented methods for distilling tall oil under vacuum from sulphate pulping waste, and developed commercially applicable processes for the manufacture of tall oil soap and the use of rosin acids from tall oil for paper sizing. Such distillations of tall oil became an important industry not only in Finland but also in many other countries. By 1939, the sales of tall oil soap conquered a third of the Finnish detergent market.[49]

In the Grand Duchy of Finland, chemical education developed considerably, and since the late 19th century it has been at a high level both at the University of Helsinki and the Helsinki Polytechnic. At the same time, several Finns also studied abroad, primarily in Germany. The number of qualified chemists was notable. One might ask whether this meant significant potential for industrial applications. Certainly, a sign of early professionalization was that Finnish chemists set up one of the first national societies for chemistry.[50] However, transforming qualified expertise into commercially thriving

---

47 Enkvist, op. cit. (1972), p. 125; Lauri Niinistö, "Kemistikilta ja vuosisadan vaihteen Suomi," *Kemia-Kemi*, **18** (1991), 1008.

48 Niinistö, Ibid., 1008; Veikko Komppa, "Professori Gust. Komppa ja Suomen kemianteollisuus," *Kemia - Kemi*, **18,** no. 11-12 (1991), 1014-1015.

49 In 1911 Hellström obtained his first patent, which was a result of cooperation with the Swedish engineer Hilding Bergström, who also specialized in the byproducts of pulping. Erkki Pakkala, "Alfons Hellström - terpeenejä, tärpättiä ja mäntysuopaa," *Tekniikka*, no. 1-2, (1974), 51-52.

50 *Activities of Chemical Societies of the World* (Tokyo: The Chemical Society of Japan, 1987); Pekka

innovations did not succeed. Both of the above-mentioned institutions of higher learning provided primarily theoretical education and lacked well-equipped laboratories. They did not have extensive research programmes and their contacts with industry were rather weak. Ironically, the Russian authorities trusted the talents of chemists at Helsinki University more than Finnish industrialists did.

Another factor that hampered the development of chemistry in Finland was the shortage of research institutions. In 1890, the Finnish government set up a Testing Institute for Building Materials *(Aineenkoetuslaitos)*, which was attached to the Helsinki Polytechnic. It was not a proper research laboratory but, nevertheless, it was the first step towards establishing an institution for scientific research. Unfortunately, the institute soon stagnated due to a shortage of funds. For two decades, it was in serious difficulties as far as modernizing its equipment and hiring more staff were concerned. However, it continued testing the quality of concrete, building stone, metals and some other products. It expanded in the 1920s and was finally thoroughly reorganized in 1942.[51]

World War I revealed various weaknesses in the Finnish economy. Finnish factories were unable to acquire consulting assistance to help overcome their technological problems from foreign experts, as they had been able to do during peacetime. Some representatives of the paper industry joined with university chemists to set up a chemical research institute. In 1916, a private company, Ab Centrallaboratorium - Keskuslaboratorio Oy, was founded. After a few years, the laboratory concentrated on testing the quality of pulp and paper dispatched by Finnish wood-processing companies. It undoubtedly contributed remarkably to the development of Finnish pulping and paper making during a critical period, but its activities did not support the rise of a mainstream chemical industry. By 1920, it had abandoned plans for extensive research in chemistry and concentrated on serving only the pulp and paper industry.[52]

During the few decades before World War I, Finnish agriculture turned its emphasis from grain growing to husbandry, especially dairy production. In 1916, in its efforts to develop dairy production, the central organization of co-operative dairies, Valio, founded a chemical and bacteriological laboratory. In 1920 Artturi Ilmari Virtanen was nominated its director. Virtanen, who had studied in Switzerland, Germany and Sweden, brought biochemistry to Finland. Within a few years he managed to develop

---

Pyykkö, "Finska Kemistsamfundet 1891-1991," *Kemia - Kemi*, **18**, no. 11-12 (1991), 1003.

51 Karl-Erik Michelsen, *Valtio, teknologia, tutkimus. VTT ja kansallinen tutkimusjärjestelmän kehitys* (Espoo: VTT, 1993), pp. 42-48.

52 Anna Grönvik, *Oy Keskuslaboratorio 50 vuotta* (Helsinki, 1966), pp. 6-40; Michelsen, op. cit. (1993), pp. 51-56.

"AIV salt" that improved the quality of Finnish butter, which was mainly exported by Valio. With his colleagues, Virtanen also invented a method for producing AIV fodder that improved the proper feeding of cattle in winter-time. His invention combined the latest knowledge of biochemistry and traditional cultivation methods. The use of AIV-fodder quickly spread all over the country in the 1930s, thus bringing about a rapid and marked increase in the milk production of Finnish farms.[53]

By selling AIV chemicals to farmers and licences for its AIV patents to other companies, Valio earned considerable income. However, the company was reluctant to be the sole financier of the research programmes drafted by A. I. Virtanen. Therefore, in 1928, the Foundation for Chemical Research *(Kemiantutkimussäätiö)* was set up, and three years later, supported by funds from Finnish agribusiness and banks, it established a new Biochemical Research Institute to carry out Virtanen's ambitious research plans. As a result, Virtanen's teams in two separate research institutes represented the most advanced chemical research in interwar Finland. Unlike his predecessors in Finnish chemistry (Ernst Qvist, Ossian Aschan and Gustav Komppa), Virtanen succeeded in combining scientific research, industrial applications and adequate funding.[54] His achievements remain unique.

**8. Conclusions**

In the early 20th century, Finland had some potential to develop a thriving chemical industry. However, various other industries competed for the same resources, such as waterpower, funding, and qualified experts and managers. The lack of higher education and relevant research facilities were weaknesses in the Finnish chemical industry. Theoretical education was not supported by experimental work in well-equipped laboratories, and, with the exception of two institutes directed by A. I. Virtanen, there were very few research facilities. In the interwar period, industrial companies seldom had their own research staff and laboratories, and R & D was, consequently, neglected and the number of professional chemists remained small. Intellectual resources were, therefore, insufficient for basic research and innovative development.

It is also important to note that, with a few exceptions, such as dynamite and chlorine production, large transnational companies were not interested in investing in Finland and constructing chemical factories there. The Finnish government did not attempt to attract foreign investors by offering any concessions or subsidies either and Finnish legislation even discouraged direct foreign investments.[55]

---

[53] Michelsen, op. cit. (1993), pp. 56-59.

[54] Ibid.; Oskari Routala, "Gustaf Komppa zum vollendeten 60. Lebensjahre," *Juhlajulkaisu,* Festschrift für Gustaf Komppa (Helsinki, 1927), pp. vii-xiv.

[55] A more detailed analysis of the Finnish case is to found in Timo Myllyntaus, "The Finnish model of technology transfer," *Economic Development and Cultural Change*, **38**, no. 3 (1990), 625-643; Timo

Finally, it is worth pointing out that the transfer of technology to enterprises in recipient countries is generally a complex process, and the transfer of processes and equipment used in chemical production, in particular, is not very easy for late industrializing countries. Melvin Kranzberg's third law states that "Technology comes in packages, big and small."[56]

As far as the chemical industry is concerned, technology is generally transferred from one country to another in large packages. Considerable elements of technological know-how concerning chemical production processes and products are protected by patents, licences and intrafirm culture. New, small firms meet a variety of problems if they attempt to obtain know-how from large chemical companies. As early as at the turn of the century, the production of standard chemicals was often controlled by international cartels that were reluctant to grant quotas to newcomers.

Prior to World War II, the thresholds and bottlenecks to be overcome in developing a large-scale chemical industry were perhaps too challenging, and the Finns consequently focused on developing other industries. It was only after 1945 that they turned their attention to chemical production

---

Myllyntaus, "Technology transfer and the contextual filter in the Finnish setting. Transfer channels and mechanisms in an historical perspective," in *Mastering Technology Diffusion - The Finnish Experience*, Synnöve Vuori and Pekka Ylä-Anttila, eds, ETLA-Series B 82 (Helsinki: The Research Institute of the Finnish Economy, 1992), pp. 195-251.

[56] Melvin Kranzberg, "One last word - technology and history: 'Kranzberg's Laws,'" in *In Context, History and the History of Technology*, Stephen H. Cutcliffe and Robert C. Post, eds, (Bethlehem, Pa.: Lehigh University Press, 1989), pp. 248-249.

# CONTRIBUTORS

**Stuart Bennett** trained as a mechanical engineer and is now a reader in the Department of Automatic Control & Systems Engineering, the University of Sheffield, where he has taught since 1968. He held a senior postdoctoral research fellowship at the Museum of American History, Smithsonian Institution, Washington DC, during the academic year 1988-89 where he worked on the history of control engineering (covering the periods 1800-1930 and 1930-1995), and has published on the history of industrial measuring instruments. His current interests relate to the impact of instruments (measuring, recording and controlling) on the organizational and operational structures of the process industries. His non-historical research is on the socio-technical modelling of manufacturing processes.

**Colin Divall** is professor of railway studies at the University of York, UK. He has written extensively on the history of engineering professions in the United Kingdom and is currently writing, with Sean Johnston, a history of chemical engineering in Britain.

**Ernst Homburg** studied chemistry in Amsterdam, and continued as researcher and assistant professor in the fields of history and chemistry and society at the universities of Groningen and Nijmegen. His Ph.D. thesis, on the rise of the profession of the industrial chemist in Germany (1790-1850), was published in 1993 (*Van beroep Chemiker*, Delft University Press). He is editor of a book series on the history of technology in the Netherlands and is at present connected to the history department of the University of Maastricht.

**Sean Johnston** has taught physics and the history and philosophy of science at the University of Leeds, and is currently a research fellow in history at the University of York, UK, where he is writing a history of chemical engineering with Colin Divall. He has studied the emergence of technical communities and intellectual bases for spectroscopy and light measurement in the 19th and 20th centuries.

**Helge Kragh** is professor at the History of Science Department, Aarhus University, Denmark. He has published on the history of technology and physical sciences and is interested in the philosophy of science. His most recent book is *Cosmology and Controversy: The Historical Development of Two Theories of the Universe* (Princeton, 1996), and he is now working on a history of physics in the 20th century.

**Anders Lundgren** is lecturer in the history of ideas and science at Uppsala University. He has a general interest in the history of science since 1700, and a special interest in the history of chemistry. He has published on the Chemical Revolution, and on the relation between mining and chemistry since the 18th century. Recently he has become interested in the rise of biochemistry as an independent discipline in Sweden, as well as the science-technology relationship, with special reference to chemistry. His publications include *Everyday Science. A History of Xylocaine* (in Swedish, 1995).

**Roy M. MacLeod** is professor of history at the University of Sydney. Educated at Harvard, the LSE, and at Cambridge, where he took his Ph.D., he has written extensively in the history of science, medicine and technology. He has held senior appointments at the universities of Sussex and London and visiting appointments at many universities in Europe and the United States. He teaches military history, nuclear history, and the history of museums in Europe, Asia and Australasia. He has written or edited sixteen books, of which the most recent are *Darwin's Laboratory: Evolutionary Theory and Natural History in the Pacific* (Honolulu: University of Hawaii Press, 1994), *Technology and the Raj: Technical Transfer and Technological Change in British India, 1780-1945* (New Delhi: Sage, 1995), and *Public Science and Public Policy in Victorian Britain* (Aldershot: Variorum, 1995). In 1996, he was Edelstein International Fellow in the History of Chemistry.

**Lothar Meinzer** studied English and history at the University of Mannheim where he subsequently completed his Ph.D. with a dissertation on the Nazi period in the Palatinate. He was a lecturer in modern history at the same university until 1985 when he was appointed to his present post as head of the BASF historical archive. He is co-author of *125 Jahre BASF: Stationen ihrer Geschichte* (Ludwigshafen: BASF, 1990).

**Peter J.T. Morris** is Senior Curator, Experimental Chemistry, Science Museum, London. His publications include the *American Synthetic Rubber Research Program* (Philadelphia: University of Pennsylvania Press, 1989), and, with Colin A. Russell, *Archives of the British Chemical Industry, 1750-1914* (Faringdon, Oxfordshire: British Society for the History of Science, 1988). Research appointments have included Royal Society-British Academy Research Fellow in the History of Science and Edelstein International Fellow in the History of Chemistry.

**Timo Myllyntaus** works as a senior lecturer at the University of Helsinki, Finland. He holds a Ph.D. in economic history from the London School of Economics. He specializes in industrial history and has published on the development of sawmilling, engineering, printing and publishing, and energy production. His main study is *Electrifying Finland. The Transfer of a New Technology into a Late Industrialising Economy* (London: Macmillan, 1991).

**Rolf Petri** is lecturer in economic history at the Martin-Luther-Universität Halle-Wittenberg. He is a graduate of the University of Marburg, and the European University Institute, Florence, where he received his Ph.D. in history and civilisation (1988). His publications include the edited volume *Venedig. Ein politisches Reisbuch* (Hamburg: VSA, 1986), *Storia di Bolzano* (Padua: Poligrafo, 1989), and *La frontiera industriale* (Milan: Angeli, 1990).

**Núria Puig** is associate professor of economic history at the Universidad Complutense de Madrid. She has published, with Santiago M. López, *Ciencia e industria en España. El Instituto Químico de Sarrià, 1916-1992* (Barcelona: Institut Químic de Sarrià, 1992), and "Chemists, engineers, and entrepreneurs. The Chemical Institute of Sarrià's impact on Spanish Industry (1916-1992)," *History and Technology*, **11** (1994), 345-359. She has published widely on the Spanish alcohol industry and is currently researching the history of synthetic dyes in Spain.

**Carsten Reinhardt** is lecturer in the history of science at the University of Regensburg. His doctoral dissertation at the Technical University Berlin (1995) dealt with the history of research in the German chemical industry, 1860-1914 ("Forschung in der chemischen Industrie. Die Entwicklung synthetischer Farbstoffe bei BASF und Hoechst, 1863-1914." Publication by Medienzentrum TU Bergakademie, Freiberg, is expected in 1997). His research interests include the history of industrial research, the history of chemistry and technology in the 19th and 20th centuries, and the history of analytical chemistry in the 20th century.

**Harm G. Schröter** studied in Hamburg and at the University of East Anglia, Norwich, and wrote his doctoral thesis on the relationship between state and industry in the interwar period. He has published widely on economic competition and concentration, particularly cartels and multinational enterprises. Since 1990 he has focused on the history of technology, mainly chemical industry. He now is guest professor of social and economic history at the University of Konstanz.

**Geert J. Somsen** studied chemistry and history of science at the Free University, Amsterdam, and has recently completed his Ph.D. in history of science at Utrecht University with a thesis on H.R. Kruyt (1882-1859), his Utrecht school of colloid chemistry (1916-1946), and his influence on chemical research in the Netherlands. Somsen has been a visiting student at the University of California, San Diego, and is Othmer Post-Doctoral Fellow at the Chemical Heritage Foundation, Philadelphia, 1997-1998.

**Knut Sogner** is research fellow at the Norwegian School of Management, Sandvika. His doctoral thesis (Oslo, 1996) dealt with the Norwegian pharmaceutical company Nycomed and its innovation of non-ionic contrast media. He has published articles in English, and is currently completing a book on the Norwegian electronics firm Simrad.

**Jakob Tanner** is professor at the Research Institute for Social and Economic History, Zurich University. His doctoral dissertation, at the same university, dealt with the economic and financial history of Switzerland in World War II. He subsequently undertook research at the department of history, Basle University, and lectured at Bern, Fribourg, Basle and Zurich. As a fellow of the Maison des Sciences de l'Homme and research director at the Ecole des Hautes Etudes en Sciences Sociales, he specialized in the history of science. During 1996/97 he was guest professor at Bielefeld.

**Anthony S. Travis** is deputy director of the Sidney M. Edelstein Center for the History and Philosophy of Science, Technology and Medicine at The Hebrew University of Jerusalem, where he is also senior researcher in the history of technology. He has published widely on the history of chemical technology and the Channel tunnel project, and is series editor of the American Chemical Society-Chemical Heritage Foundation "History of the Modern Chemical Sciences Series."

## Abbreviations used in this book

| | |
|---|---|
| ABCM | Association of British Chemical Manufacturers |
| ACNA | Azienda colori nazionali affini |
| AEG | Allgemeine Elektricitäts Gesellschaft |
| Agfa | Aktiengesellschaft für Anilinfabrikation[1] |
| AGIP | Agenzia Generale Italiana Petroli |
| AIAG | Aluminium Industrie AG |
| AIChE | American Institute of Chemical Engineers |
| AKU | Algemeene Kunstzijde Unie |
| ALCOA | Aluminum Corporation of America |
| Alusuisse | Aluminium Industrie AG |
| ANIC | Azienda Nazionale Idrogenazione Combustibili |
| ASR | Applied Scientific Research |
| Bamag | Berlin-Anhaltische Maschinenbau AG |
| BASF | Badische Anilin- & Soda-Fabrik |
| BDC | British Dyestuffs Corporation |
| B.N.R. | Blagden Noakes Resins[1,2] |
| BOC | British Oxygen Company |
| BPD | Bombrini Parodi Delfino |
| CAMPSA | Campañia Arrendatoria de Monopolio de Petroleos SA |
| CCS | Cellulosa cloro soda |
| Ciba | Chemische Industrie Basel[1] |
| CISA | Compagnia industriale società anonima viscosa[3] |
| Degussa | Deutsche Gold- und Silber-Scheideanstalt[1] |
| DSIR | Department of Scientific and Industrial Research |
| DSSF | Dansk Svovlsyre- & Superfosfatfabrik |
| Du Pont | E. I. du Pont de Nemours & Company[1] |
| DUCO | Società italiana Duco[1,4] |
| EEC | European Economic Community |
| ELH Group | Elektrobosna-Lonza-Hafslund Group[5] |
| ETH | Eidgenössische Technische Hochschule[6] |
| FIAT | Fabbrica italiana automobili Torino[1] |
| FNCE | Fabricación Nacional de Colorantes y Explosivos |
| FR Gomma | Fabbriche riunite industrie della gomma |
| GAVEG | Gasverarbeitungs-Gesellschaft m.b.H.[7] |
| HIAG | Holzverkohlungs-Industrie AG |

1 Indexed under the abbreviation
2 This is a conjecture which arose from a discussion with Daphne Cooper of Blagden Industries.
3 Originally Compagnia italiana seta artificiale
4 Presumably a contraction of Du Pont Company; a trade name for Du Pont paints in the US
5 Became Consortium für elektrochemische Industrie GmbH in 1903 and indexed under that name
6 Indexed as Federal Institute of Technology under Switzerland
7 m.b.H. = mit beschränkter Haftung (private limited company)

| | |
|---|---|
| I.C.I. | Imperial Chemical Industries Ltd |
| I.G. | I.G. Farbenindustrie AG |
| I.G. Chemie | Interessengemeinschaft Chemie[1,8] |
| I.G. Farben | I.G. Farbenindustrie AG[9] |
| IChemE | Institution of Chemical Engineers |
| IHEC | International Hydro-Engineering and Chemical Co. |
| IHP | International Hydrogenation Patents Co. Ltd |
| INA | Industria Nazionale d'Alluminio |
| IPCA | Industria piemontese colori di anilina |
| IRI | Istituto per la ricostruzione industriale |
| JASCO | Joint American Study Company[10] |
| KWG | Kaiser-Wilhelm-Gesellschaft zur Förderung der Wissenschaften |
| M&B | May & Baker Ltd |
| Norsk Hydro | Norsk Hydroelektrisk Kvaelstof A/S |
| ØK | Østasiatisk Kompagni |
| ONIA | Office National Industriel de l'Azote |
| PSR | Pure Scientific Research |
| ROMSA | Raffineria olii minerali società anonima |
| SAFFA | Società anonima fabbriche fiammiferi e affini |
| SAIGS | Società anonima italiana gomma sintetica |
| SAVA | Società alluminio Veneto anonima |
| SCI | Society of Chemical Industry |
| SCIA | Società cellulosa italiana anonima |
| SED | Sociedad Española de la Dinamita |
| SHIV | Schweizerische Handels- und Industrieverein |
| SIN | Sociedad Española del Nitrógeno |
| SIPE | Società italiana prodotti esplodenti |
| SIRI | Società italiana ricerche industriali |
| SMI | Società metallurgica italiana |
| SNIA | Società nazionale industria applicazioni viscosa[1, 11] |
| SPEA | Società prodotti esplodenti autarchici |
| Standard Oil | Standard Oil of New Jersey[12] |
| UEE | Unión Española de Explosivos |
| UNI | Ente nazionale italiano di unificazione |
| UOP | Universal Oil Products |
| VAW | Vereinigte Aluminium Werke |
| VEBA | Vereinigte Elektrizität und Bergwerks AG |
| WPA | Works Progress Administration |

---

8 The Swiss dye cartel, not to be confused with the I.G. Farben subsidiary with the same short name

9 The 'I.G.' in I.G. Farbenindustrie is not an abbreviation, see footnote 10 on p. 92

10 Not to be confused with the Japanese instrument company with the same initials

11 Originally called Società di navigazione italo-americana, because the capital for the artificial silk works originated from a former shipping company

12 Also called Jersey Standard and Esso (phonetic SO); now called Exxon (in US)

# INDEX

## —E—

## —F—

## —G—

## —H—

## —I—

## —M—

## —N—

## —O—

## —T—

## —U—

## —V—



## —W—

The following individuals and organizations are thanked for assistance provided in the preparation of the index: Bayerarchiv; James Bohning, American Chemical Society News Service; Coalite Smokeless Fuels, Ltd; Daphne Cooper, Blagden Industries, Ltd; Camille and Henry Dreyfus Foundation, Inc., New York; Goethe Institute, London; Lothar Meinzer and Ruth Fromm, Firmenarchiv BASF AG; Sheena Russell, ICI Headquarters Library; Wolfgang Metternich, Firmenarchiv Hoechst AG; Hans Michels, Imperial College, London; Manfred Rasch; Royal Society of Chemistry Information Centre and Library; David Rhees; Jane Fynn, Rhône-Poulenc Rorer, Ltd; Science Museum Library; Andrew Tait, Bayer AG, Leverkusen; the Archivist, Trinity College, Oxford; and Magda Wheatley.

# Chemists and Chemistry

1. M. Morselli: *Amedeo Avogadro* [1776–1856]. A Scientific Biography. 1984 ISBN 90-277-1624-2
2. F.W.J. McCosh: *Boussingault* [1802–1887]. Chemist and Agriculturist. 1984 ISBN 90-277-1682-X
3. D.A. Stansfield: *Thomas Beddoes, MD (1760–1808).* Chemists, Physician Democrat. 1984 ISBN 90-277-1686-2
4. Pieter Eduard Verkade (1891–1979): *A History of the Nomenclature of Organic Chemistry.* Edited by F.C. Alderweireldt, H.J.T. Bos, L. Maat, P.J. Slootmaekers and B.M. Wepster. 1985 ISBN 90-277-1643-9
5. A. Thackray, J.L. Sturchio, P.T. Carroll and R. Bud: *Chemistry in America, 1876–1976.* Historical Indicators. 1985 ISBN 90-277-1720-6; Pb 90-277-2662-0
6. G.B. Kauffman (ed.): *Frederick Soddy (1877–1956).* Early Pioneer in Radiochemistry. 1986 ISBN 90-277-1926-8
7. R.B. Seymour and Tai Cheng (eds.): *History of Polyolefins.* The World's Most Widely Used Polymers. 1986 ISBN 90-277-2128-9
8. J.T. Stock and M.V. Orna, OSU (eds.): *The History and Preservation of Chemical Instrumentation.* 1986 ISBN 90-277-2269-2
9. N.A. Peppas (ed.): *One Hundred Years of Chemical Engineering.* From Lewis M. Norton (MIT, 1888) to Present. 1989 ISBN 0-7923-0145-8
10. R.B. Seymour (ed.): *Pioneers in Polymer Science.* 1989 ISBN 0-7923-0300-8
11. H. Benninga: *A History of Lactic Acid Making.* A Chapter in the History of Biotechnology. 1990 ISBN 0-7923-0625-2
12. R. Mierzecki: *The Historical Development of Chemical Concepts.* 1991 ISBN 0-7923-0915-4
13. S. Ross: *Nineteenth-Century Attitudes: Men of Science.* 1991 ISBN 0-7923-1308-9
14. Torbern Bergman (1775) (ed.): *Chemical Lectures of H.T. Scheffer.* Translated by J.A. Schufle. 1992 ISBN 0-7923-1760-2
15. C.H. Evans (ed.): *Episodes From the History of the Rare Earth Elements.* 1996 ISBN 0-7923-4101-5
16. A.S. Travis, H.G. Schröter, E. Homburg and P.J.T. Morris (eds.): *Determinants in the Evolution of the European Chemical Industry, 1900–1939.* New Technologies, Political Frameworks, Markets and Companies. 1998 ISBN 0-7923-4890-7

KLUWER ACADEMIC PUBLISHERS – DORDRECHT / BOSTON / LONDON

| | |
|---|---|
| | |
| | |
| | |
| | |
| | |
| | |
| | |
| | |
| | |
| | |